NUCLEAR QUADRUPOLE COUPLING CONSTANTS

NUCLEAR QUADRUPOLE COUPLING CONSTANTS

E. A. C. Lucken

Cyanamid European Research Institute
and Institute of Physical Chemistry
University of Geneva
Geneva, Switzerland

ACADEMIC PRESS

London and New York • 1969

ACADEMIC PRESS INC. (LONDON) LTD
Berkeley Square House
Berkeley Square,
London, W1X 6BA

U.S. Edition published by
ACADEMIC PRESS INC.
111 Fifth Avenue,
New York, New York 10003

Library of Congress Catalog Card Number: 68–57750

PRINTED IN GREAT BRITAIN BY J. W. ARROWSMITH LTD, BRISTOL 3

PREFACE

Although nuclear quadrupole coupling constants in atoms were first observed in 1935 by Schmidt and Schuler and for the first time in molecules (D_2 and HD) by Kellog, Rabi, Ramsey and Zacharias some four years later, it was not until the discovery of pure quadrupole resonance by Dehmelt and Kruger in 1951 that nuclear quadrupole coupling constants, for a wide variety of molecules, became readily accessible experimental quantities. Through the pioneering studies of Townes and Dailey from 1949 onwards, it became clear that this parameter, in the form of the electrostatic field-gradients at the nucleus under study, provided data of considerable interest to those interested in the theory of molecular structure.

Two streams of interest are apparent. In the first, the nuclear quadrupole coupling constant is used as a sensitive test of the adequacy of a complicated self-consistent wave function for a relatively small molecule, the predicted field gradients being calculated exactly from the wave-function under study. In the second, the experimental nuclear quadrupole coupling constant is used to obtain approximate electron distributions in the neighbourhood of the nucleus in question and to correlate this information with other experimental parameters derived for example from studies of reaction rates or ultra-violet spectroscopy. The first is thus more the concern of the physicist and the second of the chemist.

In writing this book my intention was to bring to the attention of the everyday chemist, whether organic or inorganic, the information contained in nuclear quadrupole coupling constants, to survey the data now available and to correlate it by means of the simple, although admittedly approximate, methods first introduced by Townes and Dailey. Nuclear quadrupole coupling constants provide in fact striking numerical estimates of those concepts so useful to the chemist as "ionic character" "partial double-bond character" etc. An electrostatic field gradient is a classical quantity and presents no conceptual problem to the undergraduate student in his first year; nuclear quadrupole coupling constants can thus be introduced quite early as data supporting, for example, the affirmations of a lecturer in organic chemistry as to the electron distribution in organic molecules.

To the research worker too, anxious to decide between two extreme or alternate formulations of the electronic structure of a molecule the nuclear quadrupole coupling constant of an opportunely-placed nucleus can yield a decisive parameter and, although nuclear quadrupole resonance will never have the extensive utility of nuclear magnetic resonance, nuclear quadrupole coupling constant can in certain circumstances yield simple structural information.

Finally, it is hoped that this monograph may stir the theoretician to a more extensive use of nuclear quadrupole coupling constants as tests for the more-and-more exact wave functions which he is producing for more-and-more complicated molecules and at the same time explain to the nuclear physicist why the nuclear quadrupole moments which he is interested in suffer from such considerable uncertainties (it is perhaps fortunate that the theory of nuclear structure is not yet sufficiently quantitatively exact to be much worried by this!).

The book is divided into two sections. In the first, the basic theory of quadrupole coupling is presented, the methods available for the experimental measurement of the coupling constants reviewed and the relationship between them and a molecules' electronic structure discussed both from the rigorous and approximate view-points. In the second, the experimental data are reviewed and correlated by means of the approximate methods developed in the first section, particularly those of Chapter 7. I hope that in this way it will serve a variety of interests and provide a source of reference to the now considerable literature of the subject.

Geneva E. A. C. LUCKEN
November, 1968

ACKNOWLEDGEMENTS

The first link in the chain of events which led to the writing of this book was undoubtedly provided by Professor M. J. S. Dewar, F.R.S., who initially aroused my interest in nuclear quadrupole resonance, and who subsequently by his continued encouragement provided the stimulus to overcome the many difficulties which litter the path of the young and innocent chemist who strays into the territory of the physicist and the electronic engineer. May it be found worthy of such a god-father!

The American Cyanamid Company provided, in the form of the Cyanamid European Research Institute, the atmosphere in which research can be carried out and a book such as this written, as well as the excellent library with which that Institute was endowed. Mademoiselle Jacqueline Wismer took a great load from my shoulders in so carefully preparing the manuscript for publication. Above all else my wife and children deserve my special thanks for their direct support during the period when the book was written, and for the way in which they accepted the consequent demands on my spare time.

For their permission to reproduce various figures, I thank Academic Press Inc. (Fig. 3.4), The American Institute of Physics and, respectively, Professor Y. Morino (Fig. 3.7), Professor R. L. Collins (Fig. 3.16), Dr. R. M. Sternheimer (Fig. 5.3), Dr. P. Cade (Fig. 6.3), Professor R. Bersohn (Fig. 8.1), Professor N. Bloembergen (Fig. 9.1), Professor P. J. Bray (Fig. 12.2), Dr. T. E. Haas (Fig. 13.3); The Chemical Society of Japan and Professor D. Nakamura (Fig. 13.1); John Wiley and Sons Inc. and Professor E. Scrocco (Figs. 6.1 and 10.6); Springer-Verlag and Professor E. Scrocco (Fig. 6.2).

CONTENTS

Part 2: A Survey of the Experimental Data

PART 1

THE MEASUREMENT OF NUCLEAR QUADRUPOLE COUPLING CONSTANTS AND THEIR RELATIONSHIP TO MOLECULAR WAVE-FUNCTIONS

CHAPTER 1

INTRODUCTION

The development of radiofrequency spectroscopy in the last two decades has made available to the student of molecular electronic structure the values of a large number of parameters which he can use to investigate the electron distribution in molecules. These parameters are usually measures of weak interaction occurring between atomic nuclei having magnetic or electrostatic moments and the electrons, and as such yield information about the electronic structure of the molecule *in its ground state*. Furthermore, even in large complicated molecules the information obtained usually concerns only one or two nuclei and the electrons in their immediate vicinity; thus the electron distribution in such a molecule may be studied piecemeal and the whole structure built up by assembling information from different sources. With the possible exception of proton hyperfine coupling constants in the electron spin resonance spectra of conjugated organic free radicals, nuclear quadrupole coupling constants—the subject of this monograph—are the best understood of these parameters and certainly the best understood of those measurable in diamagnetic molecules. The information they provide, about the components of the electrostatic field-gradient tensor at the nucleus under study, is essentially classical in nature and gives rise to no conceptual problems. This information can thus be presented to the undergraduate student quite early in his career.

As will be seen in the following section the nuclear quadrupole coupling energy, E_Q, is given by the relationship

$$E_Q = -\tfrac{1}{6}\mathbf{Q}_{ij} \cdot \mathbf{V}^{ij}$$

where $\mathbf{Q}_{ij}$ is the nuclear quadrupole moment tensor and $\mathbf{V}^{ij}$ is the electrostatic field-gradient tensor. E_Q is thus the inner product of two symmetric second rank tensors. Such an inner product of course depends on the mutual orientation of the two axis systems to which the two tensors are referred: for $\mathbf{Q}$ the natural system is one in which one axis coincides with the spin axis of the nucleus. In this case the cylindrical symmetry of the nucleus has the consequence that the whole quadrupole moment tensor can be expressed in terms of one parameter "the" nuclear quadrupole moment, Q. Likewise since $\mathbf{V}^{ij}$ is a symmetric tensor it can be expressed in an axis system in which all off-diagonal elements are zero, i.e. it can be diagonalized. This axis system is called the principal axis system of the field-gradient tensor and in this system the only non-zero elements are the diagonal components $\partial^2 V/\partial z^2$, $\partial^2 V/\partial y^2$, $\partial^2 V/\partial x^2$, where V is the electrostatic potential at the nucleus. Furthermore, it can be shown that the field-gradient tensor is traceless, i.e.

$$\frac{\partial^2 V}{\partial x^2} + \frac{\partial^2 V}{\partial y^2} + \frac{\partial^2 V}{\partial z^2} = 0.$$

Thus only two components are independent and conventionally the field-gradient tensor is expressed in terms of the two quantities

$$q \equiv \frac{\partial^2 V}{\partial z^2} \quad \text{and} \quad \eta \equiv \frac{(\partial^2 V/\partial x^2) - (\partial^2 V/\partial y^2)}{\partial^2 V/\partial z^2}$$

The second term is called the asymmetry parameter since it measures the deviation of the field-gradient tensor from axial symmetry.

It is evident that the interest in quadrupole coupling constants to the chemical physicist lies in the different values of the field-gradient for the same nucleus in different molecules. The field-gradient tensor is, as we have seen, characterized by the two values q and η but before it can be interpreted in terms of molecular electronic structure, it is necessary to know the orientation of the field-gradient principal axis system in the molecular framework. This requires the three further values α, β and γ representing the Eulerian angles between the two axis systems. Finally the sign of the nuclear quadrupole coupling constant—and hence the sign of the field-gradient—is required. This last point is important since many of the experimental methods only reveal the magnitude of the coupling constant, and while in many cases the sign is fairly obvious on theoretical grounds in others it is essential to have an experimental value.

The plan of this monograph is thus as follows. After a chapter giving the formal theory of nuclear quadrupole coupling constants, a chapter is devoted to a survey of the various experimental methods available for the determination of quadrupole coupling constants. Particular attention is given to the extent to which each yields the values of q, η, α, β, γ, and the sign of the coupling. Next is discussed the rigorous calculation of field-gradients from various molecular wave-functions and then the more approximate inverse process of going from an observed set of field-gradients to a molecular wave-function.

In the last half of the book are surveyed the experimental data for the various nuclei using simple methods of correlating the coupling constants with actual electron distributions. The aim is to show that even with these very imperfect methods of analysis the values of the nuclear quadrupole coupling constants can provide much information of interest to the student of molecular electronic structure and to incite the theoretician to analyse the coupling constants more exactly in particular cases. Equally important, the author has tried to show that nuclear quadrupole coupling constants provide the teacher with numbers which illustrate in a very striking way the electron distributions in simple molecules and bonds and the effects of the mutual interaction of two atoms or groups attached to a common centre. It is the author's hope that the material presented here will enable both organic and inorganic chemistry to incorporate the unique information given by nuclear quadrupole coupling constants into their undergraduate courses.

CHAPTER 2

THE THEORY OF NUCLEAR QUADRUPOLE INTERACTION

The electrostatic potential energy of a nucleus of finite dimensions in the field of its surrounding electrons is given classically by the expression

$$W_E = \int_{\tau_e}\int_{\tau_n} \frac{\rho_e(\mathbf{r}_e)\rho_n(\mathbf{r}_n)\,\mathrm{d}\tau_e\,\mathrm{d}\tau_n}{|\mathbf{r}_e - \mathbf{r}_n|} \tag{1}$$

where the symbols have the significance shown in Fig. 2.1.

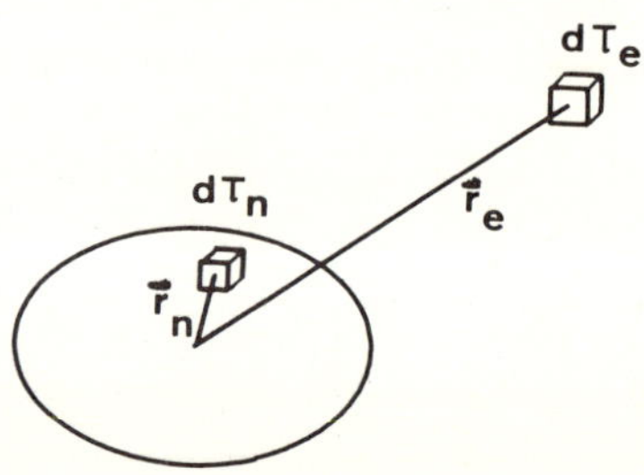

FIG. 2.1. Interaction between nuclear charge density (ρ_n, $\mathrm{d}\tau_n$) and electron charge density (ρ_e, τ_e).

The term $|\mathbf{r}_e - \mathbf{r}_n|$ can be expanded according to the well-known formula

$$\frac{1}{|\mathbf{r}_e - \mathbf{r}_n|} = 4\pi \sum_{l=0}^{\infty} \sum_{m=-l}^{m=l} \frac{1}{2l+1} \frac{r_<^l}{r_>^{l+1}} Y_l^{+m}(\theta_e, \phi_e)\, Y_l^{-m}(\theta_n, \phi_n) \tag{2}$$

where $r_<$ and $r_>$ signify the smaller and larger of the two radius vectors respectively and $Y_l^m(\theta, \phi)$ are normalized spherical harmonics of the nuclear (subscript n) or electronic (subscript e) coordinates respectively. The terms $r_e < r_n$ arise from electron penetration of the nucleus and give rise to the isotope shifts in the atomic spectra of heavy nuclei. Only s-electrons penetrate the nucleus and as will be seen later, contribute nothing to quadrupole effects, so for the present purposes the expression for electrostatic interaction may be factorized to give

$$W_E = \sum_l \sum_m N_l^m \,.\, E_l^{-m} \tag{3}$$

where

$$N_l^m = \sqrt{\frac{4\pi}{2l+1}} \int \rho_n(\mathbf{r}_n) r_n^l Y_l^m(\theta_n, \phi_n)\, d\tau_n \tag{4}$$

$$E_l^{-m} = \sqrt{\frac{4\pi}{2l+1}} \int \rho_e(\mathbf{r}_e) r_e^{-(l+1)} Y_l^{-m}(\theta_e, \phi_e)\, d\tau_e. \tag{5}$$

If the state of the nucleus, mass number A, is described by a wave function $\Psi(R_1 R_2 \ldots R_A)$ of the coordinates of its A nucleons then N_l^m is given by the expectation value of the Hamiltonian operator $\mathcal{N}_l^m$ over the nuclear wave function.

$$\mathcal{N}_l^m = \sqrt{\frac{4\pi}{2l+1}} \sum_j e_j R_j^l Y_l^m(\Theta_j . \Phi_j) \tag{6}$$

where e_j, the nucleonic charge, is e for a proton and zero for a neutron. Similarly E_l^m is the expectation value of the operator $\mathcal{E}_l^m$ over the electronic wave function where

$$\mathcal{E}_l^{-m} = e\sqrt{\frac{4\pi}{2l+1}} \sum_k r_k^{-(l+1)} Y_l^{-m}(\theta_k, \phi_k). \tag{7}$$

In these two expressions nuclear coordinates are given upper case symbols and electronic coordinates lower case symbols. The Hamiltonian for the electrostatic interaction is then

$$\mathcal{H}_E = \sum_m \sum_l \mathcal{N}_l^m . \mathcal{E}_l^{-m}. \tag{8}$$

Since the operators $\mathcal{N}_l^m$, $\mathcal{E}_l^{-m}$ transform in the same way as the spherical harmonics $Y_l^m(\theta, \phi)$ they are, by definition, tensor operators of order l. The expectation value of the operator $\mathcal{N}_l^m$ is called the nuclear multipole moment of order l.

It is well established experimentally that nuclear wave functions are either even or odd with respect to inversion. Therefore, since R_i is an even function of the nuclear coordinates the expectation value of $\mathcal{N}_l^m$ will be the integral of an odd or even function as $Y_l^m(\Theta, \Phi)$ is odd or even, i.e. as l is odd or even. Since integrals of odd functions are identically zero, only even nuclear multipole moments will be observable. In particular, nuclear electrostatic dipole moments will not occur and the first term in the summation giving rise to deviations from simple point charge interactions is the quadrupole term, $l = 2$. From the form of the operator it can be seen that the interaction energy varies approximately as $(R/r)^l$; a typical value for the nuclear radius is 10^{-12} cm, and for the radius of an electron cloud 10^{-8} cm. Since the electrostatic energy between a point charge and the electrons (the monopole term,

$l = 0$) is of the order of 10^5 cm^{-1}, the quadrupole term will be approximately 10^{-3} cm^{-1} (30 Mc/s) and the hexadecapole term ($l = 4$) 10^{-11} cm^{-1} or less than 1 c/s. It is, therefore, reasonable to concentrate our attention on the quadrupole term from now on.

The Wigner-Eckhardt theorem for tensor operators $\mathscr{A}_l^m$ shows that the matrix element between two states characterized by eigenvalues α of $\mathscr{A}_l^m$ and I and M of the angular momentum operators $\mathbf{I}^2$ and $\mathbf{I}_z$ is given by

$$(\alpha' I' M'|\mathscr{A}_l^m|\alpha I M) = A_l(I l I'|M m M') \tag{9}$$

where A_l are constants depending only on l and $(IlI'|MmM')$ are the Clebsch-Gordon coefficients for the coupling of angular momenta. From the properties of these coefficients it follows that the matrix element is zero unless $|I - I'| \leqslant l \leqslant |I + I'|$. In particular for diagonal elements, i.e. the expectation values of the operator $\mathscr{A}_l^m$ for a particular state, this condition implies that $l \leqslant 2I$ and hence multipole moments of order l are only observed if $l \leqslant 2I$. Thus quadrupole moments only occur for nuclei with spin angular momentum $\geqslant (h/2\pi)$. A similar treatment of magnetic interactions shows that only nuclei with spin angular momentum $\geqslant \frac{1}{2}(h/2\pi)$ can have a magnetic dipole moment. Hence all nuclei with quadrupole moments have magnetic moments.

It would be possible at this point to obtain the form of the quadrupole moment operator by the direct application of the Wigner-Eckhardt theorem and substitution of the appropriate Clebsch-Gordon coefficients. However, the derivation of the formula will be less esoteric if the components of $\mathscr{N}_2^m$ are written in cartesian coordinates

$$\mathscr{N}_2^{\pm 2} \equiv Q^{\pm 2} = \frac{\sqrt{6}}{4}\sum_j e_j(X_j \pm iY_j)^2 \tag{10}$$

$$\mathscr{N}_2^{\pm 1} \equiv Q^{\pm 1} = \frac{\sqrt{6}}{2}\sum_j e_j Z_j(X_j \pm iY_j) \tag{11}$$

$$\mathscr{N}_2^{0} \equiv Q^{0} = \frac{1}{2}\sum_j e_j[3Z_j^2 - (X_j^2 + Y_j^2 + Z_j^2)]. \tag{12}$$

The Wigner-Eckhardt theorem now permits the replacement of the cartesian coordinates by a constant multiple of the appropriate components of the angular momentum operator I.

$$Q^{\pm 2} \equiv Q'\frac{\sqrt{6}}{4}[\mathbf{I}_x \pm i\mathbf{I}_y]^2 \equiv Q'\frac{\sqrt{6}}{4}\mathbf{I}_\pm^2 \tag{13}$$

$$Q^{\pm 1} = Q'\frac{\sqrt{6}}{4}[\mathbf{I}_z(\mathbf{I}_x \pm i\mathbf{I}_y) + (\mathbf{I}_x \pm i\mathbf{I}_y)\mathbf{I}_z] \equiv Q'\frac{\sqrt{6}}{4}[\mathbf{I}_z\mathbf{I}_\pm + \mathbf{I}_\pm\mathbf{I}_z] \tag{14}$$

$$Q^0 = Q'\tfrac{1}{2}[3\mathbf{I}_z^2 - \mathbf{I}^2]. \tag{15}$$

Note that since the angular momentum operators do not necessarily commute we have symmetrized the expression for $Q^{\pm 1}$. The constant Q' which assimilates the summation and the values of e_j may be defined by requiring that the classical quadrupole moment ($\equiv eQ$) be equal to the expectation value of the maximum component $2Q_0$ in the quantum mechanically unrealisable situation, in which I_z lies along the spin axis.

$$eQ = Q'\langle\Psi_{I,I_z=I}|3\mathbf{I}_z^2 - \mathbf{I}^2|\Psi_{I,I_z=I}\rangle \tag{16}$$

$$= Q'[3I^2 - I(I+1)] \tag{17}$$

$$Q' = \frac{eQ}{I(2I-1)}. \tag{18}$$

Hence

$$Q^{\pm 2} = \frac{eQ}{I(2I-1)}\cdot\frac{\sqrt{6}}{4}\mathbf{I}_{\pm}^2 \tag{19}$$

$$Q^{\pm 1} = \frac{eQ}{I(2I-1)}\cdot\frac{\sqrt{6}}{4}[\mathbf{I}_z\mathbf{I}_{\pm} + \mathbf{I}_{\pm}\mathbf{I}_z] \tag{20}$$

$$Q^0 = \frac{eQ}{I(2I-1)}\cdot\tfrac{1}{2}[3\mathbf{I}_z^2 - \mathbf{I}^2]. \tag{21}$$

If the components of the electronic tensor are written in the same way it is seen for example that

$$\mathscr{E}_2^0 = \frac{e}{2}\sum_k\frac{(3Z_k^2 - r_k^2)}{r_k^6} \tag{22}$$

$$= -\frac{e}{2}\left(\frac{\partial^2 V}{\partial Z^2}\right)_{r=0} \equiv -\frac{e}{2}\mathbf{V}_{zz} \tag{23}$$

where $V(x, y, z)$ is the potential produced at the point x, y, z by the electrons. Note that $\mathbf{V}_{zz}$ is still an operator and the observable V_{zz} is its expectation value over the electronic wave function. Similarly,

$$\mathscr{E}_2^{\pm 1} - \pm\frac{e\sqrt{6}}{2}\sum_k\frac{Z_k(X_k \pm iY_k)}{r_k^5} \tag{24}$$

$$= \mp\frac{e\sqrt{6}}{6}(\mathbf{V}_{xz} \pm i\mathbf{V}_{yz}). \tag{25}$$

$$\mathscr{E}_2^{\pm 2} = \frac{e\sqrt{6}}{4}\sum_k(X_k^2 - Y_k^2 \pm 2iX_kY_k) \tag{26}$$

$$= \frac{e\sqrt{6}}{12}(\mathbf{V}_{xx}^2 - \mathbf{V}_{yy}^2 \pm 2i\mathbf{V}_{xy}). \tag{27}$$

The further development of the quadrupole coupling formula diverges into two paths depending on whether or not the atom or molecule as a whole possesses angular momentum. If it does not, for example in any phenomenon measured on a molecule in a solid, then the expectation value of $\mathbf{V}_{zz}$ becomes equal to the electronic field gradient V_{zz} in an arbitrary space-fixed co-ordinate system and the expectation values V_{zz} etc. can be directly substituted into equations (23), (25) and (27).

$$E_2^{\pm 2} = -\frac{e}{2\sqrt{6}}(V_{xx} - V_{yy} \pm 2iV_{xy}) \tag{28}$$

$$E_2^{\pm 1} = \pm\frac{e}{\sqrt{6}}(V_{xz} \pm V_{yz}) \tag{29}$$

$$E_2^0 = -\frac{e}{2}V_{zz}. \tag{30}$$

It is often convenient to have the operator expressed in Cartesian form and the following expression may be verified by multiplying out the various elements in the summation.

$$\mathscr{H}_Q = -\frac{e^2}{6}\sum_{r=x,y,z}\sum_{s=x,y,z}\left(\frac{\partial^2 V}{\partial r \partial s}\right)Q_{rs} \tag{31}$$

where

$$Q_{rs} = Q'\{\tfrac{3}{2}[\mathbf{I}_r\mathbf{I}_s + \mathbf{I}_s\mathbf{I}_r] - \delta_{rs}\mathbf{I}^2\}. \tag{32}$$

If the molecule or atom has quantized rotational angular momentum, J, then the expectation value of $\mathbf{V}_{zz}$ must be referred to the coordinate system defined by the axis of angular momentum. In this case by an argument which is exactly analogous to that used to relate the classical quadrupole moment to the quantum-mechanical quadrupole moment it may be shown that

$$\mathscr{E}_2^{\pm 2} = -\frac{\sqrt{6}}{4}\cdot\frac{eq_J}{J(2J-1)}\mathbf{J}_\pm^2 \tag{33}$$

$$\mathscr{E}_2^{\pm 1} = \pm\frac{\sqrt{6}}{4}\cdot\frac{eq_J}{J(2J-1)}[\mathbf{J}_z\mathbf{J}_\pm + \mathbf{J}_\pm\mathbf{J}_z] \tag{34}$$

$$\mathscr{E}_2^0 = -\frac{1}{2}\cdot\frac{eq_J}{J(2J-1)}[3\mathbf{J}_z^2 - \mathbf{J}^2] \tag{35}$$

where

$$q_J = \int \psi_e \left|\frac{3\cos^2\theta_J - 1}{r_e^3}\right| \psi_e^* \, d\tau$$

when the field-gradient is produced by an electron distribution described by a wave function ψ. The angle θ_J is measured with respect to the axis of angular momentum. The form of $\mathscr{H}_Q$ for a rotating molecule analogous to equation (31) for a stationary molecule may be easily verified to be

$$\mathscr{H}_Q = -\frac{1}{6}\sum_{rs} Q_{rs}(\nabla V_{rs})_J \tag{36}$$

where Q_{rs} is defined in equation (32) and

$$(\nabla V_{rs})_J = \frac{eq_J}{J(2J-1)}[\tfrac{3}{2}(\mathbf{J}_r\mathbf{J}_s + \mathbf{J}_s\mathbf{J}_r) - \delta_{rs}\mathbf{J}^2].$$

A useful form of this expression is

$$\mathscr{H}_Q = -\frac{1}{2}\frac{e^2Qq_J}{J(2J-1)I(2I-1))}[3(\mathbf{I}\,.\,\mathbf{J})^2 + \tfrac{3}{2}\mathbf{I}\,.\,\mathbf{J} - \mathbf{I}^2\mathbf{J}^2]. \tag{37}$$

It may be derived by multiplying out equation (36) and noting the following relationships

$$\sum_{rs}\mathbf{I}_r\mathbf{I}_s\mathbf{J}_r\mathbf{J}_s = \sum_r \mathbf{I}_r\mathbf{J}_r\sum_s \mathbf{I}_s\mathbf{J}_s = (\mathbf{I}\,.\,\mathbf{J})^2 \tag{38}$$

$$\sum_{rs}\mathbf{I}_r\mathbf{I}_s\delta_{rs}\mathbf{J}^2 = \sum_r \mathbf{I}_r^2\,.\,\mathbf{J}^2 = \mathbf{I}^2\mathbf{J}^2 \tag{39}$$

$$\sum_{rs}\mathbf{I}^2\,.\,\mathbf{J}_r\mathbf{J}_s\delta_{rs} = \mathbf{I}^2\mathbf{J}^2 \tag{40}$$

$$\sum_{rs}\delta_{rs}\mathbf{I}^2\mathbf{J}^2 = 3\mathbf{I}^2\mathbf{J}^2 \tag{41}$$

$$\sum_{rs}\mathbf{I}_r\mathbf{I}_s\mathbf{J}_s\mathbf{J}_r = \sum_{rs}\mathbf{I}_s\mathbf{I}_r\mathbf{J}_r\mathbf{J}_s = (\mathbf{I}\,.\,\mathbf{J})^2 + \mathbf{I}\,.\,\mathbf{J}. \tag{42}$$

All these relationships are self-evident except the last. This may be derived from the commutation rule for angular momenta

$$\mathbf{J}_r\mathbf{J}_s - \mathbf{J}_s\mathbf{J}_r = i\mathbf{J}_{r\times s'} \tag{43}$$

where $\mathbf{J}_{rxs}$ indicates the component of J perpendicular to $\mathbf{J}_r$ and $\mathbf{J}_s$, and the following manipulations

$$\sum_{rs}\mathbf{I}_r\mathbf{I}_s\mathbf{J}_s\mathbf{J}_r = \sum_{rs}\mathbf{I}_r\mathbf{I}_s\mathbf{J}_r\mathbf{J}_s + i\sum_{rs}\mathbf{I}_r\mathbf{I}_s\mathbf{J}_{r\times s} = (\mathbf{I}\,.\,\mathbf{J})^2, \tag{44}$$

but

$$\mathbf{I}_{rs} = \frac{\mathbf{I}_{rs} + \mathbf{I}_{sr} + i\mathbf{I}_{s\times r}}{2} \tag{45}$$

so

$$i\sum_{rs}\mathbf{I}_r\mathbf{I}_s\mathbf{J}_{r\times s} = \frac{1}{2}\sum_{rs}(\mathbf{I}_{rs} + \mathbf{I}_{sr})\mathbf{J}_{r\times s} - \frac{1}{2}\sum_{rs}\mathbf{I}_{r\times s}\mathbf{J}_{s\times r}. \tag{46}$$

The first term on the right-hand side is antisymmetric with respect to r and s, hence the summation vanishes, thus

$$i\sum_{rs} \mathbf{I}_r\mathbf{I}_s\mathbf{J}_{r\times s} = -\frac{1}{2}\sum_{rs} \mathbf{I}_{r\times s}\,.\,\mathbf{J}_{s\times r} = \frac{1}{2}\sum_{rs} \mathbf{I}_{r\times s}\mathbf{J}_{r\times s}$$

$$= \sum_t \mathbf{I}_t\mathbf{J}_t = \mathbf{I}\,.\,\mathbf{J}. \tag{47}$$

In the absence of strong external magnetic or electric fields the quadrupole operator (37) is diagonal in a coupled representation of the nuclear and molecular angular momenta

$$\mathbf{F} = \mathbf{I} + \mathbf{J} \tag{48}$$

$$\mathbf{F}^2 = \mathbf{I}^2 + \mathbf{J}^2 + 2\mathbf{I}\,.\,\mathbf{J} \quad (\mathbf{I} \text{ and } \mathbf{J} \text{ commute}) \tag{49}$$

i.e.

$$\mathbf{I}\,.\,\mathbf{J} = \tfrac{1}{2}\{\mathbf{F}^2 - \mathbf{I}^2 - \mathbf{J}^2\}. \tag{50}$$

The eigenvalues of the quadrupole operator are thus

$$E_Q = -\frac{e^2Qq_J}{2J(2J-1)I(2I-1)}\{\tfrac{3}{4}C(C+1) - I(I+1) - J(J+1)\} \tag{51}$$

where

$$C = [F(F+1) - I(I+1) - J(J+1)].$$

BIBLIOGRAPHY

Casimir, H. B. G., "On the Interaction between Atomic Nuclei and Electrons", Teylers Tweede Genootschap Haartem (1936); W. H. Freeman & Co., San Francisco (1963).
Ramsey, N. F., "Nucleaŕ Moments", Wiley & Sons, New York/London (1953).
Ramsey, N. F., "Molecular Beams", Oxford Univ. Press (1956).
Kopfermann, H., "Nuclear Moments", Academic Press, New York/London (1958).
Rose, M. E., "Elementary Theory of Angular Momentum", Wiley & Sons, New York/London (1957).
Tinkham, M., "Group Theory and Quantum Mechanics", McGraw-Hill (1964).

CHAPTER 3

THE EXPERIMENTAL DETERMINATION OF QUADRUPOLE COUPLING CONSTANTS

A. QUADRUPOLE COUPLING CONSTANTS IN ATOMS

1. Atomic Emission Spectroscopy

The electronic state of an atom is characterized by the value of its total angular momentum, $J\hbar$ where J can have half-integral values from zero upwards. It is also approximately characterized by the separate values $L\hbar$ and $S\hbar$ of its orbital and spin angular momentum, where L has integral values including zero and S had half-integral values including zero.

If the nucleus has a spin $I\hbar$ there is a magnetic interaction between the electronic angular momentum, J, and the nuclear magnetic moment, given by a Hamiltonian operator of the form

$$\mathscr{H} = a\,.\,\mathbf{I}\,.\,\mathbf{J}. \tag{52}$$

The first order perturbation energy produced by this operator is given by

$$E_m = \frac{a}{2}[F(F+1) - J(J+1) - I(I+1)]. \tag{53}$$

Each energy level E_J is therefore split into $(2J + 1)$ or $(2I + 1)$ components depending upon whether J or I is the smaller number.

Transitions between two electronic energy levels take place with the selection rule $\Delta F = 0, \pm 1$. From this it follows that the interval rule first proposed by Landé for the separation between atomic term multiplets is also valid here. Thus the ratio between the two separations of the triplet corresponding to a transition from a particular value of F_1 to the three allowed values of F_2 are in the ratio

$$\frac{F_1}{F_1 + 1}.$$

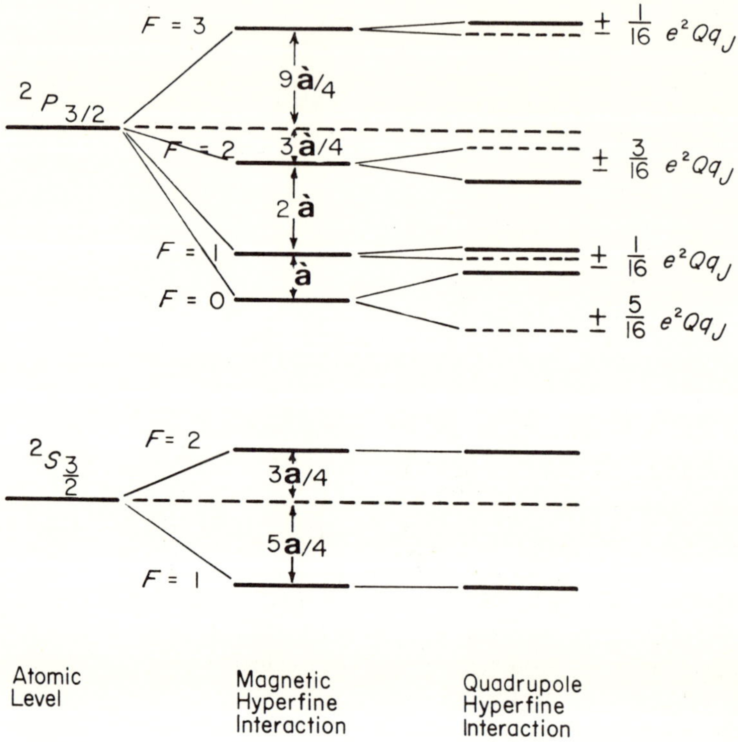

FIG. 3.1. Hyperfine structure for atomic $^2S_{\frac{1}{2}}$ and $^2P_{\frac{3}{2}}$ levels for nuclear spin $\frac{3}{2}$. The magnetic hyperfine constant, a, is assumed positive and the two possible cases e^2Qq positive (full lines) and e^2Qq negative (dotted lines) are shown. Note that the magnetic hyperfine coupling constant, A, is not in general the same in the $^2S_{\frac{3}{2}}$ state as in the $^2P_{\frac{3}{2}}$ state.

In the presence of quadrupole coupling the perturbing Hamiltonian has the form

$$\mathscr{H}_{\text{h.f.s.}} = \mathscr{H}_{I.J} + \mathscr{H}_Q \tag{54}$$

where $\mathscr{H}_Q$ is given by equation (37). Like $\mathscr{H}_{I.J}$ this is diagonalized in an F, M_F representation of electronic wave function and the first order perturbation energy becomes

$$E_{\text{h.f.s.}} = \frac{aC}{2} + \frac{e^2Qq_J}{2I(2I-1)\,.\,J(2J-1)}\left[\tfrac{3}{4}C(C+1) - I(I+1)J(J+1)\right] \tag{55}$$

where $$C = [F(F + 1) - I(I + 1) - J(J + 1)].$$

Thus in the presence of quadrupole coupling the interval rule is no longer valid and hence the quadrupole coupling can be determined from such deviations. These were first observed by Schuler and Schmidt in 1935[1] in the spectrum of Europium and constituted the first experimental demonstration of the existence of nuclear quadrupole moments. As will be seen later quadrupole coupling constants are largest for atomic $P_{3/2}$ states and the majority of quadrupole coupling measurements by atomic spectroscopy have been restricted to this case. Figure 3.1 illustrates the hyperfine structure of a level with $J = \frac{3}{2}$, $I = \frac{3}{2}$.

2. Atomic Beam Resonance Spectroscopy

The essentials of atomic and molecular beam resonance spectroscopy are the same. The brief description of the method given here will thus suffice for the needs of the present section and that of B **3** in this chapter where molecular beam resonance spectroscopy is discussed.

The essence of atomic or molecular beam experiments lies in the fact that in the absence of magnetic or electrostatic fields, molecules or atoms in the gas phase travel in straight lines between collisions. If the pressure of the gas is sufficiently low ($\sim 10^{-5}$–10^{-7} mm/Hg) the distance travelled between collisions is of the order of tens of centimetres. If the atoms or molecules have either electronic or nuclear magnetic moments their paths will be curved in the presence of an inhomogeneous magnetic field. If a molecule has a permanent electrostatic dipole moment its path will be similarly curved in an inhomogeneous electrostatic field. The original experiments of Stern and Gerlach measured the deflection of a beam of silver atoms in a magnetic field and showed that the electron spin angular momentum was quantized and that the spin angular momentum of the electron was $\hbar/2$. The actual value of the angular momentum could be obtained from a knowledge of the angle of deflection and the field strength. However, such measurements suffered from the difficulty of measuring small deflections and determining field-gradients.

A typical atomic or molecular beam resonance spectrometer is shown in Fig. 3.2.

In a sufficiently high vacuum atoms or molecules sublime from an appropriate source S and, in the absence of magnetic or electric fields, travel in straight lines without colliding with each other. Some of them will travel in a straight line from the source through the collimating slit E to a detector D. In the presence of an inhomogeneous field, A, (magnetic or electric) the molecules will follow a curved path, the degree of curvature depending on the magnetic moment (for a magnetic field) or electrostatic amount (for an electrostatic field).

It is evident that the strength and path length of the magnetic field necessary to provide a significant curvature must be much greater if the magnetic moment is purely nuclear than if it is electronic in origin. Thus for molecular beam magnetic resonance the total beam length will be 2–4 metres while for paramagnetic atomic beam magnetic resonance or polar molecular beam electric resonance the beam length is of the order of 0·5 metre.

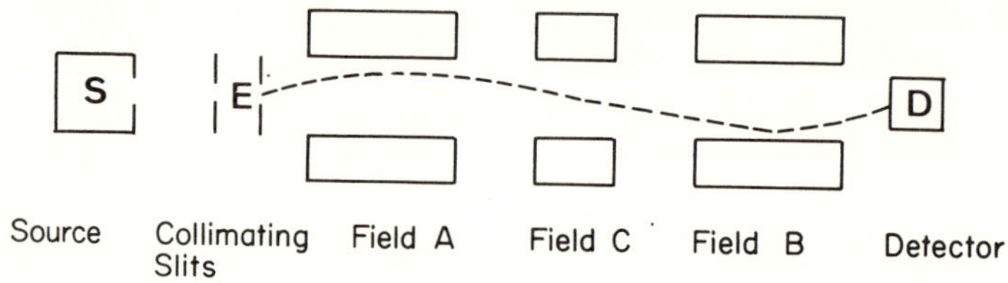

FIG. 3.2. Schematic drawing of an atomic or molecular beam resonance spectrometer.

In the presence of the inhomogeneous field, molecules leaving the source on the line *SED* will be deflected and will not pass through the collimating slit *E*. However, some molecules leaving the source at an angle to this line will be deflected so as to go through *E*. In the absence of any further field these molecules will not however reach the detector *D* but will continue in a straight line at an angle to *SED*. If now a second inhomogeneous field, a mirror image of the field *A*, is applied in the region *B* these molecules will be redeflected and, if the field strength is properly adjusted, reach the detector *D*. If the fields are exactly symmetric the condition for refocussing is that the state of the molecule is the same in the region *A* as in *B*. If in the region *C* a homogeneous field is applied, the molecule will exist in a number of states corresponding to the projection of its magnetic or electrostatic dipole on the *C*-field direction. If an oscillating magnetic or electric field, of an appropriate frequency to induce transition between these states, is now applied in the region *C* then some of the molecules will no longer be in the same state in *B* as in *A* and the number of molecules reaching the detector will decrease. A molecular or atomic beam resonance spectrum is then the plot of beam current against either frequency for a constant field strength or field strength for a constant frequency.

Atomic beam measurements are the most important means of obtaining quadrupole coupling constants of free atoms. These are of course, with perhaps the exception of the deuteron quadrupole coupling constants of light diatomic molecules, the most susceptible to exact interpretation. Indeed such measurements are usually an essential preliminary to the interpretation of coupling constants in molecules.

From what has gone before it is apparent that the interaction Hamiltonian for an atom in a magnetic field is given by

$$\mathscr{H} = g_J\beta\mathbf{J} \,.\, \mathbf{H}_0 + g_I\beta\mathbf{I} \,.\, \mathbf{H}_0 + a\mathbf{I} \,.\, \mathbf{J} + \frac{e^2Qq_J[3(\mathbf{I} \,.\, \mathbf{J})^2 + \frac{3}{2}\mathbf{I} \,.\, \mathbf{J} - \mathbf{I}^2 \,.\, \mathbf{J}^2]}{2I(2I - 1)J \,.\, (2J - 1)} \tag{56}$$

where $g_J\beta J$ = magnetic moment of the electrons, $g_I\beta I$ = magnetic moment of the nucleus, β = Bohr magneton. The first two terms in this Hamiltonian arise from the direct interaction of the electronic and nuclear magnetic moments respectively with the external magnetic field. The eigenfunctions of this Hamiltonian depend non-linearly on the strength of the magnetic field. At field strengths close to zero the nuclear and electronic spin are coupled and their resultant precesses around the field axis. Thus an F, M_F representation diagonalizes the interaction operator and the eigenvalues may be found quite straightforwardly as

$$E(F, m_F) = \beta g_F H_0 \,.\, M_F + \frac{a}{2}F(F + 1) + \frac{e^2Qq_J[\frac{3}{4}C(C + 1) - I(I + 1J(J + 1)]}{2I(2J - 1)J(2J - 1)} \tag{57}$$

where

$$g_F = \frac{1}{2F(F + 1)}\{g_I[F(F + 1) + I(I + 1) - J(J + 1)] + g_J[F(F + 1) + J(J + 1) - I(I + 1)]\}.$$

Transitions between these levels take place with the selection rule $\Delta F = 0, \pm 1, \Delta m_F = 0, \pm 1$. In general several transitions are observed and the results fitted to values of a, e^2Qq, g_I and g_J. The sign of the quadrupole coupling constant is revealed by the spectrum and in fact quadrupole coupling constants are usually determined from weak-field spectra.

At high magnetic fields the electronic angular momentum and nuclear spin precess separately about the direction of the applied field and an m, m_J representation becomes appropriate. The quadrupole Hamiltonian is not however diagonal in this representation but provided that the hyperfine splitting is large compared to the quadrupole splitting, first order perturbation theory may be used and only the diagonal elements considered. In this case we have

$$E(m, m_J) = \beta \,.\, g_I H_0 \,.\, m + \beta g_I H_0 \,.\, m_J + amm_J + \frac{e^2Qq_J}{4I(2I - 1)J(2J - 1)}[3m^2 - I(I + 1)][3m_J^2 - J(J + 1)]. \tag{58}$$

Transitions are observed with the selection rules

$$\begin{aligned} \Delta m_J &= \pm 1, \qquad \Delta m = 0 \\ \Delta m_J &= 0, \qquad \Delta m = \pm 1. \end{aligned} \tag{59}$$

In the absence of the quadrupole term the $\Delta m_J = \pm 1$ transition will for example consist of $2I + 1$ components corresponding to the different values of m, with a separation of a. The presence of the quadrupole term destroys the equality of the interval but since the quadrupole term depends on m^2 this perturbation is symmetrical about the centre of gravity of the multiplet and the sign of the coupling constant is not revealed.

3. Bibliography

Theory

Kopfermann, H., "Nuclear Moments", Academic Press, New York/London (1958).
Kuhn, H. G., "Atomic Spectra", Longmans, London (1962).
Ramsey, N. F., "Molecular Beams", Oxford University Press, (1956).

Experimental data

Lindgren, I., in "Perturbed Angular Correlations", (eds. E. Karlsson, E. Matthias and K. Siegbahn) North-Holland Publishing Co. 1964.

B. QUADRUPOLE COUPLING CONSTANTS IN GASEOUS MOLECULES

1. General Introduction

There are three general methods for the measurement of quadrupole coupling constants in gaseous molecules. The first of these, pure rotation spectroscopy, is of most general application and in favourable cases yields the complete quadrupole coupling tensor, including the sign of the components, and its orientation with respect to the molecular framework when this is not determined by the molecular symmetry. However, often only the diagonal elements of the tensor in the molecule's inertial framework are obtainable and in this case additional assumptions must be made if this tensor is to be transformed to the field-gradient principal axis system.

Molecular beam methods are mainly confined to diatomic molecules although recently there have been a few extensions to more complex systems. For diatomic molecules there is of course no problem as to the direction of the field-gradient axes and both molecular beam electric resonance and molecular beam magnetic resonance yield the sign of the coupling constant. The electric resonance method, like pure rotation spectroscopy, is confined to polar molecules; the magnetic resonance method has the advantage that homonuclear diatomic molecules can also be studied.

2. Pure Rotation Spectra

a. *General*

In classical mechanics the rotational behaviour of a three-dimensional solid body may be uniquely described in terms of that body's moments of inertia about three mutually perpendicular axes. The equations of motion have their simplest form when these axes are such that the corresponding moments of inertia have maximum values with respect to a small angular displacement of the axes: these are then the body's principal moments of inertia. Conventionally they are given the symbols $I_A < I_B < I_C$.

The rotational behaviour of a solid of molecular dimensions is also described in terms of these three quantities. From this point of view molecules may conveniently be divided into three classes: linear molecules $I_C \neq 0$, $I_A = I_B = 0$; symmetric tops having two of the moments of inertia identical and different from the third, linear molecules thus being special cases of symmetric tops; finally asymmetric tops where all three moments of inertia are different. A fourth class, the spherical top where all three moments of inertia are identical, is of little interest since all true spherical tops have no dipole moment and consequently do not exhibit rotational spectra. Accidental spherical tops can be treated as symmetric tops with the principal axis along the dipole moment axis.

The wave functions and energy levels of a molecular rotor can only be given in analytic form for symmetric tops and linear molecules. For this reason the rotational spectra of such molecules will be described first.

b. *Rotation spectra of linear molecules and symmetric tops*

The energy levels of a symmetric top (moments of inertia I_B perpendicular to the symmetry axis, I_A or I_C along the symmetry axis according to whether this moment is less than or greater than I_B) are given by the equations

$$E_{(J,K)} = hBJ(J+1) + h(C-B)K^2 \text{ (oblate top)} \tag{60}$$

$$E_{(J,K)} = hBJ(J+1) + h(A-B)K^2 \text{ (prolate top)} \tag{61}$$

where

$$A = \frac{h}{8\pi^2 I_A}, \qquad B = \frac{h}{8\pi^2 I_B}, \qquad C = \frac{h}{8\pi^2 I_C}$$

and J and K are integers, $J = 0, 1, 2 \ldots$ and $-J \leqslant K \leqslant J$.

The wave-functions corresponding to these levels have the form

$$\psi_{J,K,M} = e^{iM\phi}e^{iK\chi}\Theta_{JKM(\theta)} \qquad 62)$$

where θ, ϕ, χ are the Eulerian angles describing the position of a rotating body and $\Theta_{J,K,M(\theta)}$ is the confluent hypergeometric function. M is also an integral quantum number with values in the range $-J \leqslant M \leqslant J$ and corresponds to the projection of the total angular momentum on some external axis fixed in space. In the absence of such an axis, i.e. defined by an electrostatic or magnetic field, the energy levels are independent of M. Likewise the quantum number K corresponds to the projection of the total angular momentum on the molecular symmetry axis. Since the energy is independent of the sign of K each energy level is in all $2 . (2J + 1)$-fold degenerate.

Transitions between these levels can be induced by an oscillating electric field of the appropriate frequency according to the selection rule: $\Delta J = \pm 1$, $\Delta K = 0$. Thus the frequencies are given by

$$\nu_{J \to J+1} = 2B(J + 1). \tag{63}$$

For molecules having no permanent magnetic moment (and these of course constitute the overwhelming majority of molecules which can be studied by microwave spectroscopy) the effect of an electric field on the spectrum, the Stark effect, is useful in providing both means for the observation of microwave spectra and of assigning observed frequencies to particular transitions. We also discuss it here because of its relevance to molecular beam electric resonance spectroscopy. If a molecule has a component of its dipole moment along the direction of the angular momentum vector, and this is in general the case for symmetric tops, then application of an electrostatic field removes the degeneracy of the M levels and each level is split into $(2J + 1)$ components. In symmetric tops the component of the total dipole moment, μ, directed along the total angular momentum vector is $\mu K[J(J + 1)]^{-\frac{1}{2}}$, so that its component along the electric field is $\mu K . M[(J(J + 1))]^{-1}$ and the energy-splitting is

$$E_s' = \mu K . M . E[(J(J + 1))]^{-1} \tag{64}$$

where E is the field strength.

For spectrometers in which, as is usually the case, the electric field is applied parallel to the oscillating electric vector of the radiofrequency field, the selection rule for transitions between the levels is $\Delta J = \pm 1$, $\Delta K = 0$, $\Delta M = 0$. The frequency splitting is then given by

$$\Delta \nu' = \frac{2\mu K . M . E}{hJ(J + 1)(J + 2)}. \tag{65}$$

This is the first order Stark effect where the splitting is proportional to the field strength and of the order of 1·0 Mc/s. v^{-1} cm. Linear molecules have no component of their dipole moment along the angular momentum vector and thus do not exhibit a first order effect; however, strong fields induce an appropriate dipole and such molecules show a second order Stark splitting proportional to the square of the field strength and of the order of 0·002 Mc/s.

v^{-2} cm^2. It can be shown straightforwardly to be given by the formula

$$E_s^2 = \frac{\mu^2 E^2}{2Bh} \cdot \left\{ \frac{(J^2 - K^2)(J^2 - M^2)}{J^3(2J - 1)(2J + 1)} - \frac{[(J+1)^2 - M^2]}{(J+1)^3(2J+1)(2J+3)} \right\}. \tag{66}$$

In linear molecules $K = 0$ and this simplifies to

$$E_s^2 = \frac{\mu^2 E^2 \,.\, J(J+1) - 3M^2}{2hBJ(J+1)(2J-1)(2J+3)}. \tag{67}$$

The $\pm M$ lines remain degenerate and each line, therefore, has $(J + 1)$ components. The second order Stark effect is usually too small to be detected in the presence of a first order effect. Thus for symmetric tops it is usually seen only for levels with $K = 0$.

c. *Quadrupole effects in symmetric tops*

There is no advantage to be gained by treating the perfectly general case of quadrupole coupling in symmetric tops. The simplest case is one in which the quadrupolar nucleus lies on the symmetry axis of the molecule, in this case since a symmetric top necessarily has at least threefold symmetry about this axis the asymmetry parameter of the coupling tensor is identically zero. This choice of example has the further advantage that the quadrupole coupling in the linear molecule is then a particular case of it.

The Hamiltonian operator for quadrupole coupling, which will be treated as a perturbation, is given by equation (36) of Chapter 2. The first order quadrupole perturbation energy is thus given by equation (51) of Chapter 2. In the type of molecule under consideration the molecular symmetry axis, Z_m, coincides with the field-gradient principal axis. Thus

$$\frac{\partial^2 V}{\partial Z_m^2} = -2\frac{\partial^2 V}{\partial X_m^2} = -2\frac{\partial^2 V}{\partial Y_m^2} \tag{68}$$

and the field-gradient along the angular momentum vector, q_J, is given by

$$q_J \equiv \left\langle \frac{d^2 V}{dZ_J^2} \right\rangle = \left\langle \frac{d^2 V}{dZ_m^2} \,.\, \cos \theta M_J + \frac{d^2 V}{dX_m^2} \,.\, \sin \theta M_J \right\rangle$$

$$= \frac{d^2 V}{dZ_m^2} \left\langle \frac{3 \cos^2 \theta M_J - 1}{2} \right\rangle_{AV}. \tag{69}$$

The quantity in brackets can be obtained straightforwardly from the rotational wave function of a symmetric top (equation 62)

$$\left(\frac{3 \cos^2\theta M_J - 1}{2} \right)_{AV} = \frac{1}{2} \int \psi_{J,K,M=J}(3 \cos^2\theta - 1)\psi_{J,K,M=J}\, \mathrm{d}\tau \tag{70}$$

$$= \left[\frac{3K^2}{J(J+1)} - 1 \right] \frac{J}{2J+3}. \tag{71}$$

Thus the first order quadrupole coupling energy is given by

$$E_Q = -\frac{e^2Qq\{[3K^2/J(J+1)]-1\}}{2I(2I-1)(2J-1)(2J+3)}[\tfrac{3}{4}C(C+1)-I(I+1).J\ (J+1)]. \quad (72)$$

Transitions between quadrupole-split levels naturally normally occur without a change in the direction of nuclear spin. The selection rules are thus $\Delta J = 0, \pm 1, \Delta F = 0, \pm 1, \Delta I = 0$. Intensities are governed by the usual law for electric dipole transitions

$$\text{Intensity} \propto \left| \int \psi_L^* . e_Z . \psi_1 \, d\tau \right| \quad (73)$$

where ψ_2 and ψ_1 are the wave-functions of the two states connected by the radiation and e_Z is the electric dipole vector. In the present case this leads to the following relative intensities

Transitions

$$J - \rightarrow J$$

$$(F-1) \rightarrow F = \frac{B(J+F+I+1)(J+F+I)(J+F-I)(J+F-I-1)}{F} \quad (74)$$

$$F \rightarrow F = \frac{-B(J+F+I+1)(J+F-I)(J-F+1)(J-F-I-1)(2F+1)}{F(F+1)} \quad (75)$$

$$(F+1) \rightarrow F = \frac{B(J-F+I)(J-F+I-1)(J-F-I-1)(J-F-I-2)}{(F+1)}. \quad (76)$$

Tables of the values of the above complex expression for transitions $J \rightarrow J + 1$ for values of J from 0–10 and with nuclear spins of unity to 11/2 are given in Townes and Schawlow, Appendix 1, together with values for the relative intensities.

Figure 3 shows the energy level diagram of a linear molecule having a nucleus with spin $I = 1$ and $e^2Qq < 0$ for the levels $J = 1$ and $J = 2$ together with the theoretical spectrum for a transition between these levels. Note that the sign of the coupling is revealed by the structure of the spectrum, thus for the $J = 1$ to $J = 2$ transition a positive coupling constant would have yielded a spectrum which was the mirror image of that shown in Fig. 3.

The above expression for quadrupole coupling energy has been obtained by first order perturbation theory. When the quadrupole coupling constant is large, for example in certain iodine compounds where it can be of the order of 2000 Mc/s compared with a rotational transition of the order of 20,000 Mc/s, it is occasionally necessary to use second order theory.

$$E_Q^2 = \sum_{J'K'} \frac{\langle \psi_{IJKFM_F} | \mathscr{H}_Q | \psi_{IJ'K'FM_F} \rangle^2}{E_{JK} - E_{J'K'}}. \quad (77)$$

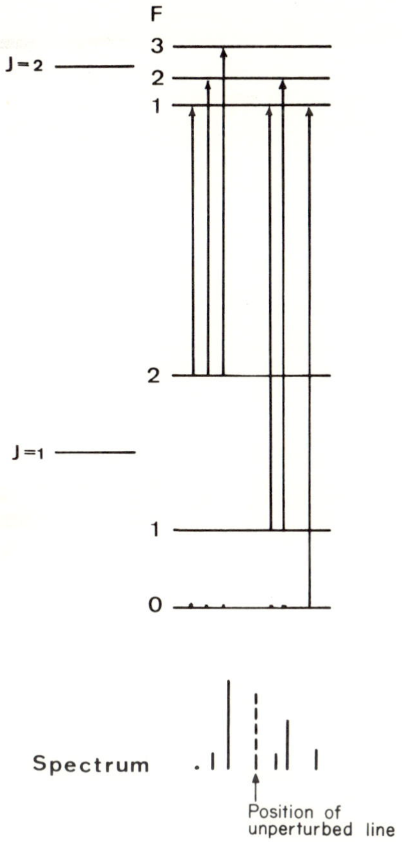

FIG. 3.3. Quadrupole fine structure in levels $J = 1$, and $J = 2$ for $I = 1$ and $e^2Qq_J < 0$.

The quadrupole coupling connects levels with $\Delta J = \pm 1, \pm 2$. The application of second order theory is quite straightforward and its effect is to modify slightly the separation between the lines. It will not be discussed further here for symmetric tops.

It has so far been assumed that the quadrupolar nucleus lies on the symmetry axis of the molecule. If it lies off the axis (e.g. $CHCI_3$) then at least two other identical nuclei are present and the evaluation of the quadrupole coupling interaction becomes much more complex.

d. *Spectra of asymmetric tops*

An asymmetric top has three distinct moments of inertia perpendicular to each other $I_A < I_B < I_C$. Even in the absence of quadrupole interaction it is not possible to give analytic expressions for the rotational wave functions and energies of the general asymmetric top.

The total angular momentum and its projection on an external axis ($J + M$) are constants of the motion, and it is convenient and conventional to designate a particular rotational state by the further "quantum numbers" K_1 and K_{-1} which are the values of K in the corresponding limiting oblate ($I_B = I_A$) and prolate ($I_B = I_C$) symmetric tops. Qualitatively the difference between symmetric and asymmetric tops is that the $\pm K$ levels are no longer degenerate, moreover all transitions ΔK_{-1} and ΔK_1 are allowed with the restriction that they cannot both be even. For slightly asymmetric prolate and oblate tops respectively, the energy levels are given by the formulae

$$E_{pr} = h\left(\frac{B + C}{2}\right)J(J + 1) + h\left(A - \frac{B + C}{2}\right)\omega \tag{78}$$

$$E_{ob} = h\left(\frac{A + B}{2}\right)J(J + 1) + h\left(C - \frac{A + B}{2}\right)\omega. \tag{79}$$

ω is given by a series expansion

$$\omega = K + c_1 b + c_2 b^2 + c_3 b^3 \ldots \tag{80}$$

where b is the rotational asymmetry parameter defined for an almost prolate and almost oblate top respectively as

$$b_{pr} = \frac{C - B}{2A - B - C} \tag{81}$$

$$b_{ob} = \frac{A - B}{2C - B - A}. \tag{82}$$

The coefficients c_1, c_2, c_3 are listed in Townes and Schawlow, Appendix III.

The spectra of asymmetric tops are thus very complex and the Stark effect is an essential tool in their analysis since it permits the assignment of K_{-1} and K_1.

Analytic expressions are not possible for the rotational wave functions of an asymmetric top, but since symmetric top wave functions form a complete set it is convenient to express the asymmetric top functions in terms of these. Moreover, since J and M are good quantum numbers for asymmetric tops, the expansion reduces to

$$\psi_{J,K_{-1},K,M} = \sum_K a_{JKM}\psi_{JKM}. \tag{83}$$

The choice of the symmetric top functions will of course be governed by whether the molecule is more closely approximated by an oblate or prolate top. Although it is possible to reduce the number of terms in this expansion even further by means of the symmetry properties of the wave functions, the evaluation of the coefficients a_{JKM} is sometimes troublesome. An analytical treatment of the asymmetric top is given by Burkhard and Brittin.[2]

e. *Quadrupole coupling in asymmetric tops*

The first order quadrupole interaction energy in an asymmetric top is of course still given by equation (51) of Chapter 2. In the absence of an analytical form for the rotational wave functions, however, the problem lies in the evaluation of q_J,

$$q_J = \int \psi_{J,K_{-1},K,M}\left(\frac{\partial^2 V}{\partial Z_J^2}\right)\psi^*_{J,K_{-1},K,M}\,\mathrm{d}\tau. \tag{84}$$

Moreover, in general the field gradient principal axes do not coincide with the axes of the principal moments of inertia. If $\psi_{J,K_{-1},K,M}$ is expanded in a series of symmetric top functions, then it can be shown by a straightforward extension of the methods used in deriving equation (69) that

$$q_J = \frac{q_m}{(J+1)(2J+3)}\sum_K a_{JK}^2[3K^2 - J(J+1)] - \eta_M \sum_K a_{J,K}\cdot a_{J,K+2}\{[J^2-(K+1)^2][(J+1)^2-(K+1)^2]\}^{\frac{1}{2}} \tag{85}$$

where

$$q_m = \frac{\partial^2 V}{\partial Z_m^2} \tag{86}$$

$$\eta_M = \frac{(\partial^2 V/\partial X_m^2) - (\partial^2 V/\partial Y_m^2)}{(\partial^2 V/\partial Z_m^2)} \tag{87}$$

and Z_m is the principal moment of inertia axis which most nearly approximates the symmetry axis of the corresponding symmetric top from which the functions $\psi_{J,K_{-1},K_1,M}$ have been derived. Note the difference between the field-gradient asymmetry parameter defined here and the one appropriate to the axes of the field-gradient coordinates system, η, which necessarily lies within the limits $0 \leqslant \eta \leqslant 1$. This distinction is unfortunately not always made clear in the earlier literature.

Other methods of evaluation q_J have been employed and are discussed in Townes and Schawlow. All these methods, however, show that the first order quadrupole coupling energy depends on the components of the field-gradient along the axes of the principal moments of inertia. A computer programme for the diagonalization of the complete rotational plus quadrupole Hamiltonian has been described[3] together with formulae for the various matrix elements.

In the absence of further information it is impossible to determine from the coefficients q_a, q_b, q_c the values of the components q_z, q_x, q_y. When, however, the quadrupolar nucleus is singly bonded to the molecule, e.g. the halogens, experience has shown that the principal axis of the field-gradient system coincides almost exactly with the bond direction. This assumption then allows the determination of q_z by an appropriate rotation of coordinates.

After this rotation, it will either be found that $q'_x = q'_y$, in which case the field-gradient x and y axes are indeterminate, or the molecular symmetry, e.g. vinyl chloride, will enable an appropriate choice of the x and y axes to be made. When multiply bonded nuclei are under investigation similar assumptions are also sometimes possible.

In favourable cases, it is possible to determine the directions of the field-gradient axes by isotopic substitution. For example[4] in the molecules $HONO_3$ and $DONO_3$ the field-gradient axes lie in the same direction and the components q_z, q_x, q_y are essentially identical. The directions of the axes of the principal moments of inertia are, however, rather different and can be calculated from the known isotopic masses, bond lengths and angles. The components of the field-gradient along these axes, as yielded by the quadrupole hyperfine structure, will also therefore be different and the combination of these two sets of data allows the direct determination of q_z and η as well as the orientation of the field-gradient axes in the molecular framework. It will be appreciated, however, that this method is only successful if isotopic substitution produces an appreciable rotation of the inertial framework.

The second order quadrupole interaction energy is given by

$$E_Q^2 = \sum_{J',K'_{-1},K_1} \frac{\langle \psi_{I,J,K_{-1},K_1,F} | \mathscr{H}_Q | \psi_{I,J',K'_{-1},K'_1,F} \rangle^2}{W_{JK_{-1},K_1} - W_{J'K'_{-1}K'_1}}. \tag{88}$$

These matrix elements are non-zero only if $J = J' \pm 1$ or $J' \pm 2$. They have been evaluated by Bragg[5] and it can be shown that second order quadrupole interaction involves terms off-diagonal in the field-gradient referred to the inertial axes, i.e. $(\partial^2 V/\partial X_a \partial Y_b)$ etc.

If these terms are known the appropriate rotation to the principal axis system, i.e. off-diagonal elements are zero, can then be made and q_z, η and the orientation of the principal axes in the molecular framework determined. However, the determination of the off-diagonal terms necessitates a significant contribution from second-order effects in the quadrupole coupling. This method was used by Flygare and Gwinn[6] to show that in methylene dichloride the z principal axis follows the C—Cl bond to within $\pm 1°$.

f. *Summary*

In the most favourable cases all the necessary information about the quadrupole coupling tensor can be obtained from the quadrupole hyperfine structure of the pure rotational spectrum. It should be noted, however, that the precision with which the components can be measured is much smaller than that obtainable by quadrupole resonance. However, as will be seen later, this apparent superiority of the pure quadrupole resonance method is illusory.

3. Molecular Beam Resonance Spectroscopy

a. *General*

The principle of the experimental method of molecular beam resonance spectroscopy is essentially the same as that of atomic beam resonance spectroscopy (A.**2**) but since molecules, unlike atoms, can have permanent electrostatic dipole moments the deflecting fields can be electrostatic as well as magnetic. There are thus two types of molecular beam resonance spectra, i.e. magnetic resonance and electric resonance, and they will be discussed separately. Both methods are of course concerned with isolated molecules in the gaseous state, consequently the rotational wave-functions described above are an appropriate starting point for the development of this subject. Since however molecular beam resonance methods are capable of detecting much smaller energy differences than pure rotation spectroscopy a number of additional small terms which contribute to the Hamiltonian operator leading to the wave-function of a rotating molecule must also be introduced. Indeed it is this increased resolving power, leading to the more accurate determination of hyperfine coupling constants, which constitutes one of the principal interests of these methods.

Since a molecule is made up of charged particles, molecular rotation generates a magnetic field. Of course rotation of the positive nuclei produces a field opposite in sign to that produced by the electrons, but even in non-polar molecules the average distance of the electrons from the molecular center of gravity is not usually the same as the average distance of the nucleons so that there is usually a resultant small but finite magnetic field. Thus a rotating molecule, even if it is diamagnetic in the usual sense, is associated with a small magnetic moment, μ_R, both of which of course will be directed along the axis of rotational angular momentum and proportional to the rotational quantum number. Thus

$$\begin{aligned} \mu_J &= \mu_R \mathbf{J} \\ &\equiv g_J \mu_N \mathbf{J}. \end{aligned} \tag{89}$$

This latter step is merely a convenient change of scale to show the relationship between nuclear moments and rotational moments. For the lightest diatomic molecule, H_2, g_J is approximately unity but decreases markedly as the nuclear masses increase. If the molecule contains a magnetic nucleus there is an interaction between the rotational magnetic moment and the nuclear magnetic moment which will have the form

$$\mathscr{H}_{IJ} = \mathrm{c}\,.\,I\,.\,J. \tag{90}$$

In addition to this term, if the molecule possesses more than one magnetic nucleus there is the possibility of coupling between them. This of course is the spin–spin coupling familiar to nuclear magnetic resonance. It consists of two parts: a direct dipole–dipole interaction which is anisotropic but which does *not* in general average to zero over rotational motion in a well-defined rotational state, and the isotropic electron-coupled interaction. Usually this last term is an order of magnitude greater than the rotation-averaged anisotropic term and both of them are negligible unless the two nuclei under study are directly attached to each other by a chemical bond.

The presence of these terms has the unfortunate result of making the formulae for the energy levels extremely bulky and cumbersome although their inclusion is quite straightforward from the theoretical point of view.

b. *Molecular beam electric resonance spectra*

The discussion of these spectra follows naturally upon that of pure rotation spectra since the phenomenon observed is usually a direct transition between the Stark-levels of the pure rotation spectrum without a change in rotational state, i.e. $\Delta J = 0$, although occasionally $\Delta J = \pm 1$ transitions are also studied. The method is complementary to that of pure rotation spectroscopy since in the study of such molecules as the diatomic alkali halides whose vapour pressure is so low the more conventional method is not sensitive enough. Indeed, for reasons which will become clear shortly, its application is almost entirely limited to diatomic molecules.

For a diatomic polar molecule of dipole moment μ, having two nuclei with spin I_1 and I_2 respectively, the complete Hamiltonian operator describing its behaviour in an electrostatic field, E, may be written

$$\mathscr{H} = B\mathbf{J}^2 - \mu\,.\,\mathbf{E} - e^1q_1Q_1\frac{[3(\mathbf{I}_1\mathbf{J})^2 + \frac{3}{2}(\mathbf{I}_1\mathbf{J}) - (\mathbf{I}_1^2\mathbf{J}^2)]}{2I_1(2I_1 - 1)(2J - 1)(2J + 3)}$$
$$- e^2q_2Q_2\frac{[3(\mathbf{I}_2\,.\,\mathbf{J})^2 + \frac{3}{2}(\mathbf{I}_2\,.\,\mathbf{J}) - (\mathbf{I}_2^2\,.\,\mathbf{J}^2)]}{2I_2(2I_2 - 1)(2J - 1)(2J + 3)} + c_1\mathbf{I}_1\,.\,\mathbf{J} + c_2\mathbf{I}_2\,.\,\mathbf{J}$$
$$+ \frac{g_1g_2\mu_N^2}{R^3}\frac{[3(\mathbf{I}_1\,.\,\mathbf{J})(\mathbf{I}_2\,.\,\mathbf{J}) + 3(\mathbf{I}_2\,.\,\mathbf{J})(\mathbf{I}_1\,.\,\mathbf{J}) - 2(\mathbf{I}_1\,.\,\mathbf{I}_2)\mathbf{J}^2]}{(2J - 1)(2J + 3)} + c_3\mathbf{I}_1\,.\,\mathbf{I}_2. \tag{91}$$

The origin of these terms will be clear from the discussion of quadrupole coupling in linear molecules in section B.2b and the foregoing discussion on magnetic interactions. Thus the penultimate term is the rotationally-averaged dipole-dipole interaction while the final term is the isotropic spin–spin interaction.

It is clear that the diagonalization of this Hamiltonian is not something which can be undertaken lightly. For the present purposes, however, where we only wish to illustrate the use of the method in determining quadrupole coupling constants, we may limit our attention to the case when only one nucleus has a quadrupole moment and where all the magnetic interaction terms are neglected. This naturally results in a very considerable simplification and in the absence of an electrostatic field the Hamiltonian is diagonalized by the usual coupled representation and the energy levels given by equation (51) of Chapter 2. The coupling is of course between the rotational angular momentum and the spin of the quadrupolar nucleus. Since we are dealing with linear molecules, which have no first order Stark effect, the electrostatic field splits these levels into doubly-degenerate sets corresponding to the value of $|m_F|$; at high field strengths the rotational angular momentum is completely uncoupled from that of the quadrupolar nucleus and both angular momentum vectors precess independently about the direction of the applied field. For the electrostatic fields used in electric resonance experiments it is usually necessary to include third and fourth order terms in the perturbation calculation of eigenfunctions of $\boldsymbol{\mu} . \mathbf{E}$. The effect of these terms is to make the energy of terms $m_J \neq 0$ decrease as the field strength increases rather than the uniform increase predicted by second order theory (equation 66). At sufficiently high field strengths even the $m_J = 0$ levels begin to level off and ultimately decrease. When the rotational and nuclear angular momenta are precessing independently the levels fall into groups corresponding to the various values of $|m_J|$. In the high field limit the levels with $|m_J| \neq L$ are split into sublevels by the quadrupole hyperfine interaction corresponding to the various values of $|m|$. The levels $m_J = 1$ have twice as many substrates, essentially owing to the fact that the quadrupole Hamiltonian connects levels with $\Delta M = 2$. Thus for the level $J = 1$ for a molecule having a quadrupolar nucleus with spin $I = \frac{3}{2}$ and $e^2Qq > 0$, the energy-level diagram has the form shown in Fig. 4 (weak field).

At high fields the selection rule is

$$\Delta(m + m_J) \equiv \Delta M = 0, \pm 1 \tag{92}$$

and the high-field spectrum is shown in Fig. 4 (strong field). For a negative coupling constant the spectrum would be the mirror image of that shown in Fig. 4 (strong field). Thus the molecular beam electric resonance method reveals both the sign and magnitude of the coupling constant. Since for the diatomic molecules to which it is applied the directions of the field-gradient axes are defined by the molecular symmetry, the field-gradient tensor is here now completely determined. Formulae for the energy levels of the Hamiltonian of equation (91) may be found in Hughes and Grabner[7] as well as in the monographs by Ramsey and by Kopfermann (see Bibliography).

In addition to the radiofrequency spectra between Stark-split levels described above, i.e. $\Delta I = 0$, molecular beam microwave spectra, $\Delta J = 1$, are also investigated. These measurements yield the moment of inertia of the molecule usually with greater accuracy than is provided by simple rotational spectroscopy.

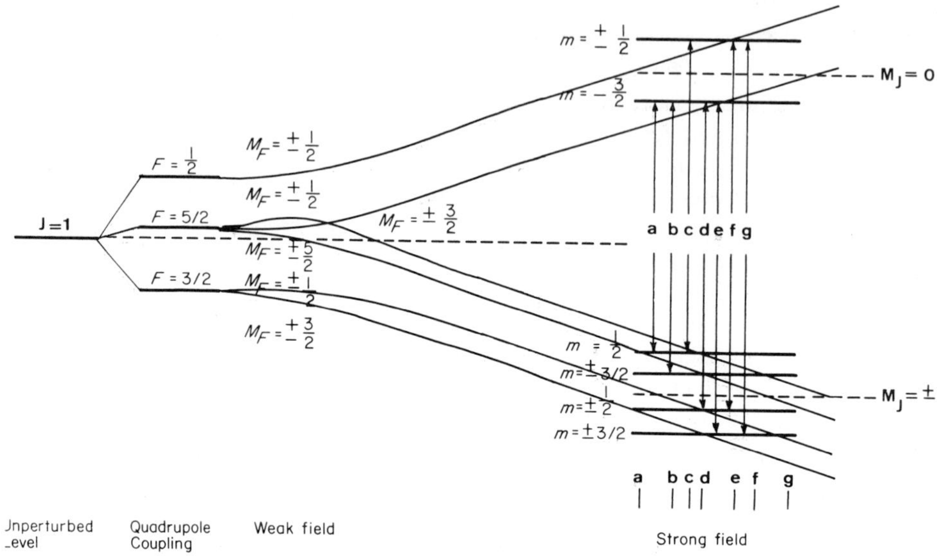

FIG. 3.4. Stark splitting of a level $J = 1$ for a linear molecule and the corresponding molecular beam electric resonance transitions.

The resolution of molecular beam spectrometers is such that in general data are obtained for each populated vibrational state of the molecule. In particular the quadrupole coupling constants of the alkali halides often vary markedly with vibrational quantum number and of course any analysis of these parameters must account for this phenomenon.

A particular advantage of the electric resonance method over the magnetic resonance method is that since molecular electrostatic dipole moments are quite large a suitable arrangement of the deflecting fields, collimating slits and stops in the spectrometer can usually lead to the selection of just one rotational substate, $(J, |m_J)$ for study. This possibility has been turned to advantage in the so-called *Maser* where rotationally selected molecules undergo induced transitions to a lower energy state with emission of microwave radiation of the corresponding frequency. With suitable arrangements of the stimulating and emitted vibrations this leads either to amplification of the stimulating frequency or sustained oscillations at that frequency. Apart from its considerable intrinsic interest this phenomenon is of importance in the present connection since the method has been adapted to the determination of the very small coupling constants of deuterium atoms in simple molecules.[8] For the details of this, however, the reader is referred to the original papers.

c. *Molecular beam magnetic resonance*

(1) *General Introduction.* As we saw above all molecules possess rotational moments and all those molecules of interest to us here have nuclear moments. They are thus all deflected by magnetic fields but since the magnetic moments are small the deflections are small and even with the pathlengths of one or more meters which are common in molecular beam magnetic resonance spectrometers, it is not usually possible to select particular rotational states for observation. For light molecules such as H_2 this is no problem since the magnitude of the rotational quantum is sufficiently great that only the lowest rotational states are appreciably populated at room temperature. For heavy molecules, however, where in addition high temperature may have to be used to achieve appreciable molecular concentrations in the gas phase, the rotational states up to $J = 100$ or more may be populated and the resulting spectra represent averages over the rotational states. The outstanding advantage of the magnetic resonance method is however that it yields values of the hyperfine interactions in non-polar molecules which are of course not accessible to study by pure rotation spectroscopy or molecular beam electric resonance spectroscopy. Oddly enough the molecular beam magnetic resonance method historically preceded the electric resonance method and was originally used mainly in the study of heteronuclear molecules such as the alkali halides. Since however the electric resonance method gives results of a much higher precision its use nowadays is mainly confined to the homoatomic case.

(2) *Molecular beam resonance in diatomic molecules.* The Hamiltonian operator for an isolated diatomic molecule in a magnetic field is given by

$$\begin{aligned}\mathscr{H} = {} & B\mathbf{J}^2 - e^2q_1Q_1\left\{\frac{3(\mathbf{I}_1 \, . \, \mathbf{J})^2 + \frac{3}{2}(\mathbf{I}_1 \, . \, \mathbf{J}) - \mathbf{I}_1^2 \, . \, \mathbf{J}^2}{2I_1(2I_1 - 1)(2J - 1)(2J + 3)}\right\} \\ & - e^2q_2Q_2\left\{\frac{3(\mathbf{I}_2 \, . \, \mathbf{J})^2 + \frac{3}{2}(\mathbf{I}_2 \, . \, \mathbf{J}) - \mathbf{I}_2^2 \, . \, \mathbf{J}^2}{2I_2(2I_2 - 1)(2J - 1)(2J + 3)}\right\} \\ & + \mu_N g_1 \mathbf{H} \, . \, \mathbf{I}_1 + \mu_N g_2 \mathbf{H} \, . \, \mathbf{I}_2 + \mu_N g_J \mathbf{H} \, . \, \mathbf{J} + c_1 \mathbf{I}_1 \, . \, \mathbf{J} + c_2 \mathbf{I}_2 \, . \, \mathbf{J} \\ & + \frac{g_1 g_2 \mu_N^2}{R^3} \frac{[3(\mathbf{I}_1 \, . \, \mathbf{J})(\mathbf{I}_2 \, . \, \mathbf{J}) + 3(\mathbf{I}_2 \, . \, \mathbf{J})(\mathbf{I}_1 \, . \, \mathbf{J}) - 2(\mathbf{I}_1 \, . \, \mathbf{I}_2)\mathbf{J}^2]}{(2J - 1)(2J + 3)} + c_3 \mathbf{I}_1 \, . \, \mathbf{I}_2. \end{aligned} \tag{93}$$

If a homonuclear molecule is under study then $\mathbf{I}_1 = \mathbf{I}_2$ and although the above Hamiltonian retains its validity it is convenient to express it in terms of

$\mathbf{I}_R = \mathbf{I}_1 + \mathbf{I}_2$. With this transformation the quadrupole term in the Hamiltonian above becomes

$$\mathscr{H}_Q = \frac{e^2Qq}{2I(2I-1)}\left[1 - \frac{I_R(I_R+1) + 4I(I+1)}{2I_R(2I_R+3)}\right] \left[\frac{3(\mathbf{I}_R \cdot \mathbf{J})^2 + \frac{3}{2}(\mathbf{I}_R \cdot \mathbf{J}) - \mathbf{I}_R^2 \cdot \mathbf{J}^2}{(2J-1)(2J+3)}\right] \tag{94}$$

and the rotationally-averaged anisotropic spin–spin coupling becomes

$$\mathscr{H}_{\text{aniso}} = \left(\frac{g_I^2\mu_N^2}{r^3}\right)\left[\frac{I_R(I_R+1) + 4I(I+1)}{2I_R(2I_R+3)}\right]\left[\frac{3(\mathbf{I}_R \cdot \mathbf{J})^2 + \frac{3}{2}(\mathbf{I}_R \cdot \mathbf{J}) - \mathbf{I}_R^2 \cdot \mathbf{J}^2}{(2J-1)(2J+3)}\right]. \tag{95}$$

Both these terms are formally similar, having the same dependence on $\mathbf{I}_R$ and $\mathbf{J}$, hence they are not distinguished and only their sum can be derived in a molecular beam magnetic resonance experiment.

With the present-day knowledge of interatomic distances and nuclear magnetic moments it is of course possible to calculate the anisotropic spin–spin coupling term quite exactly and hence deduce the quadrupole coupling. In fact historically it was because the measurements on H_2 and D_2 did not give corresponding results for the spin–spin coupling term that the quadrupole moment of the deuteron was first discovered.[9] In most other molecules, however, the spin–spin coupling term is much less than that due to the quadrupole coupling and it is usually neglected.

As already stated most molecules other than H_2 in a molecular beam exist in all rotational states up to quite high quantum numbers. In this case and in the limit of high magnetic fields the two identical nuclei of a homonuclear molecule effectively precess independently about the magnetic field vector and the spectrum of a homonuclear molecule becomes essentially the same as that of a heteronuclear molecule. In high field resonance transitions are usually observed corresponding to $\Delta m = \pm 1$ with no corresponding change in M_J. Attention may therefore be here confined to that part of the Hamiltonian dependent on m and the following expression for the m-dependent energy levels may be obtained:

$$E_{(m)} = ng_1\beta H - e^2Qq\,\frac{(3m_J^2 - J(J+1))[3m^2 - I(I+1)]}{4I(2I-1)(2J+3)(2J-1)}. \tag{96}$$

This of course neglects the term due to internuclear coupling. For high values of J this expression is approximated by

$$E_{(m)} = mg_I\beta H - \frac{e^2Qq(3\cos^2\alpha - 1)(3m^2 - 1)}{16I(2I-1)} \tag{97}$$

where $\cos\alpha = [m/J(J+1)]$. For a particular value of $\cos\alpha$ there are thus $2I$ transitions with $\Delta M_I = \pm 1$. For example, with $I = \frac{3}{2}$ there is a central line at $g_I\beta H$ flanked by two satellites separated from the central resonance by an energy difference $(3\cos^2\alpha - 1)/8 \,.\, e^2Qq$. Since there are a large number of different values of J and M_I, $\cos\alpha$ may be treated as a continuous variable. In this way the envelope of all transitions may be calculated and can be shown[10] to have maxima in the region of $\cos\alpha = 1$. Thus for $I = \frac{3}{2}$ and $I = 1$ the line shapes are shown in Fig. 5. The actual line shapes are of course broadened by the spin–spin and spin-rotation terms and the discontinuities at the wings are not observed. These line-shapes are essentially the same as those observed in the nuclear magnetic resonance spectra of powders exhibiting quadrupole coupling.

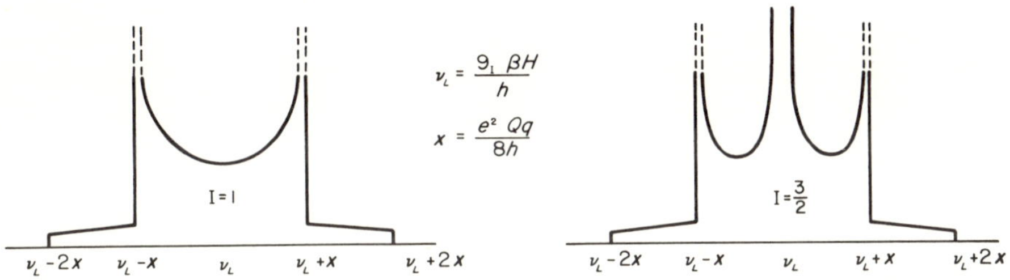

FIG. 3.5. Line shapes of molecular beam magnetic resonances.

(3) *The sign of the coupling constant.* Since the quadrupole-broadened spectrum is symmetrical about $g_I\beta H$ it would appear at first sight that the sign of the quadrupole coupling constant could not be revealed by a molecular beam magnetic resonance experiment. To determine the sign it is necessary to determine which satellite is due to the transition $\frac{3}{2} \to \frac{1}{2}$ and which to the transition $-\frac{3}{2} \to -\frac{1}{2}$. In a molecular beam magnetic resonance experiment it is possible to do this mechanically.[11] Figure 6 shows schematically the paths of molecules with positive and negative magnetic moments through the spectrometer. Insertion of an asymmetric obstacle thus cuts out molecules with positive or negative moment, depending on the position of the obstacle and the directions of the magnetic fields. Consequently the satellite corresponding to the molecules with the sign of moment which has been eliminated from the beam will be removed. Since rotational magnetic moments are small the sign of the molecular moment is determined by the sign of the nuclear moments within it. The nuclear moment is given by

$$\mu = mg_I\beta. \tag{98}$$

Hence this experiment gives the relative values of g_I and e^2Qq and thus the absolute value of e^2Qq if g_1 is known, as it usually is, from a separate source.

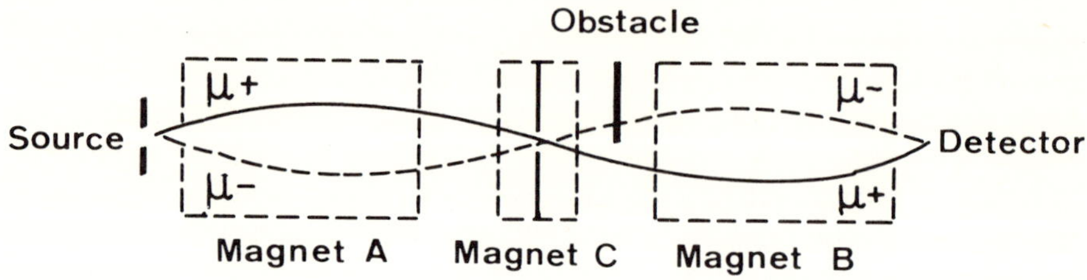

FIG. 3.6. Determination of the sign of the quadrupole coupling constant in molecular beam magnetic resonance.

It should be noted that the above theory has only been developed for $e^2Qq \ll g_I\beta H$. If this condition is not met it is impossible to observe strong-field magnetic resonance spectra. At weak fields tending to zero it should be possible to observe what amount to pure quadrupole transitions but this does not so far appear to have been attempted.

d. *Summary*

Molecular beam magnetic resonance yields both the sign and magnitude of the coupling constants of diatomic homoatomic molecules provided that these are not too large. The inherent precision of the method for heavy molecules is not very great.

4. Bibliography

Theory

Ramsey, N. F., *Molecular Beams*, Oxford University Press 1956.

Sugden, T. M., and Kenney, C. N., *Microwave Spectroscopy of Gases*, Van Nostrand, London 1965.

Townes, C. H., and Schawlow, A. L., *Microwave Spectroscopy*, McGraw Hill, New York 1955.

Kopfermann, H., *Nuclear Moments*, Academic Press, New York/London 1958.

Wollrab, J. E., *Rotational Spectra and Molecular Structure*, Academic Press, New York/London 1967.

Experimental Data

Starck, B., *Molecular Constants from Microwave Spectroscopy*, Landolt-Börnstein, New Series, Group II, Vol. 4, Springer Verlag 1967.

C. QUADRUPOLE COUPLING IN SOLIDS

1. General Introduction

There are four principal methods for determining quadrupole coupling constants in solids; nuclear quadrupole resonance, nuclear magnetic resonance, electron paramagnetic resonance and Mössbauer spectroscopy. The first two methods are in fact closely related and are complementary rather than separate methods. For the magnetic fields practically employed to date the Larmor precession frequencies of most nuclei is less than 15 Mc/s. If the quadrupole coupling constant is of the same order as or greater than this then it will usually be impossible to observe a magnetic resonance signal* and pure quadrupole resonance, i.e. zero field magnetic resonance, is the appropriate method of determining the coupling constant. If however the coupling constant is much less than the practical Larmor frequency then a magnetic resonance spectrum will exhibit quadrupole fine structure yielding the value of the coupling constant. The disadvantage of both these methods is that they require rather large quantities of material (about one cubic centimetre) and while this is not usually a problem with studies on polycrystalline material, it is not usually easy to make single crystals (necessary for a complete determination of the quadrupole coupling tensor) of these dimensions.

Electron spin resonance is much less demanding in this respect since crystals of the order of magnitude of around 10 mm^3 can be employed. In this case, however, studies on polycrystalline materials are usually not possible and of course the method is limited to paramagnetic compounds.

Mössbauer spectroscopy is of course confined to nuclei exhibiting the Mössbauer effect. As we will see below, however, it has the advantage that quadrupole effects are always present (unless the field-gradients are zero) and this renders accessible the measurement of coupling constants for nuclei which in their ground state have nuclear spin less than or equal to one half.

These four methods can potentially provide coupling constants for some isotope or another of most elements in the periodic table.

2. Nuclear Quadrupole Resonance

a. *General*

In nuclear quadrupole resonance, direct transitions between quadrupole levels are observed. This is only possible in the rigid crystalline state where the field-gradient axes are fixed in space. In a liquid and even in a solid just below its melting point molecular vibrations modulate the electronic field-gradient and induce transitions between quadrupole levels to such an extent that the life-time of a quadrupole state becomes so short that uncertainty broadening makes the line too broad to be detectable. In a gas at low pressure where the time between collisions is of the order of the reciprocal of the natural line width pure quadrupole transitions should again be detectable, however the measurement of quadrupole coupling constants in the gas phase is more easily achieved by the detection of hyperfine structure in microwave pure-rotation spectra.

* See however the last section of III.Cb.

Nuclear quadrupole resonance is observed then in crystalline solids, the component molecules of which are usually diamagnetic and have closed shell electronic structures. No electronic orbital degeneracy exists and the expectation values of $\mathbf{V}_{jk}$ is taken over a single wave function. The expectation values of $\mathbf{V}_{jk}$ are then classical electrostatic field-gradients and the energy levels arise from the different orientations of the nucleus in this framework.

It is convenient at this stage to choose an electronic coordinate system such that the off-diagonal elements $V_{xy} = V_{yz} = V_{zx} = 0$, i.e. the axes lie along the directions of maximum field-gradient. Then

$$V^0 = \tfrac{1}{2}V_{zz} \tag{99}$$

$$V^{\pm 1} = \frac{1}{2\sqrt{6}}(V_{zx} \pm iV_{yz}) = 0 \tag{100}$$

$$V^{\pm 2} = \frac{1}{2\sqrt{6}}(V_{xy} - V_{yy} \pm 2iV_{xy}) = \frac{V_{xx} - V_{yy}}{2\sqrt{6}}. \tag{101}$$

If we define $eq = V_{zz}$, $\eta = (V_{xx} - V_{yy})/V_{zz}$ (note that two terms describe the tensor completely since its trace is zero) then the Hamiltonian operator for pure quadrupole resonance becomes

$$\mathscr{H}_Q = A\left\{3\mathbf{I}_z^2 - \mathbf{I}^2 - \frac{\eta}{2}(\mathbf{I}_+^2 + \mathbf{I}_-^2)\right\}$$

$$A \equiv \frac{e^2Qq}{4I(2I - 1)}. \tag{102}$$

b. *Quadrupole spectra with axial field-gradients*

Electronic field-gradients often have axial symmetry, i.e. $\eta = 0$, and in this case the theory is particularly simple. $\mathscr{H}_Q$ is given by

$$\mathscr{H}_Q = A\{3\mathbf{I}_z^2 - \mathbf{I}^2\} \tag{103}$$

and this is diagonal in an m representation. The eigenvalues of this operator are given by

$$E_m = A\{3m^2 - I(I + 1)\}. \tag{104}$$

Thus the levels $\pm m$ are degenerate.

Transitions between quadrupole levels are brought about in practice by the coupling of the nuclear magnetic dipole moment with an oscillating magnetic field of the appropriate frequency. Here the time-dependent Hamiltonian is given by

$$\mathscr{H}_{M(t)} = -\gamma\hbar[H_{x(t)}\mathbf{I}_x + H_{y(t)}\mathbf{I}_y + H_{z(t)}\mathbf{I}_z] \tag{105}$$

where γ is the gyromagnetic ratio of the nucleus under study and $H_{x(t)}$,

$H_{y(t)}$ and $H_{z(t)}$ are the components of the linearly polarized magnetic field $2H_1 . \cos 2\pi\nu t$. The well-known matrix elements between angular momentum states are

$$\langle m|\mathbf{I}_z|m'\rangle = m\delta_{mm'} \tag{106}$$

$$\langle m|\mathbf{I}_x \pm i\mathbf{I}_y|m'\rangle = [(I \pm m)(I \mp m + 1)]^{\frac{1}{2}}\delta_{m\mp 1,m''}. \tag{107}$$

Thus the x and y components of the magnetic field produce transitions between levels with $\Delta m = \pm 1$, the z component a change $\Delta m = 0$. The latter involves no change in energy and thus have no further interest here.

The frequency which induces the transition between two levels $|m|$ and $|m + 1|$ is given by equation (104) and the relationship $\nu = \Delta E/h$

$$\nu = \frac{3|A|}{h}(2|m| + 1). \tag{108}$$

Thus there are I distinct transitions for integral spins and $(I - \frac{1}{2})$ for half-integral spins. The magnitude of the coupling constant can thus be determined directly from the frequencies in the quadrupole spectrum. Its sign is not however afforded since the same spectrum is obtained if the quadrupole levels are ordered in the sense $E_{|m|=I} > E_{|m|=I-1} > E_{|m|=I-2} \ldots$ or in the sense $E_{|m|=I} < E_{|m|=I-1} < E_{|m|=I-2} \ldots$. The same situation prevails when we are dealing with spectra having non-axial field-gradients (see section c) or with quadrupole spectra the presence of small magnetic fields (see sections d and e). We will return briefly to this point in the following chapter when we deal with nuclear magnetic resonance.

The intensity of a pure quadrupole transition depends upon the orientation of the linearly oscillating magnetic field to the axis of the cylindrically symmetric field-gradient, being a maximum when the field lies in a plane perpendicular to the symmetry axis and zero when it is parallel to the axis. Obviously for a polycrystalline sample an average intensity is obtained; for a single crystal measurement of intensities as a function of orientation can provide in principle the direction of the symmetry axis in the crystal and hence, if the crystal structure is known, the orientation of the symmetry axis in the molecule. Since, however, most substances crystallize with more than one molecule in the unit cell, the observed intensity variation is a superposition of two or more out-of-phase variations and the positions of maxima and minima are not well defined. A more accurate method lies in examining the behaviour of the pure quadrupole spectrum under the influence of a small magnetic field (the nuclear quadrupole Zeeman effect) and this is discussed in section d.

c. *Pure quadrupole spectra with non-axial field-gradients*

When the field-gradient tensor does not have axial symmetry, i.e. $\eta \neq 0$, then equations (102) and (107) yield for the off-diagonal matrix elements between the angular momentum eigenfunctions

$$\langle m'|\mathscr{H}_Q|m\rangle = \frac{A\eta}{z}[(I \mp m - 1)(I \mp m)(I \pm m + 1)(I \pm m + 2)]^{\frac{1}{2}}\delta_{m',m\pm 2}. \quad (109)$$

Thus mixing occurs between states having $\Delta m = 2$ and the quadrupole interaction energy must be obtained by diagonalization of the interaction matrix. It is more convenient to treat non-integral and integral spins separately.

(1) *Half-integral spins.* The secular equations for various values of the nuclear spin are given in Table 1. Only for $I = \frac{3}{2}$ can the solution be obtained in analytical form yielding

$$E_m = E_{|\frac{3}{2}|} = 3A\left(1 + \frac{\eta^2}{2}\right)^{\frac{1}{2}} \quad (110)$$

$$E_m = E_{|\frac{1}{2}|} = -3A\left(1 + \frac{\eta^2}{3}\right)^{\frac{1}{2}}. \quad (111)$$

For other values of nuclear spin the equations must be solved numerically for different values of η and solutions have been tabulated.[12] For $I = \frac{5}{2}$ an approximate solution, valid for $\eta \leqslant 0{\cdot}1$, is given by

$$E_{|\frac{5}{2}|} = A\left(10 + \frac{\eta^2}{9}\right) \quad (112)$$

$$E_{|\frac{3}{2}|} = A(-2 + 3\eta^2) \quad (113)$$

$$E_{|\frac{1}{2}|} = A\left(-8 - \frac{32\eta^2}{9}\right). \quad (114)$$

TABLE 3.1. Secular equations for pure quadrupole interaction

I	Secular equation	Units of E
$\frac{3}{2}$	$E^2 - 3\eta^2 - 9 = 0$	A
$\frac{5}{2}$	$E^3 - (3\eta^2 + 21)E^2 - 20(1 - \eta^2) = 0$	$2A$
$\frac{7}{2}$	$E^4 - 42\left(1 + \frac{\eta^2}{3}\right)E^2 - 64(1 - \eta^2) + 105\left(1 + \frac{\eta^2}{3}\right)^2 = 0$	$3A$
$\frac{9}{2}$	$E^5 - 33\left(1 + \frac{\eta^2}{3}\right)E^3 - 44(1 - \eta^2)E^2 + 44\left(1 + \frac{\eta^2}{3}\right)^2 E + 144\left(1 + \frac{\eta^2}{3}\right)(1 - \eta^2) = 0$	$6A$

It will be noted that the $\pm m$ degeneracy is not removed by asymmetry. For spin $\frac{3}{2}$ therefore, where only one transition is observed, the pure quadrupole spectrum does not yield separable values for e^2Qq and η. The frequency of the transition is given by

$$\nu = \frac{e^2Qq}{2h}\left(1 + \frac{\eta^2}{3}\right)^{\frac{1}{2}} \tag{115}$$

and η can only be determined by observation of the nuclear quadrupole resonance Zeeman effect (see sections d and e). For half-integer spins higher than $\frac{3}{2}$ at least two transitions can be observed and e^2Qq and η can both be determined. Thus for $I = \frac{5}{2}$, $\eta < 0{\cdot}1$.

$$\nu(\tfrac{5}{2} \rightarrow \tfrac{3}{2}) = \frac{3e^2Qq}{10h}(1 - \tfrac{11}{54}\eta^2) \tag{116}$$

$$\nu(\tfrac{3}{2} \rightarrow \tfrac{1}{2}) = \frac{3e^2Qq}{20h}(1 + \tfrac{5}{54}\eta^2). \tag{117}$$

(2) *Integral spins.* Integral non-zero nuclear spins are rare for stable isotopes and the only elements having isotopes of integral spin with a relatively large natural abundance are to be found among the lighter nuclei. Thus Li^6 ($I = 1$), B^{10} ($I = 3$), N^{14} ($I = 1$). Deuterium ($I = 1$) has a moment too small for pure quadrupole resonance to have ever been detected even in isotopically pure specimens. Since there exists more abundant stable isotopes of Lithium (Li^7, $I = \frac{3}{2}$, 92·6% abundance) and of Boron (B^{11}, $I = \frac{3}{2}$, 81·2% abundance) of half-integral spin, it is usually more convenient to study the resonances from these. Thus only N^{14} (abundance 99·635%) is of any significant interest to pure quadrupole resonance and therefore the discussion given here will be limited to spin of unity.

It is evident that asymmetry has no effect on the level $m = 0$ (there are no levels $m = \pm 2$ with which it can mix). The secular equation for the diagonalization of the $m = \pm 1$ submatrix then becomes

$$(A - E_{\pm})^2 = A^2\eta^2 \tag{118}$$

therefore

$$E_{\pm} = A(1 \pm \eta). \tag{119}$$

Since the ground state energy, E_0, is $-2A$, the resonance line is split into a doublet

$$\nu = \frac{3e^2Qq}{4h}\left(1 \pm \frac{\eta}{3}\right). \tag{120}$$

A low frequency transition ($\Delta m = 0$) is also possible between $E+$ and $E-$ and has been observed in cases where η is sufficiently large.

For unit spin, therefore, the asymmetry parameter can be determined directly from the spectrum. For a monocrystalline sample the X and Y directions can be determined in principle from intensity measurements, since the high-frequency line has maximum intensity when the oscillating magnetic field lies in the X direction, the low frequency line having here zero intensity. The situation is reversed when the field lies in the Y direction. The Z direction is obviously perpendicular to the X–Y plane. This method suffers from the disadvantages mentioned above for half-integral spins, however, as will be seen shortly, it is the only one available for nuclei of unit spin.

d. *The nuclear quadrupole resonance Zeeman effect: Axial field-gradients*

The application of a constant magnetic field, H_0, in a direction with polar coordinates θ, ϕ in the field-gradient tensor axis system, introduces an additional term into the Hamiltonian operator given by

$$\mathscr{H}_M = \mu H_0[\mathbf{I}_z \cos\theta + \mathbf{I}_x \sin\theta \cos\phi + \mathbf{I}_y \sin\theta \sin\phi] \tag{121}$$

where μ is the nuclear magnetic moment (for axial symmetry the X and Y axes and hence the origin for the angle ϕ are of course arbitrary). In the present circumstances the magnitude of the field is such that $\mu H_0/I \ll e^2Qq$. The opposite case, $\mu H_0/I \gg e^2Qq$, will be treated in a subsequent section where the effect of quadrupole coupling constants on nuclear magnetic resonance spectra is considered.

For states with $|m| > \frac{1}{2}$ the presence of the magnetic field removes the degeneracy between the $\pm m$ levels and first order perturbation theory yields

$$E_{\pm m} = A[3m^2 - I(I+1)] \mp m\frac{\mu}{I}\,.\,H_0 \cos\theta. \tag{122}$$

For nuclei with half-integral spins the situation for the levels $|m| = \frac{1}{2}$ is somewhat different since the off-diagonal elements under $\mathscr{H}_m$, $\langle\frac{1}{2}|\mathscr{H}_M|-\frac{1}{2}\rangle$ are non-zero, having the value $-\mu H_0/4(2I + 1/I)\sin\theta$. Diagonalization of the $\mathscr{H}_M$ submatrix for $|m| = \frac{1}{2}$ then yields two new states

$$\psi_+ = \psi_{+\frac{1}{2}} \sin\alpha + \psi_{-\frac{1}{2}} \cos\alpha \tag{123}$$

$$\psi_- = \psi_{-\frac{1}{2}} \sin\alpha - \psi_{+\frac{1}{2}} \cos\alpha \tag{124}$$

where

$$\tan\alpha = \left[\frac{(f+1)}{(f-1)}\right]^{\frac{1}{2}}$$

$$f = [1 + (I + \tfrac{1}{2})^2 \tan^2\theta]^{\frac{1}{2}}.$$

The energies of these states $E_{\pm}$ are given by

$$E_{\pm} = A[\tfrac{3}{4} - I(I+1)] \mp \frac{f}{2}\frac{\mu . H_0}{I}\cos\theta. \tag{125}$$

For nuclei with integral spins the lowest level, $m = 0$, is non-degenerate and its energy is not affected by the magnetic field.

The selection rule $\Delta m = \pm 1$ is still valid, hence for transitions between levels not involving the mixed states $\psi_{\pm}$ each resonance is symmetrically split to yield a doublet of separation $(2\mu/hI)H_0 \cos\theta$. However four transitions are possible between the mixed states and the neighbouring $|m| = \frac{3}{2}$ levels of frequencies given by the expression

$$\nu = \frac{1}{h}\left[6A \pm \tfrac{1}{2}(3 \pm f)\frac{\mu}{I} H_0 \cos\theta\right]. \tag{126}$$

Thus the original line is split into a symmetrical quadruplet, the outer pair being conventionally referred to as the α doublet and the inner as the β doublet. The intensities of these doublets are not identical but can be readily seen from the form of the mixed states ψ_+ and ψ_- (equations 123 and 124) to be in the ratio $\alpha:\beta = (f+1):(f-1)$. The α doublet is thus always the more intense but the intensity varies with orientation since f is a function of θ.

If resonances not involving transitions to or from the mixed states are being observed it is evident that zones of equal splitting lie on a cone of half-angle θ about the symmetry axis and that the splitting is a maximum when θ is 0° and zero when $\theta = 90°$. It is thus here relatively easy to locate the principal axes of the crystalline field-gradients. The behaviour of the α and β transitions is more complex and it is convenient to consider some particular orientations:

(i) $\theta = 0°$, $f = 1$ and the β doublet has zero intensity. The splitting of the α doublet is given by

$$\Delta\nu = \frac{2\mu}{Ih} H_0$$

(ii) $\theta = 90°$; all lines are of equal intensity. The α and the β lines coincide and the resultant spectrum is a doublet with separation

$$\Delta\nu = \frac{(I + \frac{1}{2})}{I}\mu H_0$$

(iii) When $f = 3$, given by $\theta = \tan^{-1}[2\sqrt{2}/(I + \frac{1}{2})]$ the α lines coincide and the separation of the β lines is

$$\Delta\nu = \frac{6\mu H_0}{I}$$

The intensity of the central single α-line is four times that of each β line.

Thus the most convenient way of determining the z axis from transitions involving the mixed states (and this is inevitable if $I = \frac{3}{2}$) is to determine the locus of non-splitting of the α component which, when $I = \frac{3}{2}$, is a right circular one of semi-angle $\theta = \tan^{-1}\sqrt{2} = 54° 44'$.

e. *Zeeman effect on nuclear quadrupole resonance with non-axial field-gradients*

(1) *Half-integral spins.* The Hamiltonian operator for a constant magnetic field H_0 (equation 121) connects states differing by $\Delta m = \pm 1$ while the asymmetry terms connect states with $\Delta m = \pm 2$. The secular equation may be diagonalized in the normal way but an analytic solution for any values of the magnetic field and asymmetry parameter is only possible for $I = \frac{3}{2}$. The energy levels are given by an expression of the form:

$$E_{m\pm} = E_{m(0)} \pm \frac{\mu H_0}{2I}[a_m^2 \cos^2\theta + (b_m^2 + c_m^2 + 2b_m c_m \cos 2\phi)\sin^2\theta]$$

$$\equiv E_{m(0)} \pm \frac{\mu H_0}{2I} \,.\, f(m, \theta, \phi) \qquad (127)$$

where $E_{m(0)}$ are the energy levels in the absence of a magnetic field (equations 110, 111). A general solution for the coefficients a_m, b_m, c_m, correct to the first order in η, is given by

$$|a_m| = |2_m|; \qquad b_{\frac{1}{2}} = (I + \tfrac{1}{2}), \qquad b_{m > \frac{1}{2}} = 0;$$

$$c_{\frac{1}{2}} = -c_{\frac{3}{2}} = -(I - \tfrac{1}{2})(I + \tfrac{1}{2})(I + \tfrac{3}{2})\frac{\eta}{6}; \qquad c_{m > \frac{3}{2}} = 0. \qquad (128)$$

For $I = \frac{3}{2}$ the exact solution is given by

$$a_{\frac{3}{2}} = -1 - 2/\rho, \qquad a_{\frac{1}{2}} = -1 + 2/\rho$$

$$b_{\frac{3}{2}} = 1 - 1/\rho, \qquad b_{\frac{1}{2}} = 1 + 1/\rho$$

$$c_{\frac{3}{2}} = -c_{\frac{1}{2}} = \eta/\rho, \qquad \rho = (1 + \eta^2/3)^{\frac{1}{2}}. \qquad (129)$$

Tabulated values for the coefficients for $I = \frac{5}{2}, \frac{7}{2}$ and $\frac{9}{2}$ are available.[12]

It is seen that once again the presence of a magnetic field removes the degeneracy between the states $\pm m$. Moreover because of the mixing between states with $\Delta m = \pm 2$ caused by the asymmetry all four transitions between the two pairs of adjacent levels become possible, as in the general case for the $m \pm \frac{1}{2} \leftrightarrow m \pm \frac{3}{2}$ transitions. The frequencies are given by

$$\nu = \nu_0 \pm \frac{\mu H_0}{2Ih}[f(m_1, \theta, \phi) \pm f(m_2, \theta, \phi)] \qquad (130)$$

where ν_0 is the frequency in the absence of the magnetic field. The inner or α pair are the most intense while the β pair are weak and dependent on the value of η.

It is useful to consider the locus (θ_0, ϕ_0) about the Z axis of no splitting of the α-pair as was done previously for the transitions $m \pm \frac{1}{2} \leftrightarrow m \pm \frac{3}{2}$ where the electrostatic field had axial symmetry. It is possible to show that for the various transitions the loci are given by the following formulae

$m \pm \frac{1}{2} \to m \pm \frac{3}{2}$:

$$\sin^2 \theta_0 = 8\left[8 + (I + \tfrac{1}{2})^2 - \frac{\eta}{3}(I + \tfrac{1}{2})^2(I - \tfrac{1}{2})(I + \tfrac{3}{2}) \cos^2 \phi_0\right]^{-1} \tag{131}$$

$m \pm \frac{3}{2} \to m \pm \frac{5}{2}$:

$$\sin^2 \theta_0 = 16\left[16 + \frac{\eta^2}{36}(I - \tfrac{1}{2})^2(I + \tfrac{1}{2})^2(I + \tfrac{3}{2})^2\right]^{-1} \tag{132}$$

$m \pm \frac{5}{2} \to m \pm \frac{7}{2}$ and higher transitions: $\sin^2 \theta_0 = 1.$ (133)

Thus the first locus is a right elliptical cone from which we can derive the asymmetry parameter η by the relationship

$$\eta = \left\{\frac{3[8 + (I + \frac{1}{2})^2]}{(I + \frac{1}{2})^{\frac{1}{2}}(I - \frac{1}{2})(I + \frac{3}{2})}\right\}\left\{\frac{\sin^2 \theta_0(\max) - \sin^2 \theta_0(\min)}{\sin^2 \theta_0(\max) + \sin^2 \theta_0(\min)}\right\}. \tag{134}$$

The second locus is a rather flat right circular cone. The asymmetry parameter is given by the relationship

$$\eta = \frac{2}{3} \cdot \frac{\cot \theta}{(I - \frac{1}{2})(I + \frac{1}{2})(I + \frac{3}{2})}. \tag{135}$$

All subsequent loci only determine the X–Y plane. Although the principal axis (Z) of the field-gradient tensor may be located accurately and easily by the zero-splitting locus method, a corresponding accuracy for the directions of X and Y for moderate values of η, and for the value of η itself, can only be achieved with careful design and construction of the apparatus used for rotating the crystal and for providing a moderate but homogeneous magnetic field (up to 100 gauss is usually sufficient). Even with such care the straightforward method can rarely achieve an accuracy of much better than $\pm 0{\cdot}01$ in η since the width of the resonance lines makes precise location of a non-split orientation difficult. Bucci, Cecchi and Scrocco[13] have devised a method which is essentially a combination of electronic sharpening of absorption lines combined with a least-squares computer analysis of the data which it is claimed gives an accuracy of $\pm 0{\cdot}001$ in η.

Rehn has abandoned the zero-splitting in locus method entirely and achieves an accuracy of $\pm 0{\cdot}001$ in η by measuring the splitting of the α and

β components when the magnetic field is perpendicular to the Z-axis.[14] Inspection of equation (130) for this case (i.e. $\eta = 0$) shows that if $\eta = 0$ the α and β lines coincide (see section d). Rehn shows that when the asymmetry is non-zero the splitting between the α and β components is given by the approximate relationship

$$\nu_\alpha - \nu_\beta \cong \frac{\mu H_0 \eta}{Ih}\left[1 + \frac{\eta}{6}(\cos 2\phi - \eta)\right] \tag{136}$$

If $\eta \leqslant 0{\cdot}25$ the term in ϕ may be neglected entirely to yield

$$\nu_\alpha - \nu_\beta = \frac{\mu H_0}{Ih} \,.\, \eta. \tag{137}$$

Thus in general a preliminary investigation is required to determine the direction of the principal axis and, eventually, the X and Y axes. A simple frequency measurement with the crystal in the appropriate orientation then sufficies to yield η to a considerable accuracy.

All these methods for the determination of η of course require monocrystalline samples and for resonances of ^{35}Cl or other nuclei in the frequency range 10–50 Mc/s this implies single crystals with a volume of at least one cubic centimetre. This condition is of course far from being generally realizable and in these circumstances when only a polycrystalline sample is available, a method due to Morino and Toyama[15] may be employed to yield the magnitude of η though not of course the orientation of the field-gradient axes. These authors showed that if a small magnetic field is applied parallel to the direction of the radiofrequency field in a pure quadrupole resonance measurement on a polycrystalline solid then the resultant broad absorption line has the shape illustrated in Fig. 7 below. Since however the application of the magnetic field considerably reduces the signal intensity a very sensitive spectrometer is required but given this an accuracy of 0·02 in. η can be achieved without too much difficulty.

(2) *Integral spins.* Again only the case of $I = 1$ will be considered. First order perturbation theory shows that the $m = 0$ level is unperturbed and the $m = \pm 1$ levels are given by

$$E_\pm = \frac{e^2Qq}{4} \pm \frac{\mu^2 H_0^2}{I^2}\cos^2\theta - \left(\frac{e^2Qq \,.\, \eta}{4}\right)^2. \tag{138}$$

A dependence on ϕ only arises in the second order and is usually too small to be detected. Thus although Zeeman splitting studies of single crystals containing nuclei of unit spin reveal the direction of the principle axis, the X and Y axes must be located by intensity measurements.

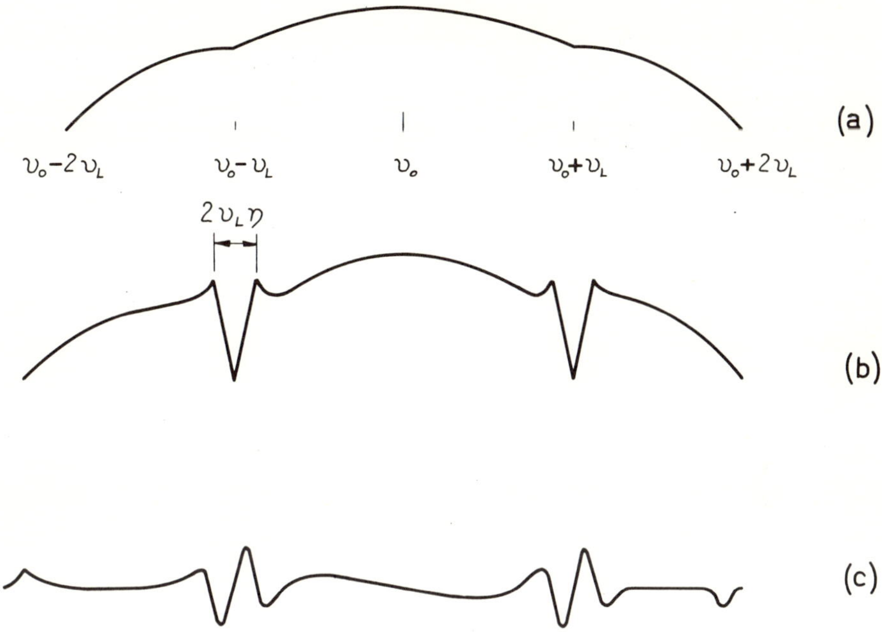

FIG. 3.7. Envelope shape for $I = \frac{3}{2}$ with external magnetic field and radiofrequency fields parallel. ν_0 is the position of the unperturbed line; ν_L is the Larmor frequency of the nucleus in the external magnetic field. (a) $\eta = 0$, (b) $\eta \neq 0$, (c) derivative form of (b).

f. *Summary*

It has been shown that pure quadrupole resonance does not reveal the sign of the coupling constant. For all nuclei except those of $I = \frac{3}{2}$ the spectrum of the polycrysytalline powder yields both the coupling constant and the asymmetry parameter. For nuclei with $I = \frac{3}{2}$ Zeeman studies are necessary for the determination of η and such studies, combined with a knowledge of the crystal structure, are in any case necessary for the location of the field-gradient axes in the molecular framework. For nuclei with $I = 1$ the location of the X and Y axes cannot be carried out with much precision.

3. Nuclear Magnetic Resonance

a. *General*

The Hamiltonian for the interaction of a nuclear magnetic dipole with an externally applied homogeneous magnetic field, H_0, is given by

$$\mathscr{H}_M = -\frac{\mu H_0}{I} \cdot \mathbf{I}_z. \tag{139}$$

The form of this Hamiltonian is such that it is assumed that the nuclear magnetic dipoles are free to align themselves in a magnetic field applied in an arbitrary direction (z) and not already tied to a coordinate system. This is the distinction between the Hamiltonian employed here and that used above for the description of the Zeeman effect in pure quadrupole resonance (equation 121).

Application of this Hamiltonian to the eigenfunctions of nuclear angular momentum shows that a system of equally spaced energy levels is produced, given by the expression

$$E_m = m \, . \frac{\mu H_0}{I}. \tag{140}$$

Transitions between levels $\Delta m = \pm 1$ can be induced by a linearly oscillating magnetic field in the X–Y plane of frequency

$$\nu_m = \frac{\mu H_0}{hI}. \tag{141}$$

Equation (141), valid for an isolated nucleus, implies a sharp resonance frequency. For the macroscopic assembly of nuclear spins with which nuclear magnetic resonance is concerned, dipole–dipole interactions between neighbouring nuclei result in considerable line-broadening. Experimentally nuclear magnetic resonance is usually observed by subjecting the sample to an oscillating magnetic field of fixed frequency and sweeping the magnetic field through the value at which resonance occurs according to equation (141). Line widths are, therefore, usually expressed in gauss but it will be realized that for a particular nucleus there is a bilateral correspondence between field and frequency. For solids these dipole interactions usually result in line widths of several gauss. Since, however, nuclear magnetic moments are not of themselves strongly tied to the molecular framework, nuclear magnetic resonance, unlike nuclear quadrupole resonance, is also possible in the liquid state or in solution, for the nucleus can take up a fixed orientation in space while the molecule vibrates and rotates around it. Indeed, this molecular motion has the beneficial effect of averaging out dipole–dipole or other anisotropic interactions and line widths in the liquid state are of the order of 0·1 milligauss, thus permitting the observation of separate resonance lines due to the same nuclei in different chemical environment within the same molecule. If however the nucleus has a quadrupole moment or is magnetically coupled to another nucleus having a quadrupole moment the consequent coupling of the nucleus to the molecular coordinate system induces transitions between nuclear magnetic states and, according to the uncertainty principle, increases the resonance line width. Measurements of line-widths in such circumstances can be used to estimate the values of quadrupole coupling constants (see section c).

b. *Quadrupole effects in the nuclear magnetic resonance spectra of solids*

In the derivation which follows it is assumed for the sake of simplicity that the asymmetry parameter η is zero. Then if Z is the principal axis of the field-gradient tensor and z is parallel to the direction of the magnetic field H_0 and θ is the angle between Z and z

$$\mathbf{I}_Z = \mathbf{I}_z \cos\theta + \mathbf{I}_x \sin\theta. \tag{142}$$

The Hamiltonian operator is given by

$$\mathscr{H} = \mathscr{H}_M + \mathscr{H}_Q \tag{143}$$

$$\mathscr{H}_Q = A[3\mathbf{I}_Z^2 - \mathbf{I}^2]$$

and from equation (142) $\mathscr{H}_Q$ becomes

$$\mathscr{H}_Q = A[3\mathbf{I}_Z^2\cos^2\theta + 3\mathbf{I}_x^2\sin^2\theta + 3\sin\theta\cos\theta(\mathbf{I}_z\mathbf{I}_x + \mathbf{I}_x\mathbf{I}_z) - \mathbf{I}^2] \tag{144}$$

which by making use of the easily verified relationships

$$\begin{aligned}\mathbf{I}_z\mathbf{I}_x &= \tfrac{1}{2}\mathbf{I}_z(\mathbf{I}_+ + \mathbf{I}_-)\\ \mathbf{I}_x\mathbf{I}_z &= \tfrac{1}{2}(\mathbf{I}_+ + \mathbf{I}_-)\mathbf{I}_z\\ \mathbf{I}_x^2 &= \mathbf{I}^2 - \mathbf{I}_z^2 - \mathbf{I}_y^2\\ &= \tfrac{1}{2}(\mathbf{I}_+^2 + \mathbf{I}_-^2) + \mathbf{I}^2 - \mathbf{I}_z^2\end{aligned} \tag{145}$$

may be reduced to

$$\mathscr{H}_Q = \frac{A}{2}\{(3\cos^2\theta - 1)(3\mathbf{I}_z^2 - \mathbf{I}^2) + 3\sin\theta\cos\theta$$

$$\times [\mathbf{I}_z(\mathbf{I}_+ + \mathbf{I}_-) + (\mathbf{I}_+ + \mathbf{I}_-)\mathbf{I}_z] + \tfrac{3}{2}\sin^2\theta(\mathbf{I}_+^2 + \mathbf{I}_-^2)\}. \tag{146}$$

In this form the diagonal and off-diagonal matrix elements are at once apparent.

The effect of $\mathscr{H}_Q$ is usually treated by perturbation theory and it may be readily seen that in first order each level E_m^0 is shifted by an amount E_m^1

$$E_m^1 = \frac{A}{2}[3\cos 2\theta - 1][3m^2 - I(I+1)]. \tag{147}$$

Since the shift depends on m and the magnetic resonance lines arise from transitions $\Delta m = \pm 1$, the originally single line is split into $2I$ lines corresponding to the different values of m

$$\nu_{m\to m+1} = \nu_m^0 + \frac{3A}{h}(m + \tfrac{1}{2})(3\cos^2\theta - 1). \tag{148}$$

For nuclei with half-integral spin the central line ($m = -\tfrac{1}{2} \to m = \tfrac{1}{2}$) is unshifted. If the ratio $3A\,.\,I/4\mu H_0$ is not insignificant then the second order perturbation term must be taken into account and in particular it may be seen that the central line is shifted by an amount

$$\Delta\nu^2_{-\frac{1}{2}\to\frac{1}{2}} = \frac{9A^2 I}{16h\mu H_0}[4I(I+1)-3][9\cos^2\theta - 1][\cos^2\theta - 1]. \tag{149}$$

The extension of the above theory to the general case of non-axial field-gradient tensor is straightforward. The first order change in energy is given by

$$E^1_m = A[3m^2 - I(I+1)]V_{zz}. \tag{150}$$

If the polar angles of the z axis with respect to the crystalline coordinates XYZ are θ, ϕ

$$V_{zz} = V_{ZZ}\cos^2\theta + V_{XX}\sin^2\theta\cos^2\phi + V_{YY}\sin^2\theta\sin^2\phi \tag{151}$$

$$= V_{ZZ}\left[\cos^2\theta - \tfrac{1}{2}\sin^2\theta + \frac{\eta}{2}\sin^2\theta\cos 2\phi\right].$$

Hence

$$E^1_m = \frac{A}{2}[3\cos^2\theta - 1 + \eta\sin^2\theta\cos 2\phi][3m^2 - I(I+1)] \tag{152}$$

and the resonance line splits up into $2I$ components whose shift in frequency from the unperturbed line is given by

$$\nu^1_m - \nu^0_m = \frac{3A}{h}[3\cos^2\theta - 1 + \eta\sin^2\theta\cos 2\phi][m + \tfrac{1}{2}]. \tag{153}$$

If the sample under study is a single crystal it is apparent that a study of the splitting pattern as a function of the orientation of the crystal in the magnetic field will reveal the values of the coupling constant and the asymmetry parameter. If the crystal structure is known, the orientation of the field-gradient tensor in the molecular framework can also be determined. If, however, only a polycrystalline sample is available, then the effect of quadrupole coupling is averaged over all orientations. Figure 8 illustrates the idealized line-shapes to be expected for $I = \frac{3}{2}$ and $I = \frac{5}{2}$ when $\eta = 0$ and first order theory applies. When η is appreciable further structure appears in the spectrum and Fig. 9 shows the line shape expected for $I = \frac{3}{3}$, $\eta \neq 0$.

In the introduction to this section it was remarked that nuclear magnetic resonance spectra cannot usually be observed in the presence of large quadrupole effects. This is certainly true for the transitions at or around the value of the Larmor frequency usually employed. As shown in equations (123) and (124) the presence of a magnetic field mixes the states $m = \pm\frac{1}{2}$ to give two new states $\psi_\pm$. In sections **C2d** and *e* we were concerned with transitions between these states and the neighbouring $m = \pm\frac{3}{2}$ levels, but direct transitions between the $\psi_\pm$ levels are also possible with a frequency given by (see

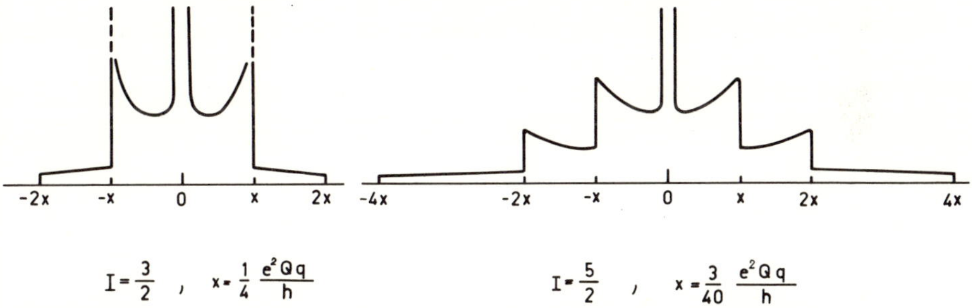

FIG. 3.8. Idealized line shapes for polycrystalline powders of substances having axially symmetric quadrupole coupling constants.

equation 125) approximately

$$\nu = \frac{f}{2hI} \mu H_0 \cos \theta. \tag{154}$$

When θ, the angle between the magnetic field vector and the field-gradient z-axis, is equal to 90° this reduces to

$$\begin{aligned} \nu &= \frac{2\mu H_0}{hI} \cos \theta \\ &= 2\nu_L \end{aligned} \tag{155}$$

where ν_L is the normal Larmor precession frequency. This expression is derived essentially from perturbation theory since it was assumed that the quadrupole energy splitting was much greater than the magnetic splitting. However for $I = \frac{3}{2}$ the complete secular determinant may be solved fairly

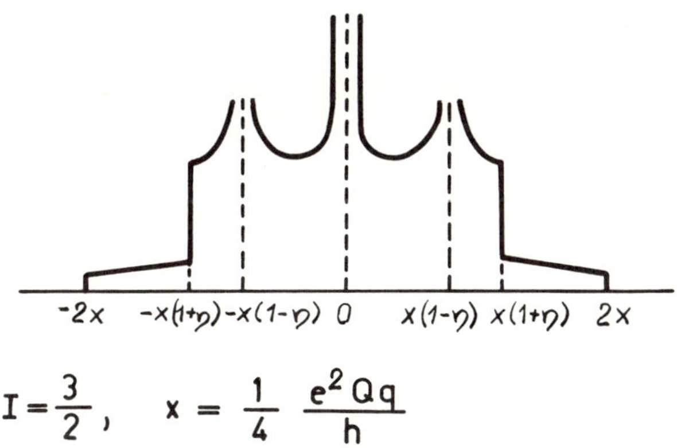

FIG. 3.9. Idealized line shape for a polycrystalline powder of a substance having asymmetric field-gradients.

straightforwardly and, as shown by Segel and Barnes,[16] the frequency of this transition ($\theta = 90°$, $\eta = 0$) is given by

$$\nu = \nu_L + \tfrac{1}{2}[(\nu_Q + 2\nu_L)^2 - 2\nu_Q\nu_L]^{\frac{1}{2}} \\ - \tfrac{1}{2}[(\nu_Q - 2\nu_L)^2 + 2\nu_Q\nu_L]^{\frac{1}{2}} \tag{156}$$

where ν_Q is the pure quadrupole resonance frequency. For $\nu_Q \gg \nu_L$ this reduces to

$$\nu = 2\nu_L\left[1 - \frac{3}{4}\left(\frac{\nu_L}{\nu_Q}\right)^2\right]. \tag{157}$$

These transitions are particularly advantageous to study because although in a powder all possible angles, θ, are present the resultant powder line shape presents a sharp discontinuity at ν which can often be observed. Furthermore, the quadrupole coupling constant only affects ν in second order, and consequently undetectably-wide pure quadrupole resonance lines can be made to manifest their presence in this way. All that is required is to study ν as a function of magnetic field and to fit the results to equation (156) or (157).

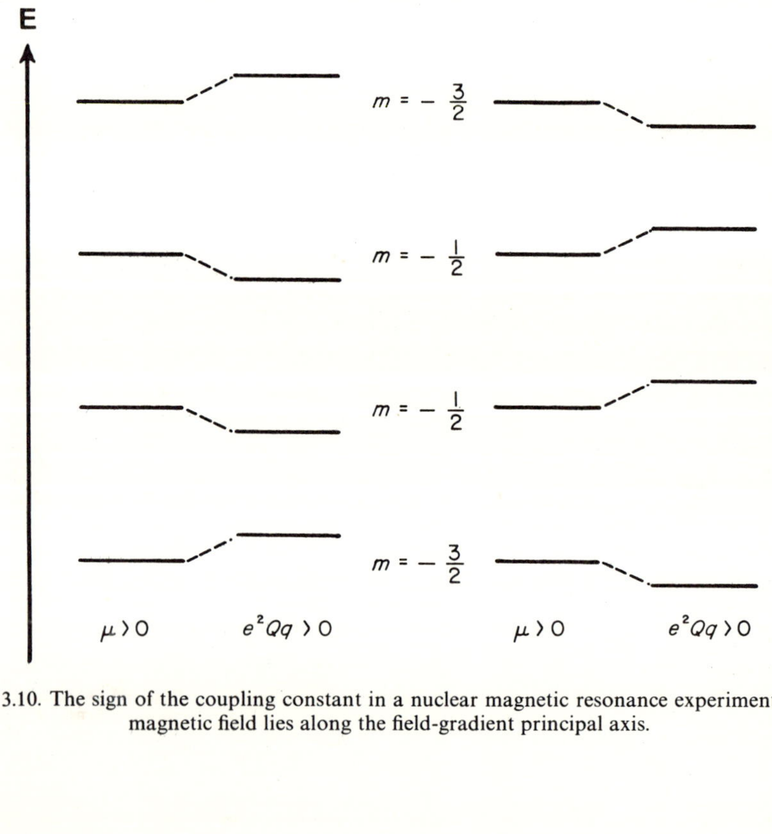

FIG. 3.10. The sign of the coupling constant in a nuclear magnetic resonance experiment. The magnetic field lies along the field-gradient principal axis.

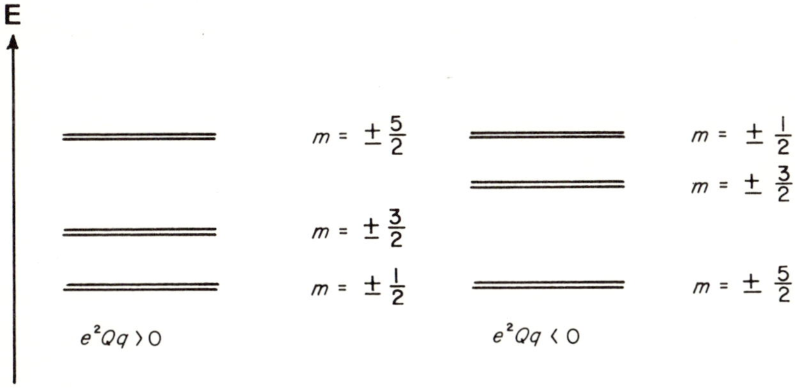

FIG. 3.11. The sign of the coupling constant in a pure quadrupole resonance experiment.

Since the splitting produced by quadrupole effects is symmetric about the unperturbed resonance line, the positions of the quadrupole satellites do not reveal the sign of the coupling constant. It will be recalled that an essentially similar situation prevails for pure quadrupole resonance. In order to obtain the sign of the coupling constant we must be able to distinguish between the systems of energy levels illustrated for $I = \frac{3}{2}$ in Fig. 10 for nuclear magnetic resonance and for $I = \frac{5}{2}$ in Fig. 11 for pure quadrupole resonance respectively (note that in the nuclear magnetic resonance experiment the sign of the nuclear magnetic moment can be determined experimentally, the order of the levels corresponding to different values of m is thus known and illustrated in the figure for the usual case where the nuclear magnetic moment is positive). As we have seen the *positions* of the resonance lines will not do this but there remains the possibility that the intensities of the lines might yield the relevant information.

At ordinary temperatures (the meaning of ordinary is defined below) the population differences between adjacent magnetic energy levels, separated by an energy $h\nu$, is given by $h\nu/kT$. This statement is true as long as $h\sum_i \nu_i \ll kT$, where $\sum_i \nu_i$ is the sum of the energy differences of adjacent levels, and this condition defines "ordinary" temperatures. The intensity of an absorption between two such levels is of course proportional to the population difference between them and in these circumstances it is evident that at ordinary temperatures the relative intensities of, say, the two transitions $m = \pm\frac{5}{2} \leftrightarrow m = \pm\frac{3}{2}$ and $m = \pm\frac{3}{2} \leftrightarrow m = \pm\frac{1}{2}$ of Fig. 11 are independent of whether it is the level $m = \pm\frac{5}{2}$ or $m = \pm\frac{1}{2}$ which has actually the lowest energy. Only when the temperature is so low (or the energy separations so high) that the Boltzmann factor must be expressed by additional terms in the expansion

$$\exp\left(\frac{\Delta E}{kT}\right) = 1 - \frac{\Delta E}{kT} + \frac{1}{2}\frac{\Delta E^2}{k^2T^2} - \frac{1}{6}\frac{\Delta E^3}{k^3T^3} \qquad (158)$$

will the absolute populations of the levels affect the relative intensities and distinguish between the orders of the various levels, thus yielding the sign of the coupling constant. To the author's knowledge, at least, such experiments have yet to be performed. However Abragam[17] has determined the sign of a nuclear quadrupole coupling constant by creating an "artificial" population difference by a double-resonance method.

c. *Quadrupole effects in the nuclear magnetic resonance spectra of liquids*

The spin-lattice relaxation time of a nuclear magnetic resonance absorption of a nucleus of spin $I = \frac{1}{2}$ in a pure liquid is determined entirely by the details of the magnetic dipole–dipole interactions between the various magnetic nuclei within the molecule and between nuclei of neighbouring molecules. The details of such interactions and their quantitative evaluation are sufficiently complex to form the subject of an entire monograph, but in general magnetic resonance of nuclei with spin $I = \frac{1}{2}$ in solutions of low viscosity are sharp, with widths of 1 milligauss or less corresponding to relaxation times of the order of seconds. When however a nucleus of $I > \frac{1}{2}$ is under study, and this nucleus is in a site of non-cubic symmetry, the dominant term determining the spin-lattice relaxation time is usually the modulation of the quadrupole coupling by the molecular rotation and vibration. It can be shown that the quadrupole relaxation term, $(1/T_1)_{\text{quad}}$, is given by

$$\left(\frac{1}{T_1}\right)_{\text{quad}} = \frac{3}{40} \cdot \frac{2I + 3}{I^2(2I - 1)} \left(1 + \frac{\eta^2}{3}\right) \left(\frac{e^2Qq}{\hbar}\right)^2 . \tau_c \tag{159}$$

where τ_q is the correlation time for the molecular motion, i.e. an intramolecular term and all other symbols have their usual significance. In general this correlation time is proportional to the viscosity of the solution divided by the absolute temperature and if T_c is estimated in this way then an order of magnitude of the nuclear quadrupole coupling constant may be obtained.

A more exact evaluation of the quadrupole coupling constant may be made if data are available for two isotopes of the same nucleus, only one of which has a quadrupole moment. Such a pair is Hydrogen and Deuterium and examples of the use of this method in this case are given in Chapter 6. The spin lattice relaxation time of an isotopically pure liquid, having only one magnetic nucleus (A) and no quadrupole relaxation, can be expressed as

$$\left(\frac{1}{T_1}\right) = \left(\frac{1}{T_{1AA}}\right)_{\text{inter}} + \left(\frac{1}{T_{1A}}\right)_{\text{intra}} \tag{160}$$

The double label on the intermolecular term implies interaction between two like nuclei in different molecules.

For the quadrupolar nucleus, again as an isotopically pure liquid, the quadrupolar relaxation given by equation (159) must be added:

$$\left(\frac{1}{T_1}\right) = \left(\frac{1}{T_{1BB}}\right)_{\text{inter}} + \left(\frac{1}{T_{1B}}\right)_{\text{intra}} + \left(\frac{1}{T_{1B}}\right)_{\text{quad}}. \tag{161}$$

It may be shown that the ratio of the intermolecular relaxation times is given by

$$\left(\frac{T_{1BB}}{T_{1AA}}\right) = \left(\frac{\gamma_A}{\gamma_B}\right)^4 \left[\frac{I_A(I_A+1)}{I_B(I_B+1)}\right] \tag{162}$$

where γ_A, γ_B are the nuclear gyromagnetic ratios. In a mixture of the isotopically pure liquids the relaxation time of the non-quadrupolar nucleus, mole fraction x, is given by

$$\left(\frac{1}{T_1}\right)_A = \left(\frac{1}{T_{1A}}\right)_{\text{intra}} + x.\left(\frac{1}{T_{1AA}}\right)_{\text{inter}} + (1-x)\left(\frac{1}{T_{1AB}}\right)_{\text{inter}}. \tag{163}$$

The last term in this expression corresponds to relaxation of nucleus A by magnetic dipole interaction with nucleus B in another molecule. It can be shown that it is given by the expression

$$\left(\frac{T_{1AB}}{T_{1AA}}\right)_{\text{inter}} = \frac{3}{2}.\left(\frac{\gamma_A}{\gamma_B}\right)^2 \frac{[I_A(I_A+1)]}{[I_B(I^B+1)]} \tag{164}$$

Studies of such mixtures may then be used to obtain $(1/T_{1A})_{\text{intra}}$. The theoretical expression for $(1/T_1)_{\text{intra}}$, for an isotopically pure molecule, is

$$\left(\frac{1}{T_{1A}}\right)_{\text{intra}} = \tfrac{3}{2}\hbar^2\gamma_A^4\left(\sum_{i\neq j} r_{ij}^{-6}\right)\tau_d \tag{165}$$

where τ_d is an intramolecular correlation time and where r_{ij} is the distance between the various pairs of magnetically identical nuclei in the molecule. This formula allows $(1/T_{1B})_{\text{intra}}$ to be calculated from the value of $(1/T_{1A})_{\text{intra}}$, and hence to obtain $(1/T_{1B})_{\text{quad}}$ from equation (161).

There remains the question of τ_q. For a rigid spherical molecule $\tau_d = \tau_q$ and substitution of this value in equation (159) leads to a value of the quadrupole coupling constant. When the molecules are not rigid spheres this relationship is no longer exact but very probably holds fairly well and it has been applied in most work to date.

d. *Summary*

In the most favourable case when large crystalline samples of solids having known internal structure are available, then e^2Qq, η and the directions of the field-gradient axes may all be determined, although once again the sign of

e^2Qq is not normally revealed. Even for powdered samples it is usually possible to measure e^2Qq and obtain a good estimate of η as long as the quadrupole coupling energy is not comparable with the nuclear magnetic resonance energy. The advent of super-conducting magnets may well make it possible to overcome even this proviso.

Values of quadrupole coupling constants derived from the measurement of either line-widths or relaxation times in solution are of course approximate but have the merit of being easily obtainable. Where closely related molecules of similar size have nuclei in axial field-gradients, the relative values of the coupling constants within the series should be quite accurate. Such measurements are often useful as a preliminary to a search for a pure quadrupole resonance of the nucleus in question.

4. Electron Spin Resonance

a. *General*

It is beyond the scope of this monograph to discuss the most general case for the electron spin resonance spectrum of a completely asymmetric molecule having several unpaired electrons with both spin and orbital momentum, containing a variety of nuclei with different spins, magnetic dipole moments and electric quadrupole moments, and indeed such a discussion would obscure the essential features of quadrupole coupling. Therefore an axially symmetric molecule is considered, having just one unpaired electron and one magnetic and quadrupolar nucleus lying on the symmetry axis. The axial symmetry will be supposed to be at least trigonal, hence the asymmetry of the field-gradient is zero. This choice has the advantage that it covers many of the spectra of transition metal and rare-earth complexes for which experimental data are available.

The Hamiltonian for the interaction between a magnetic field and a system having both orbital (L) and spin (S) angular momentum is

$$\mathscr{H}_E = \beta H_0(\mathbf{L} + g_s\mathbf{S}) \tag{166}$$

where β is the Bohr magneton and g_s the magnetogyric ratio of the free electron. When only one unpaired electron with no orbital momentum is present this reduces to

$$\mathscr{H}_E = g_s\beta H_0\mathbf{S}_z \tag{167}$$

which has eigenvalues

$$E_E^\circ = M_s g_s\beta H_0 \equiv \pm\tfrac{1}{2}g_s\beta H_0 \tag{168}$$

where M_s is the electron spin magnetic quantum number which for the case considered here can have values of $\pm\frac{1}{2}$. Transitions between these two levels may be induced by an oscillating magnetic field in the xy plane of frequency

$$\nu_E = \frac{g_s \beta H_0}{h}. \tag{169}$$

For free electrons $g = 2{\cdot}0023$ which, for fields of 10·000 gauss corresponds to a resonance frequency of 27,900 Mc/s. Most electron spin resonance measurements are done at frequencies of 10,000 or 25,000 Mc/s. since measuring equipment in these frequency ranges is readily available commercially.

The electron spin resonance spectrum of any system having one unpaired electron with both spin and orbital momentum may be described phenomenologically by the Hamiltonian

$$\mathscr{H}_E = \beta(g_x \mathbf{H}_x \mathbf{S}_x + g_y \mathbf{H}_y \mathbf{S}_y + g_z \mathbf{H}_z \mathbf{S}_z) \tag{170}$$

i.e. by ignoring the presence of the orbital momentum operator but replacing the scalar g_s by a tensor quantity $\mathbf{g}$. For paramagnetic transition metal and rare-earth ions, it is usually possible to correlate rather exactly the observed components of $\mathbf{g}$ with electronic structure of the ion and where considerable contributions from orbital momenta are possible, the three components can have very different values. Thus the resonance of the lowest state of Ce^{3+} in a matrix of lanthanum ethyl sulphate has $g_x = g_y = 2{\cdot}185$, $g_z = 0{\cdot}955$. For the system considered here $g_x = g_y \equiv g_\perp$ and $g_z \equiv g_\parallel$, and the eigenvalues of the electron spin resonance Hamiltonian operator for an orientation θ of the symmetry axis, z, to the magnetic field H_0 is given by

$$E_E^0 = M_s g \beta H_0 \tag{171}$$

where

$$g = \sqrt{g_\parallel^2 \cos^2 \theta + g_\perp^2 \sin^2 \theta}.$$

An excellent introduction to this subject is given by Carrington and Longuet-Higgins[18] where references to more detailed discussion may also be found.

b. *Quadrupole effects in the* e.s.r. *spectra of solids*

In a perfectly general way it can be seen that the presence of a magnetic and quadrupolar nucleus in the paramagnetic spectrum will add a term $\mathscr{H}_{FS}$ to the Hamiltonian operator, given for the specific case considered here by

$$\begin{aligned}\mathscr{H}_{FS} &= \mathscr{H}_M + \mathscr{H}_Q \\ &= a\,.\,\mathbf{S}\,.\,\mathbf{I} + A(3\mathbf{I}_z - \mathbf{I}^2) - \frac{\mu_N}{I}\,.\,H_0\mathbf{I}.\end{aligned} \tag{172}$$

The first term represents the interaction between the nuclear and electronic spins; the second the familiar quadrupole interaction and the third the direct action between the nuclear spin and the external magnetic field.

Interaction between nuclear and electronic spins may arise in two ways. In the first place the magnetic field at the nucleus is the sum of the external field and the average magnetic field produced by the unpaired electron. Electrons move rapidly compared with nuclei, so the field produced by the electron can be replaced by the field produced by one or more magnetic dipoles situated at particular points on the molecular framework. Since the nuclear dipole and the equivalent electron dipole are always parallel to the external field, the field produced by the electron at the nucleus depends on the angle(s) between the vector joining the nucleus to the equivalent dipole(s). The average of this interaction over all directions in space is zero, therefore this term does not contribute to electron resonance spectra of substances in the gas or liquid phases. The second mechanism for electron nuclear magnetic interaction arises if there is a spin imbalance at the nucleus in question, i.e. if the wave function describing the motion of the unpaired electron has a finite value at the nucleus. This term is obviously independent of orientation and is responsible for the hyperfine structure of the spectra of gaseous or liquid substances.

The factor representing the sum of these two effects is a tensor quantity and for the particular case considered here is uniquely given by its two components $a_{\parallel}$ and $a_{\perp}$, parallel and perpendicular to the molecular symmetry axis respectively. In the absence of quadrupole coupling the eigenvalues of the fine-structure Hamiltonian, $\mathscr{H}_{FS}$, are given, to first order, by

$$E_{FS} = aM_s m - \frac{\mu_n H_0 m}{I}\left[\frac{a_{\parallel}g_{\parallel}}{ag}\cos^2\theta + \frac{a_{\perp}g_{\perp}}{ag}\sin^2\theta\right] \tag{173}$$

where

$$a = \sqrt{\frac{a_{\parallel}^2 g_{\parallel}^2}{g^2}\cos^2\theta + \frac{a_{\perp}^2 g_{\perp}^2 \sin^2\theta}{g^2}}.$$

Transitions between these levels normally take place with no change in nuclear magnetic quantum number, i.e. $\Delta m = 0$. In these circumstances the second term of equation (173) contributes nothing to the hyperfine structure and the electron spin resonance line is, therefore, split into $(2I + 1)$ equally spaced components of equal intensity whose position in the magnetic field for an exciting frequency ν is given by

$$H_m = H_0 + \frac{a\,.\,m}{\beta g} \tag{174}$$

where

$$H_0 = \frac{h\nu}{\beta g}.$$

It is perhaps worth remarking at this point that a necessary requisite for the observation of quadrupole effects in electron spin resonance spectra is that the nuclear electronic magnetic moments be coupled, i.e. that the tensor a be non-zero. There is no direct interaction between a nuclear quadrupole and an electronic magnetic dipole so that if a quadrupolar nucleus in a molecule is in a position far removed from the paramagnetic site, the electron spin resonance spectrum will reveal nothing about its quadrupole coupling constant.

The presence of quadrupole interaction introduces a further term into the expression for the eigenvalues of $\mathscr{H}_{FS}$ given by

$$\begin{aligned} E_Q &= A\left[\frac{3m^2 - I(I+1)}{2}\right]\left[\frac{3a_{\|}^2 g_{\|}^2 \cos^2\theta}{a^2 g^2} - 1\right] \\ &\equiv A[3m^2 - I(I+1)]f'(\theta). \end{aligned} \tag{175}$$

In the derivation of this expression it is assumed that the quadrupole interaction energy is much smaller than the magnetic interaction of the nucleus with the electron spin and external field, i.e. the axis of precession of the nuclear spin remains unchanged.

It is at once apparent that for the transitions $\Delta m = 0$ which have been considered heretofore, and in the limit of first order theory, quadrupole coupling produces no effect on the spectrum. However, as we have previously seen, the quadrupole operator connects levels with $\Delta m = \pm 2$ and transitions corresponding to these differences become allowed for directions of the magnetic field at an angle to the symmetry axis. The transitions $\Delta m = \pm 1$ are also allowed when the magnetic coupling tensor a is highly anisotropic and when the magnetic field is in directions neither perpendicular nor parallel to the symmetry axis. The effect of quadrupole coupling is illustrated for a nucleus with $I = \frac{3}{2}$. The hyperfine energy levels are then given by

$$\begin{aligned} E_{+\frac{3}{2}} &= \pm\frac{3a}{4} - \frac{3}{2}\frac{\mu_N H_0}{I}f(\theta) + \tfrac{3}{2}Af'(\theta) \\ E_{+\frac{1}{2}} &= \pm\frac{1a}{4} - \frac{1}{2}\frac{\mu_N H_0}{I}f(\theta) - \tfrac{3}{2}Af'(\theta) \\ E_{-\frac{1}{2}} &= \mp\frac{1a}{4} + \frac{1}{2}\frac{\mu_N H_0}{I}f(\theta) - \tfrac{3}{2}Af'(\theta) \\ E_{-\frac{3}{2}} &= \mp\frac{3a}{4} + \frac{3}{2}\frac{\mu_N H_0}{I}f(\theta) + \tfrac{3}{2}Af'(\theta). \end{aligned} \tag{176}$$

The upper signs in the formulae give the position of the energy levels relative to $\frac{1}{2}g\beta H_0$ for $M_s = +\frac{1}{2}$ while the lower signs give the corresponding values relative to $-\frac{1}{2}g\beta H_0$ for $M_s = -\frac{1}{2}$.

The energy level diagram is given in Fig. 12a and the corresponding spectrum for the various transitions in Fig. 12b.

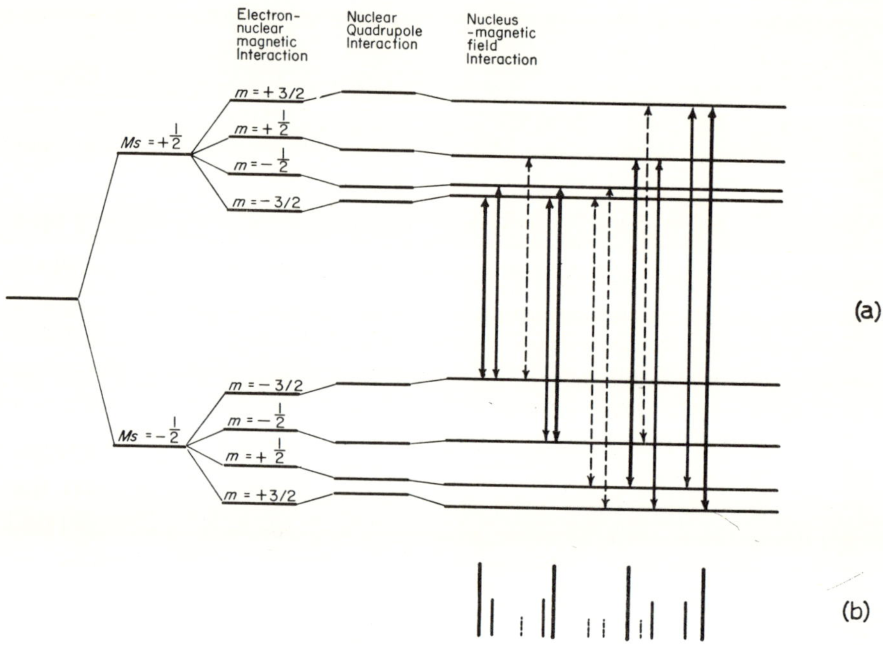

FIG. 3.12. Energy levels (a) and spectrum (b) of a paramagnetic molecule, $s = \frac{1}{2}$, $I = \frac{3}{2}$. Heavy full lines are transitions $\Delta m = 0$, light full lines $\Delta m = \pm 1$, dotted lines $\Delta m = \pm 2$.

A consideration of equation (176) and the figure shows readily that the electron spin resonance experiment reveals at once the relative sign of **a** and e^2Qq. It is often possible to obtain the probable absolute sign of **a** on the theoretical grounds so that the absolute sign of e^2Qq is then revealed.

The first order theory so far presented is usually quite adequate for electron spin resonance experiments at frequences of 25,000 Mc/s which have the additional advantage that the term $\mu H_0/I$ is larger so that the asymmetry in the high field and low field $\Delta m = \pm 1$, $\Delta m = \pm 2$ transitions is more marked. At 10,000 Mc/s however second order terms become appreciable and have to be taken into account in the analysis of the spectrum. These terms are discussed by Bleaney *et al.*[19] who gives a good account of the effect of quadrupole coupling on the electron spin resonance spectra of paramagnetic complex ions.

The extension of the above theory to compounds having an asymmetry parameter, η, so that as a consequence all three components of the tensor a are also different is straightforward and is given by Bleaney. The directions of the axes and the value of η are given directly by the spectrum. The general case in which none of the directions of the principal axes of g, a and A coincide does not appear to have been treated theoretically, but in any case an analytical solution to this problem would not be possible.

c. *Summary*

In common with the methods of pure quadrupole resonance and nuclear magnetic resonance, e^2Qq, η and the directions of the field-gradient axes can all be determined by electron spin resonance measurements conducted on single crystals. In favourable cases the absolute sign of the coupling constant can be determined. The method finds its most general application in the determination of the quadrupole coupling constants of paramagnetic complexes of the transition metals and the lanthanons.

5. Mössbauer Spectroscopy

a. *General*

In Mössbauer spectroscopy the phenomenon that is observed is the resonant absorption of a γ-photon by an atomic nucleus which is thereby raised from its ground state to an excited state. This excitation is always accompanied by a change of at least unity in the nuclear spin angular momentum, I; as a consequence even if the nuclear ground state has zero spin the excited state will have a spin of unity and in general quadruple coupling effects can be observed. The details of Mössbauer spectroscopy need not concern us here; it will suffice to say that the observation of Mössbauer spectra requires, by the very nature of the effect, solid and in general crystalline materials, and very often temperatures as low as that of liquid helium. Mössbauer spectroscopy is confined to nuclei in which the first excited state lies at not too great an energy and at present the practical limit lies at about 200 keV. Table 2 shows a list of some of the more important nuclei in which the Mössbauer effect has been observed.

It is convenient to illustrate the discussion of Mössbauer spectroscopy with the example of the ^{57}Fe nucleus for which by far the largest number of Mössbauer spectra are available. In the ground state this nucleus has a spin $I = \frac{1}{2}$ so that the spin of the excited nucleus, Fe^{57}, is $I = \frac{3}{2}$. If the nucleus is in a non-cubic environment therefore the energy level diagram is that shown in

TABLE 3.2. Nuclei exhibiting the Mössbauer effect. This list is by no means exhaustive.

Isotope	Gamma ray energy (keV)	Lifetime of excited state (nanoseconds)	Nuclear spin		Quadrupole moment (barns)		Magnetic moment (nuclear magnetons)	
			Ground state	Excited state	Ground	Excited	Ground	Excited
K^{40}	29·4	—	4	3	−0·09	—	−1·298	—
Fe^{57}	14·0	100·0	$\frac{1}{2}$	$\frac{3}{2}$	0	+0·3	+0·09057	−0·15521
Ni^{61}	67·4	5·2	$\frac{3}{2}$	$\frac{5}{2}$	—	—	+0·746	+0·34
Sn^{119}	23·8	18·0	$\frac{1}{2}$	$\frac{3}{2}$	0	−0·07	−1·046	+0·76
I^{127}	59·0	—	$\frac{5}{2}$	$\frac{7}{2}$	−0·79	−0·71	+2·8091	—
I^{129}	26·8	—	$\frac{7}{2}$	$\frac{5}{2}$	−0·55	−0·68	+2·617	—
Xe^{129}	40·0	1·0	$\frac{1}{2}$	$\frac{3}{2}$	0	−0·41	−0·777	—
Cs^{133}	81·0	6·0	$\frac{7}{2}$	$\frac{5}{2}$	−0·3	—	+2·579	+3·3
Eu^{151}	21·6	3·0	$\frac{7}{2}$	$\frac{5}{2}$	+0·95	+1·2	+3·464	+2·5
Tm^{169}	8·41	8·0	$\frac{1}{2}$	$\frac{3}{2}$	0	−1·6	+0·59	+0·6
Yb^{170}	84·2	1·6	0	2	—	—	—	—
W^{183}	46·5	0·15	$\frac{1}{2}$	$\frac{3}{2}$	—	—	+0·117	—
	99·1	0·57	$\frac{1}{2}$	$\frac{5}{2}$	—	—	+0·117	—
Os^{188}	155·0	0·73	0	2	—	—	+0·49	—
Ir^{193}	73·0	—	$\frac{3}{2}$	$\frac{1}{2}$	1·5	0	+0·17	—
Pt^{195}	98·9	0·16	$\frac{1}{2}$	$\frac{3}{2}$	—	—	+0·61	—
Au^{197}	77·3	1·9	$\frac{3}{2}$	$\frac{1}{2}$	+0·56	0	+0·145	+0·37

Fig. 13. It is clear that the Mössbauer spectrum corresponding to this energy-level scheme is just a doublet with a separation, $\Delta\nu$, equal to that of the separation between the $m = \pm\frac{1}{2}$ and the $m = \pm\frac{3}{2}$ levels.

The extension to other nuclei is straightforward. It should be noted that since in the absence of magnetic fields the axis of quantization of the nuclear spin system is just the field-gradient principal axis, the situation is exactly the same as in pure quadrupole resonance and polycrystalline samples give sharp resonance spectra. However, for an excited nucleus with $I = \frac{3}{2}$, just as in pure quadrupole resonance spectroscopy, the quadrupole splitting is a

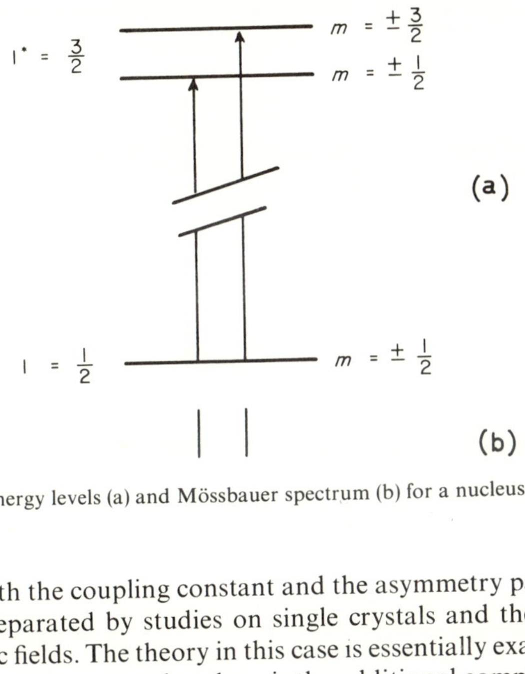

FIG. 3.13. Energy levels (a) and Mössbauer spectrum (b) for a nucleus $I = \frac{1}{2}$. $I^* = \frac{3}{2}$.

function of both the coupling constant and the asymmetry parameter. These can only be separated by studies on single crystals and the application of small magnetic fields. The theory in this case is essentially exactly the same as for pure quadrupole except that there is the additional complication that the degeneracy of the $m = \pm\frac{1}{2}$ level of the ground state is also lifted by the magnetic field.

b. *Spectra for axially-symmetric field-gradients*

When the field-gradient tensor has axial symmetry the quadrupole Hamiltonian is diagonal in the nuclear spin functions, as was seen in section C.2b. For $I = \frac{3}{2}$ we then have four states which we may label by their values of I and m, thus $|\frac{3}{2}, \frac{3}{2}\rangle$, $|\frac{3}{2}, \frac{1}{2}\rangle$, $|\frac{3}{2}, -\frac{1}{2}\rangle$ and $|\frac{3}{2}, -\frac{3}{2}\rangle$. The energy separation of these levels is of course given by $\Delta E = \frac{1}{2}e^2Qq$. The states of the unexcited Fe^{57} nucleus with $I = \frac{1}{2}$ may be similarly designated $|\frac{1}{2}, \frac{1}{2}\rangle$ and $|\frac{1}{2}, -\frac{1}{2}\rangle$.

Mössbauer spectra are usually observed using a beam of unpolarized γ-photons and if the field-gradient principal axis is at an angle θ to the direction of incidence of the γ-rays (see Fig. 14) transitions between the levels of Fe^{57} and Fe^{57*} take place according to the selection rule $\Delta m = 0, \pm 1$. The relative intensities of these transitions depend upon θ in the manner set out in Table 3. For a power, of course, an average must be made over the angle θ and, provided that the probability of the Mössbauer transition is itself independent of θ, both peaks of the quadrupole have equal intensity. It does, however, appear that in some cases the Mössbauer transition does not have an angular dependence and in this case the spectrum is asymmetric. This point has some importance since in the earlier literature the appearance of asymmetric doublets led to claims that the splitting was due not to quadrupole coupling but to the presence of nuclei in two chemically-different environments.

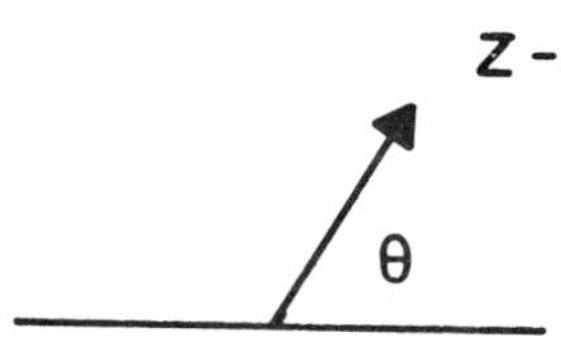

FIG. 3.14. Direction of incident beam of γ-protons.

The selection rules of Table 3 show that the principal axis of the field-gradient tensor may be determined from single-crystal studies by measuring the intensities of the two peaks as a function of θ. Thus in general the ratio of the intensity of the peak corresponding to the transitions to the $m = \pm\frac{3}{2}$ levels to the intensity of the peak for the transition to the $m = \pm\frac{1}{2}$ level is given by

$$\frac{i_{\frac{3}{2}}}{i_{\frac{1}{2}}} = \frac{1 + \cos^2\theta}{\frac{2}{3} + \sin^2\theta}. \tag{177}$$

TABLE 3.3

Transition		Δm	Relative intensity
$\lvert\frac{1}{2}, \frac{1}{2}\rangle$	$\lvert\frac{3}{2}, \frac{3}{2}\rangle$	$+1$	$(1 + \cos^2\theta)$
$\lvert\frac{1}{2}, \frac{1}{2}\rangle$	$\lvert\frac{3}{2}, \frac{1}{2}\rangle$	0	$\frac{4}{3}\sin^2\theta$
$\lvert\frac{1}{2}, \frac{1}{2}\rangle$	$\lvert\frac{3}{2}, -\frac{1}{2}\rangle$	-1	$\frac{1}{3}(1 + \cos^2\theta)$
$\lvert\frac{1}{2}, -\frac{1}{2}\rangle$	$\lvert\frac{3}{2}, \frac{1}{2}\rangle$	$+1$	$\frac{1}{3}(1 + \cos^2\theta)$
$\lvert\frac{1}{2}, -\frac{1}{2}\rangle$	$\lvert\frac{3}{2}, -\frac{1}{2}\rangle$	0	$\frac{4}{3}\sin^2\theta$
$\lvert\frac{1}{2}, -\frac{1}{2}\rangle$	$\lvert\frac{3}{2}, -\frac{3}{2}\rangle$	-1	$(1 + \cos^2\theta)$

In addition to yielding the direction of the z axis, such studies allow the determination of the sign of the coupling constant since the relative intensities of the two lines allow them to be associated with the appropriate values of m. Thus if the $m = \pm\frac{3}{2}$ levels occur at higher energies than the $m = \pm\frac{1}{2}$ levels the coupling constant is positive, while if the reverse is true the coupling constant is negative. A number of determinations of this kind have indeed been carried out but their feasability depends upon the availability of large single crystals which can be cut into thin plates.

Mössbauer spectroscopy thus offers the distinct advantage that the sign of the coupling constant can be easily determined. This parameter is indeed of considerable importance in the multiply-coordinated atoms since it is often difficult to be sure of the sign of the field-gradient on purely theoretical grounds. Even when only polycrystalline samples are available the sign can be determined by a method which depends upon the different behaviour of the $m = \pm\frac{3}{2}$ and $m = \pm\frac{1}{2}$ levels in a magnetic field.[20,21] Consider the energy level diagram of Fig. 14. If a magnetic field is applied along the field-gradient principal axis the various energy levels are split to yield the level diagram of Fig. 15. In Fig. 15 the example of the Fe^{57} nucleus has again been taken: here it is known that the magnetic moment of the Fe^{57} nucleus is negative while that of the Fe^{57*} nucleus is positive. Furthermore the ratio of the gyromagnetic ratios of the Fe^{57} and Fe^{57*} nuclei is 1·715. Under the influence of the magnetic field the levels split, the separation of the $|\frac{3}{2}, \frac{3}{2}\rangle$ and $|\frac{3}{2}, -\frac{3}{2}\rangle$ levels is three times that of the $|\frac{3}{2}, \frac{1}{2}\rangle$ and $|\frac{3}{2}, -\frac{1}{2}\rangle$ levels. Owing to the fact that the transitions with $\Delta m = 2$ are forbidden, however, the single absorption line, which previously corresponded to the transition to the $m = \pm\frac{3}{2}$ level, splits into a symmetrical doublet while the line corresponding to the transition to the $m = \pm\frac{1}{2}$ has four components. In a polycrystalline powder the situation is rather more complex than this since the magnetic field has all possible orientations to the magnetic field. It can however be shown that the corresponding diagram, i.e. with the splittings averaged over all angles θ, is given by Fig. 16. This figure is of course for sharp transitions but since the lines are broad the spectrum will appear as a triplet (the transitions to the $|\frac{3}{2}, \pm\frac{1}{2}\rangle$ levels) and a doublet (the transition to the $|\frac{3}{2}, \pm\frac{3}{2}\rangle$ levels) so that the origin of each peak of the quadrupole spectrum is determined and the sign of the coupling constant thus revealed. Figure 16 shows an example of such a spectrum. In ferromagnetic materials the internal field provides a similar splitting from which the sign of the coupling constant may be obtained. In this case, however, the direction of the internal field is determined by the orientation of the individual domains and a complete magnetic hyperfine spectrum results even for a polycrystalline powder.

c. *Spectra for non-axial field-gradients*

As we saw in section C.2c the presence of an asymmetry parameter does not remove the degeneracy of the $\pm m$ levels but simply mixes states with

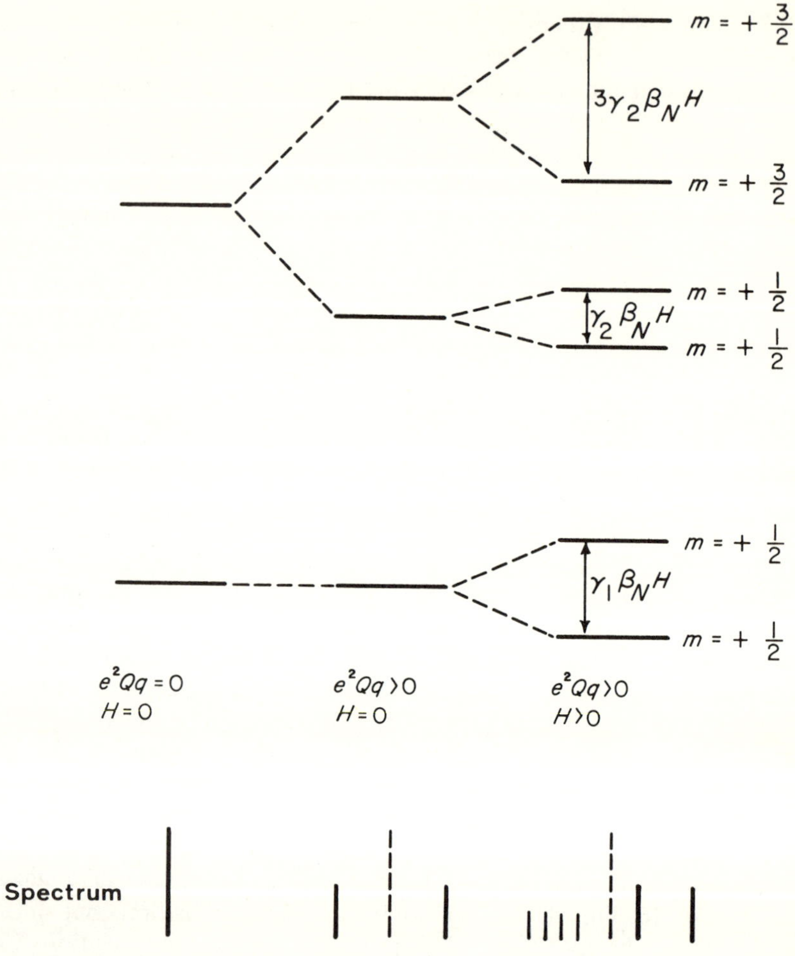

FIG. 3.15. The effect of magnetic fields on the energy levels of Fig. 13 together with the corresponding spectrum.

$\Delta m = \pm 2$. For $I = \frac{3}{2}$ the separation between the two quadrupole levels is given by

$$\Delta \nu = \frac{e^2Qq}{h}\left(1 + \frac{\eta^2}{3}\right)^{\frac{1}{2}}. \tag{178}$$

The only way in which the asymmetry parameter can be determined in this case is by studies on single crystals.

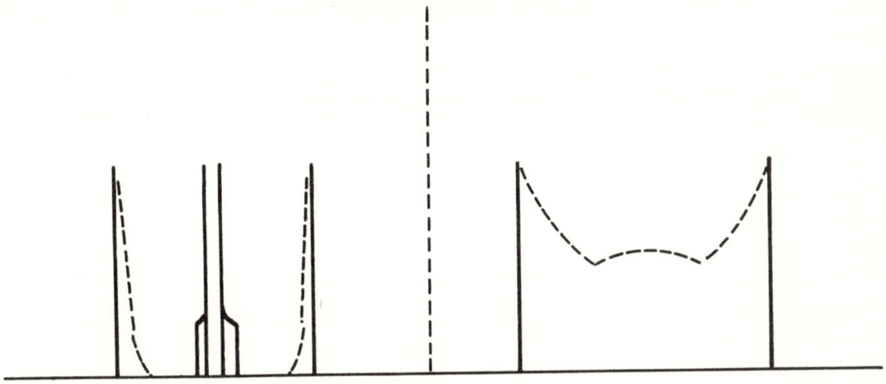

FIG. 3.16. Mössbauer spectrum of polycrystalline powder of a nucleus $I = \frac{1}{2}$, $I^* = \frac{3}{2}$ under the influence of a magnetic field.

Single crystal studies of course also reveal the sign of the coupling constant. If the asymmetry parameter is not too large ($\langle \sim 0{\cdot}6$) the magnetic perturbation method retains its validity. At high asymmetries, however, the powder spectrum simply shows a pair of doublets: this can be seen qualitatively by a consideration of the limiting case $\eta = 1{\cdot}0$ where the field-gradient components are given by $q_z = -q_y, q_x = 0$, i.e. the sign of the coupling constant is indeterminate.

d. *Summary*

Mössbauer spectra can in principle yield all the parameters which characterize a quadrupole coupling tensor, although of course a crystal structure determination is necessary if the orientation of the field-gradient axes in the molecular framework is also required. There is one additional point which has not been brought out in the preceding discussion but which will be evident from a consideration of actual Mössbauer spectra; this is that since quadrupole splittings are comparable with the width of the spectral lines the accuracy with which they can be measured is often rather low.

The discussion above has been confined in its quantitative details to a spin for the excited nuclear state of $I = \frac{3}{2}$; this is the case for Fe^{57*} and Sn^{119*} for which a large amount of data is available. If the spin of the ground state is greater than or equal to unity the ground state levels are also split when the nucleus is in a non-cubic environment and many separate transitions are observed. It is however a straightforward matter to work out the various possibilities and there is no point considering them in detail here.

6. Bibliography

Theory

Nuclear Quadrupole Resonance

Das, T. P. and Hahn, E. L., "Nuclear Quadrupole Resonance Spectroscopy", Academic Press, New York/London 1958.

Nuclear Magnetic Resonance

Cohen, M. H. and Reif, F., "Solid State Physics" 5, p. 322. Academic Press, New York/London 1957.

Slichter, C. P., "Principles of Magnetic Resonance", Harper and Row, New York 1963.

Abragam, A., "The Principles of Nuclear Magnetism", Oxford University Press 1961.

Electron Spin Resonance

Kopfermann, H., "Nuclear Moments", Academic Press, New York/London 1957.

Ingram, D. J. E., "Spectroscopy at Radio and Microwave Frequencies", Butterworths-London 1955.

Altschuler, S. A. and Kozyrer, B. M., "Electron Paramagnetic Resonance", Academic Press, New York/London 1964.

Mössbauer Spectroscopy

Goldanskii, V. I., "The Mössbauer Effect and its Applications in Chemistry". Consultants Bureau, New York 1964.

Duncan, J. F. and Golding, R. M., *Q. Rev.* **19**, 36, 1965.

Frauenfelder, H., *The Mössbauer Effect*, Benjamin, New York 1962.

Wertheim, G. K., *Mössbauer Effect*, Academic Press, New York 1964.

Experimental Results

Lindgren, I., *In* "Perturbed Angular Correlations," (eds. E. Karlsson, E. Matthias and K. Siegbahn) North-Holland Publishing Co. 1964.

Starck, B., "Molecular Constants from Microwave Spectroscopy," Landolt-Börnstein, New Series, Group II, Vol. 4, Springer Verlag 1967.

Segel, S. L. and Barnes, R. G., "Catalog of Nuclear Quadrupole Interactions and Resonance Frequencies in Solids," Ames Laboratory, Research and Development Report U.S.A. E.C. Part I, 1962, Part II, 1965.

Magnetic Properties of Coordination and Metal Organic Compounds of the Transition Elements

Landolt-Börnstein, New Series, Group II, Vol. 2, Springer Verlag 1966.

Muir, A. H., Ando, K. J. and Coogan, H. M., Mössbauer Effect Data Index, 1958–1965, Interscience 1967.

REFERENCES

1. H. Schuler and T. Schmidt, *Z. Phys.* **94**, 457 (1935).
2. D. G. Burkhard and W. E. Brittin, *J. Mol. Spec.* **18**, 87–109 (1965).
3. H. P. Benz, A. Bauder and Hs.H. Gunthard, *J. Mol. Spec.* **21**, 156 (1966).
4. D. J. Millen and J. R. Morton, *J. Chem. Soc.* 1523 (1960).
5. J. K. Bragg, *Phys. Rev.* **74**, 533 (1948).
6. W. H. Flygare and W. D. Gwinn, *J. Chem. Phys*, **36**, 787 (1962).
7. V. Hughes and L. Grabner, *Phys. Rev.* **79**, 829 (1950).
8. P. Thaddeus and L. C. Krisher, *Rev. Sci. Ins.* **32**, 1083 (1961); also P. Thaddeus, L. C. Krisher and J. H. N. Loubser, *J. Chem. Phys.* **40**, 257 (1964).

9. J. M. B. Kellog, I. I. Rabs, N. F. Ramsey and J. R. Zacharias, *Phys. Rev.* **55**, 728 (1936); also *Phys. Rev.* **57**, 677 (1940).
10. B. T. Feld and W. E. Lamb, *Phys. Rev.* **67**, 15 (1945); also H. M. Foley, *Phys. Rev.* **71**, 747 (1947).
11. R. A. Logan, R. E. Coté and P. Kusch, *Phys. Rev.* **86**, 280 (1952).
12. M. H. Cohen, *Phys. Rev.* **96**, 1278 (1954).
13. P. Bucci, P. Cecchi and E. Scrocco, *Ric. Sci.* **34**, 129 (1964).
14. V. Rehn, *J. Chem. Phys.* **38**, 749 (1963).
15. Y. Morino and M. Toyama, *J. Chem. Phys.* **35**, 1289 (1961).
16. S. Segel and R. G. Barnes, *Phys. Rev. Lett.* **15**, 886 (1965).
17. A. Abragam and M. Chapellier, *Phys. Lett.* **11**, 207 (1967).
18. A. Carrington and H. C. Longuet-Higgins, *Q. Rev.* **14**, 427 (1960).
19. B. Bleaney, K. D. Bowers and D. J. E. Ingram, *Proc. R. Soc.* **228**A, 147 (1955).
20. S. L. Ruby and P. A. Flinn, *Rev. Mod. Phys.* **36**, 351 (1964).
21. R. L. Collins, *J. Chem. Phys.* **42**, 1072 (1965).

CHAPTER 4

NUCLEAR QUADRUPOLE COUPLING CONSTANTS IN ATOMS

A. GENERAL

The problem which will be discussed in the next two chapters is the calculation of field-gradients from atomic or molecular wave-functions. If the system under study has angular momentum there are two problems: namely to calculate first the field-gradients with respect to the angular momentum axis or axes and then to go from there to the space-fixed axis system. The first problem is simply the evaluation of the expectation value of the field-gradient operator q_J

$$\mathscr{H}_{q_J} = \frac{3\cos^2\theta_J - 1}{r^3} \tag{1}$$

over the wave-function Ψ_{JJ} (the subscript is to emphasize that the system has angular momentum and that we evaluate it in the substate $m_J = J$). Thus

$$q_J = \int_\tau \psi_{JJ} \frac{(3\cos^2\theta_J - 1)}{r^3} \psi_{JJ}^* \, \mathrm{d}\tau. \tag{2}$$

The next step is to perform the appropriate rotations of coordinates to convert this to the space-fixed system and the details of this vary from system to system. For molecules the problem has been discussed in Chapter 3, section B. For atoms the problem is usually fairly straightforward and is discussed in the present section.

A discussion of the evaluation of field-gradients in atoms forms an essential preliminary to a discussion of field-gradients in molecules. In the first place we rely mainly on measurements on atoms for reliable values of nuclear quadrupole moments. Secondly most molecular wave functions are expressed as linear combinations of atomic wave functions so that it is essential to understand the coupling in various atomic states before we can go on to the molecular problem.

B. HYDROGEN-LIKE ATOMS

It is instructive to consider initially the value of the field-gradient in a hydrogen-like atom for which of course the wave functions have a simple analytic form, at least in the non-relativistic approximation and neglecting electron spin. Hydrogen-like wave functions may be separated into radial and angular parts

$$\Psi = R^{nl}_{(r)} \cdot Y^{l,m}(\theta, \phi) \tag{3}$$

where

$$R^{nl}_{(r)} = -\sqrt{\frac{8}{n^3 a_0^3} \frac{(n-l-1)!}{2n[(n+l)!]^3}} \cdot \left(\frac{2r}{na_0}\right)^l \cdot l^{-r/na_0} L^{2l+1}_{n+l}\left(\frac{2r}{na_0}\right)$$

and n, l, m are the principal, azimuthal and magnetic quantum numbers respectively,* a_0 is the Bohr radius, $L^{2l+1}_{n+l}(2r/na_0)$ is associated Laguerre polynomial. The functions $Y^{l,m}(\theta, \phi)$ are the normalized spherical harmonics. Substitution of this wave function in equation (2) and evaluation over the angular coordinates yields

$$q_J = q_{nl,m=l} = -\frac{2l}{2l+3}\left\langle\frac{1}{r^3}\right\rangle. \tag{4}$$

Evaluation of the radial part of the integral yields

$$\left\langle\frac{1}{r^3}\right\rangle = \frac{1}{l(l+\frac{1}{2})(l+1)} \cdot \frac{Z^3}{n^3 a_0^3} \tag{5}$$

where Z is the atomic charge.

These two formulae serve to illustrate the trends we will expect throughout the atomic table. Firstly the field-gradient depends on Z^3 so that heavy atoms will have much larger field-gradients than light ones. Of course the field-gradients do not increase as dramatically as formula (5) suggests since for a polyelectronic atom the outer electrons are shielded from the nucleus by the electronic core. More significant is the effect of l on the field-gradient. In order to compare these quantities it is preferable to evaluate q along the symmetry axis of the orbital in question, i.e. $q_{nl,m=0}$. Manipulation of the usual formulae for spherical harmonics yields

$$q_{nlm} = \left[\frac{l(l+1) - 3m^2}{l(l+1)}\right] q_{nl,m=0} \tag{6}$$

* The magnetic quantum number m used here should not be confused with the nuclear magnetic quantum number m of previous sections.

i.e.

$$q_{nl,m=0} = \frac{2l(l+1)}{(2l-1)(2l+3)}\left\langle\frac{1}{r^3}\right\rangle. \tag{7}$$

In Table 1 are set out the results of the calculation of the field-gradient in hydrogen-like atoms by application of equations (5) and (7) for different values of l for the same principal quantum number n. For s states, i.e. $l = 0$, the system has spherical symmetry and naturally the field-gradients are zero. The most interesting thing to note, however, is the considerable decrease in field-gradient in going from p-electrons, $l = 1$, to d and f electrons, $l = 2$ and 3. This dominance of p-electrons is important in the approximate methods which have been evolved for the correlation of coupling constants with molecular electronic structure and will be illustrated experimentally below for polyelectronic atoms.

TABLE 4.1. Field-gradients in hydrogen-like atoms.

l	$q_{nl,m=0}\left[\text{units of}\left(\frac{Z}{n^*a_0}\right)^3\right]$
0	0
1	4/15
2	4/105
3	4/315

A further quite general point brought out by equation (5) is the dependence of $\langle 1/r^3\rangle$ on n^{-3}. For atoms, where often the quadrupole coupling constant is measured in an excited state, this means that the lower the energy of the excited state then in general the higher will be the value of the field-gradient.

Even for a hydrogen-like atom with just one electron the above formulae are incorrect because of our neglect of electron spin and of relativistic effects. In the presence of electron spin the state of the atom is characterized exactly only by its total angular momentum, J. Equation (4) then becomes

$$q_{nl,m=0} = -\frac{2J-1}{2J+2}\cdot\left\langle\frac{1}{r^3}\right\rangle. \tag{8}$$

A further complication is the presence of a relativistic correction. Casimir[1] has shown that the solution of the usual Dirac equations for an electron in a central field yields the following relationship

$$q_{nl,m=l} = \frac{2J-1}{2J+2}\cdot R_r(J,l)\left\langle\frac{1}{r^3}\right\rangle \tag{9}$$

where $R_r(J, l)$ the relativistic correction is given by

$$R_r(J, l) = \frac{l(l+1)(2l+1)}{\rho(\rho^2 - 1)(2\rho^2 - 1)}[3k(k+1) - \rho^2 + 1] \tag{10}$$

where $k = l \pm 1$ for $J = l \pm \frac{1}{2}$, $\rho = \sqrt{k^2 - Z^2\alpha^2}$, Z = atomic number of nucleus, α = fine structure constant = $e^2/\hbar c$. This function is plotted in Fig. 1 for the configurations $l = 1$, $j = \frac{3}{2}(^2P_{\frac{3}{2}})$; $l = 2, j = \frac{3}{2}(^2D_{\frac{3}{2}})$; $l = 2$, $j = \frac{5}{2}(^2D_{\frac{5}{2}})$. It is clear that it is only important for heavy nuclei. Casimir's

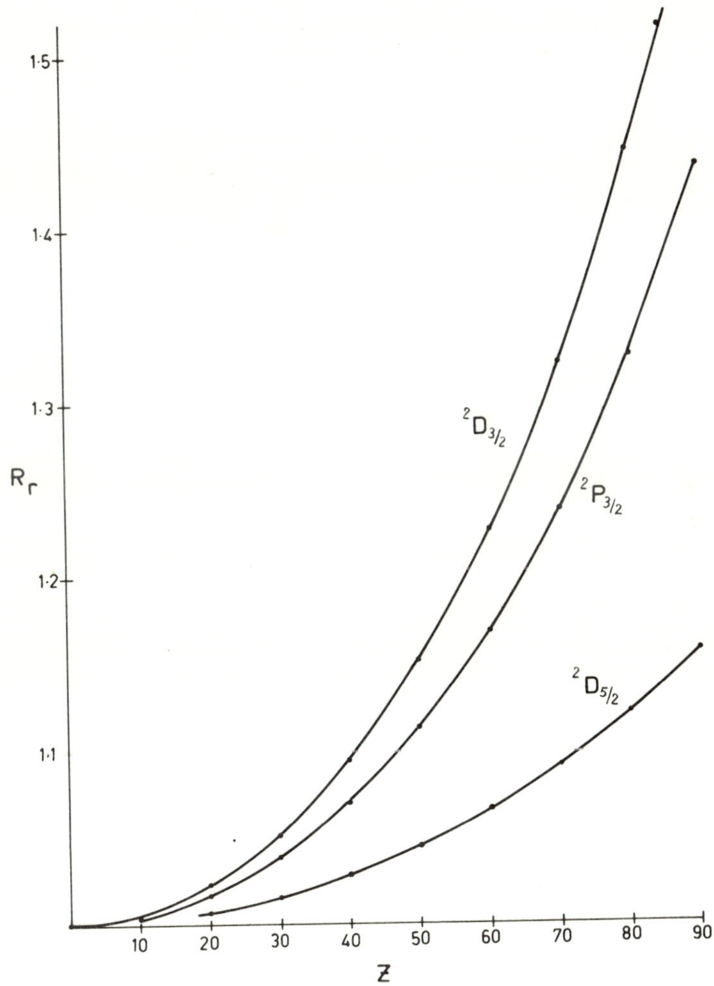

FIG. 4.1. The relativity correction, R_r, for various configurations.

formula was in fact derived for an unbound electron, i.e. $n = \infty$; for a hydrogen atom the correction for $n = 2$ is somewhat different[2] from that given by Casimir. We return to this point in the discussion of field-gradients in polyelectronic atoms with one valence electron.

C. ATOMS HAVING ONE VALENCE ELECTRON

It is usual in most calculations of atomic structure to express the basis functions, in terms of which the final atomic structure is expressed, in the form

$$\psi_{n,l,m,s} = \chi_{(r)}^{n.l} \cdot Y^{lm}(\theta, \phi)\phi^{s} \tag{11}$$

where $\chi_{(r)}^{nl}$ is a radial function constant for all given values of n and l. $Y^{lm}(\theta, \phi)$ are spherical harmonics and ϕ^s are spin functions independent of n, l, m. The wave function of the complete atom is then written as a Slater determinant of the basis functions

$$\Phi = |\psi^1\psi^2 \ldots \psi^N|.$$

A consequence of this form of the basis function is that closed shells have spherical symmetry and the field-gradient at the nucleus arises only from the valence electrons. The errors introduced by this restriction are discussed in Chapter 5; for the present attention is concentrated on the field-gradients arising only from valence electrons. When the atom has just one valence electron it is clear that the assumed form of the basis functions leads to the analogue of equation (8) for the field-gradient

$$q_J = -\frac{2J-1}{2J+2} \cdot R\left\langle\frac{1}{r^3}\right\rangle_{\text{valence}}. \tag{12}$$

If the atom is one electron short of a closed shell structure, e.g. Cl^-, the expression is just the same but with a positive sign. The field-gradient can be thought of as arising from a positive hole in a completely filled shell. In the present case, however, $\langle 1/r^3 \rangle$ is not given by the hydrogenic function, equation (5), but by the average over the radial function $\chi_{(r)}^{v}$ (the superscripts l and m have been dropped and replaced by v indicating the radial function of the valence electron). Calculations of $\langle 1/r^3 \rangle$ can be made from the appropriate Hartree–Fock atomic orbitals. For the neutral atoms He to Ru a tabulation of these values is available.[3] Likewise the relativity correction is not given by equation (10) but by an analogous form in which Z is replaced by an effective charge, Z_i.

Fortunately for the student of nuclear structure, whose main interest lies in an accurate value for the nuclear quadrupole moment, there exists several experimental measures of $\langle 1/r^3 \rangle_{\text{valence}}$. For hydrogen-like orbitals the energy-difference between configurations having the same orbital and spin angular momenta but different total angular momentum, e.g. the difference between the configurations $^2P_{\frac{3}{2}}$ and $^2P_{\frac{1}{2}}$, is given by the relationship

$$E_\lambda = H_r R_\infty \alpha^2 a_0^3 Z(l + \tfrac{1}{2}) . \left\langle \frac{1}{r^3} \right\rangle \tag{13}$$

where R_∞ is the Rydberg constant for infinite nuclear mass and H_r is a relativity correction given by

$$H_r = \frac{2l(l+1)}{\alpha^2 Z^2}(\rho' - \rho'' - 1)$$

ρ' and ρ'' are the values of ρ corresponding to $j = l + \frac{1}{2}$ and $j = l - \frac{1}{2}$ respectively (see equation (10)). For atoms with one valence electron the equation has a similar form in which Z is replaced by an effective charge Z_i. For p-electrons ($l = 1$) this is approximately the nuclear charge minus 4; more accurate values have been tabulated.[4]

Substituting the value of $\langle 1/r^3 \rangle_{\text{valence}}$ derived from the empirical value of λ and the modified form of equation (13)

$$q_J = -\frac{2l\lambda}{Z_i . R_\infty . a_0^3 . (l + \frac{1}{2})(2l + 3)} . \frac{R_r}{H_r} . \tag{14}$$

Another experimental value for $\langle 1/r^3 \rangle_{\text{valence}}$ is provided by the magnetic coupling between the nucleus and the electron. For electrons having orbital momentum in hydrogen-like atoms this is given by

$$a = F . \frac{2\mu_I \mu_S}{IJ} . \frac{l(l+1)}{J(J+1)} \left\langle \frac{1}{r^3} \right\rangle \tag{15}$$

where μ_I and μ_S are the nuclear and electron magnetic moments respectively and F is a relativity correction given by

$$F = \frac{4J(J + \frac{1}{2})(J + 1)}{\rho(4\rho^2 - 1)} .$$

This behaves similarly to R_r (see Fig. 1). For atoms with one valence electron this expression has the same form where the term $\langle 1/r^3 \rangle$ is taken only over the valence electron. Substitution of the empirical value of a to yield a value of $\langle 1/r^3 \rangle_{\text{valence}}$ from equation (15) then yields

$$q_J = \frac{aI . J(2J - 1)}{4\mu_I \mu_S l(l + 1)} . \frac{R}{F} . \tag{16}$$

It is from sources such as these that the most accurate values of q_J (and hence of Q from the observed coupling constants) have been obtained. Both equations (14) and (16) contain ratios of the relativity corrections. Schwartz[2] has shown that although Casimir's formulae gives incorrect values for the individual corrections, the ratio R/F is given fairly accurately. The ratio R/H is however too small.

Barnes and Smith[4] have tabulated experimental values of $\langle 1/r^3 \rangle_{\text{valence}}$ for p-electrons derived from the above interpolating values for some atoms where the experimental data are lacking; these values are tabulated in Table 2. They were calculated without taking account of relativistic correction factors. Also shown in Table 2 are values of $\langle 1/r^3 \rangle$ calculated from

TABLE 4.2. Values of $\langle 1/r^3 \rangle_{\text{valence}}$ for atoms having p-electrons. The "experimental" values are those given in Barnes and Smith,[4] while the "theoretical" values derive from Froese.[3]

Atom	Configuration	State	$\left\langle \frac{1}{r^3} \right\rangle$ (atomic units)	
			"experimental"	"theoretical"
Li	$2p$	^2P	0·0387	
Be	$2s2p$	^3P	0·173	
B	$2p$	^2P	0·608	0·7756
C	$2p^2$	^3P	1·23	1·6618
N	$2p^3$		(2·46)	3·0205
O	$2p^4$	^3P	4·30	4·9490
F	$2p^5$	^2P	6·56	7·5451
Na	$3p$	^2P	0·244	
Mg	$3s3p$	^3P	0·77	
Al	$3p$	^2P	1·28	1·0883
Si	$3p^2$	^3P	2·31	2·0275
P	$3p^3$		(3·48)	3·2667
S	$3p^4$	^3P	5·02	4·8364
Cl	$3p^5$	^2P	7·16	6·7688
K	$4p$	^2P	0·438	
Ca	$4s4p$	^3P	1·13	
Ga	$4p$	^2P	3·50	4·8908
Ge	$4p^2$	^2P	5·75	4·7344
Ag	$4p^3$		(7.51)	6·8542
Se	$4p^4$	^3P	9·62	9·2715
Br	$4p^5$	^2P	13·55	11·9994
Rb	$5p$	^2P	0·845	
Sr	$5s5p$	^3P	2·00	
In	$5p$	^2P	5·72	4·4572
Sn	$5p^2$	^3P	11·2	6·7468
Sb	$5p^3$		(13·0)	9·2313
Te	$5p^4$	^3P	15·0	11·9366
I	$5p^5$	^2P	18·0	14·8724

TABLE 4.3. Values of $\langle 1/r^3 \rangle$ (atomic units) calculated from Slater radial functions.

Atom	Configuration	S	Z	$Z-S$	n^*	$\left\langle \frac{1}{r^3} \right\rangle$
Li	$2p$	1·7	3	1·3	2·0	0·0915
Be	$2s2p$	2·05	4	1·95	2·0	0·2638
B	$2s^22p$	2·40	5	2·60	2·0	0·7323
C	$2s^22p^2$	2·75	6	3·25	2·0	1·4303
N	$2p^3$	3·10	7	3·90	2·0	2·4716
O	$2p^4$	3·45	8	4·55	2·0	3·9248
F	$2p^5$	3·80	9	5·20	2·0	5·8587
Na	Ne $3p$	8·80	11	2.20	3.0	0·0263
Mg	Ne $3s3p$	9·15	12	2·85	3·0	0·0571
Al	Ne $3s^23p$	9·50	13	3·50	3·0	0·1059
Si	Ne $3s^23p^2$	9·85	14	4·15	3·0	0·1765
P	$3p^3$	10·20	15	4.80	3·0	0·2731
S	$3p^4$	10·55	16	5·45	3·0	0·3997
Cl	$3p^5$	10·90	17	6·10	3·0	0·5604
K	Ar $4p$	16·8	19	2·20	3·7	0·0066
Ca	Ar $4s4p$	17·15	20	2·85	3·7	0·0143
Ga	Ar $3d^{10}4s^24p$	26·0	31	5·0	3·7	0·0772
Ge	$4p^2$	26·35	32	5·65	3·7	0·1114
As	$4p^3$	26·70	33	6·30	3·7	0·1544
Se	$4p^4$	27·05	34	6·95	3·7	0·2073
Br	$4p^5$	27·40	35	7·60	3·7	0·2711

Hartree–Fock functions and it is clear that Hartree–Fock functions constitute a reasonably satisfactory basis for the calculation of field-gradients in atoms. Comparisons of theoretical and experimental values for a number of anions and cations as well as neutral atoms have been presented by Malli and Fraga.[5]

It is interesting at this point to discuss the value of $\langle 1/r^3 \rangle$ obtained from Slater orbitals, not because anyone would think of using them to obtain atomic field-gradients, but because Slater orbitals have in the past been used as bases for molecular calculations which have then been used to calculate field-gradients at the different atoms. Obviously, the success or otherwise of such calculations must depend on the adequacy or otherwise of Slater functions in yielding values of $\langle 1/r^3 \rangle$. Normalized Slater radial functions, $\chi(r)_{\text{Slater}}$, for an atom of nuclear charge Z, are given by the expression

$$\chi(r)_{\text{Slater}} = \left\{ \frac{1}{\Gamma(2n^*+1)} \left[\frac{2(Z-S)}{n^* a_0} \right]^{(2n^*+1)} \right\}^{\frac{1}{2}} r^{n^*-1} . e^{-(Z-S)r/n^* a_0} \tag{17}$$

where n^* and S are calculated according to certain recipes. The value of

$\langle 1/r^3 \rangle$ for this function may be straightforwardly evaluated as

$$\left\langle \frac{1}{r^3} \right\rangle_{\text{Slater}} = \frac{4(Z-S)^3}{(n^*)^4(2n^*-1)(2n^*-2)} \cdot \frac{1}{a_0^3}. \tag{18}$$

Values of $\langle 1/r^3 \rangle_{\text{Slater}}$ calculated from equation (18) are set out in Table 3. For elements of the first short period the field-gradients thus predicted are in fairly good agreement with those shown in Table 1. For heavier elements, however, the values of $\langle 1/r^3 \rangle_{\text{Slater}}$ are too small by an order of magnitude. Consequently any calculation of field-gradients at such atoms in molecules is bound to fail if the basis set consists only of simple Slater orbitals.

D. ATOMS HAVING MORE THAN ONE VALENCE ELECTRON

The field-gradient operator, $\mathscr{H}_q$, is a single-particle operator so that when more than one electron gives rise to a field-gradient the total field-gradient is simply a summation of single-particle terms:

$$q_{\text{total}} = \sum_i^n q_i. \tag{19}$$

The wave function for an atom of total angular momentum $J\hbar$ may be written as a linear combination of single-particle atomic states:

$$\psi_J = \sum_{nlm} A_{nlm} \phi_{nlm}. \tag{20}$$

Then the field-gradient corresponding to ψ_J is given simply by

$$q_J = \sum_{nlm} A_{nlm}^2 \cdot \left[\frac{3m^2 - l(l+1)}{l(2l+3)} \right] \cdot \frac{2l}{2l+3} \cdot \left\langle \frac{1}{r^3} \right\rangle_{nlm} . R. \tag{21}$$

The problem is thus to determine the coefficients A_{nlm} and the radial averages $\langle 1/r^3 \rangle_{nlm}$. R is as before a relativity correction. If a particular state may be thought of as arising from just one configuration then in the limit of either Russel–Saunders coupling (spin-orbit interactions negligible) or $j\,.\,j$ coupling (spin-orbit interactions dominant) the coefficients A_{nlm} may be simply obtained from the Clebsch-Gordon coupling coefficients for angular momenta. The actual coefficients differ from these by amounts which may be usually determined from an analysis of the atomic spectrum[6].

The field-gradients arising from the configuration *sl* are particularly easy to treat since there is only one orbital giving rise to a field-gradient. It is also important experimentally and will serve as an illustration of the principles involved. This configuration gives, in the limit of Russel-Saunders coupling, four terms ${}^3L_{l+1}$, ${}^3L_l$, ${}^3L_{l-1}$ and ${}^1L_l$. Spin orbit coupling only mixes states having the same value of l, hence the field-gradients in the terms ${}^3L_{l+1}$ and $3L_{l-1}$ are independent of the coupling. Substitution of the well-known Clebsch-Gordon coefficients in equation (21) yields

$$q_{J=l+1} = -\frac{2l}{2l+3}\left\langle\frac{1}{r^3}\right\rangle . R \tag{22}$$

$$q_{J=l-1} = -\frac{2(l-1)(l+1)(2l-3)}{l(2l-1)(2l+1)}. \tag{23}$$

The Russel-Saunders states ${}^3L_l$ and ${}^1L_l$ are, however, mixed by spin orbit coupling and the extent of mixing can in general be determined by the departures of the corresponding spectral intervals from the Landé interval rule. It can be shown that the field-gradient for the ${}^3L_l$ term for intermediate coupling is given by

$$q_{J=l} = \left\langle\frac{1}{r^3}\right\rangle \left\{C_1^2\frac{(2l-1)(l+2)2l}{(l+1)(2l+1)(2l+3)} . R' + C_2^2\frac{2(l-1)}{2l+1}R'' - \frac{12C_1C_2\sqrt{l(l+1)}}{(2l+1)(2l+3)}S\right\} \tag{24}$$

where R', R'' and S are relativity corrections discussed above and $C_1 = \sin(\theta_0 - \theta)$, $C_2 = \cos(\theta_0 - \theta)$, $\theta_0 = \tan^{-1}\sqrt{l/l+1}$, $\sin^2\theta = \Delta/D$, D is the separation between the terms ${}^3L_l$ and ${}^1L_l$ while Δ is the deviation of the term ${}^3L_l$ from the position it would have had in the case of pure Russel-Saunders coupling.

From the above it is clear that in the case of complex spectra from poly-electronic atoms or ions a very complete analysis of the spectrum is necessary before the field-gradients may be analysed. Expressions analogous to equations (14) and (16) may be derived to yield "empirical" values of the term $\langle 1/r^3\rangle_{\text{valence}}$ in polyelectronic atoms and in this way satisfactory values of the nuclear quadrupole moments may be obtained.

E. SUMMARY

It is usually possible to obtain empirical values of the term $\langle 1/r^3\rangle$ from other spectroscopic parameters. In this way the observed coupling constants in atoms yield reliable values of the nuclear quadrupole moments which are an essential prerequisite of a complete understanding of the coupling constants in molecules. Insofar as they have been investigated, Hartree-Fock wave functions appear to yield reliable values of atomic field-gradients; Slater orbitals on the other hand are completely unreliable for atoms other than those belonging to the first short period.

REFERENCES

1. H. B. G. Casimir, "On the Interactions between Electrons and Nuclei", Teylers' Tweede Genooptehap, Haartem 1936; Freeman, San Francisco 1963.
2. C. Schwartz, *Phys. Rev.* **105**, 173 (1957).
3. C. Froese, *J. Chem. Phys.* **45**, 1417 (1966).
4. R. G. Barnes and W. V. Smith, *Phys. Rev.* **93**, 95 (1954).
5. G. Malli and S. Fraga, *Theor. Chim. Acta* **6**, 54 (1966).
6. E. U. Coudon and G. H. Shortly, "The Theory of Atomic Spectra", p. 72. Cambridge University Press 1935.

CHAPTER 5

STERNHEIMER SHIELDING

A. GENERAL

In section C of Chapter 4 it was remarked that the assumption that the field-gradient in atoms is due only to the valency electrons gives rise to certain errors. This is because the valence electrons polarize the originally spherically symmetrical inner shells of electrons which since they have now lost their spherical symmetry contribute to the field-gradient at the nucleus. The inner shells may be said to shield the nucleus from the field-gradient of the valence electron and the whole phenomenon is often referred to as Sternheimer shielding after R. M. Sternheimer who, in a series of papers starting in 1950 and extending to the present day, was the first to attempt to examine this phenomenon quantitatively.[1]

This chapter will be concerned with Sternheimer shielding not only in atoms (where in fact it is an artefact of our inability or unwillingness to calculate atomic structures rigorously) but also in ions situated at points in crystal lattices where the field-gradients from the other ions are non-vanishing. In these circumstances (it is supposed that the ion in question has a closed-shell structure) the field-gradient is not just given simply by a lattice sum, since the field-gradients induced by the polarization of the originally spherically symmetric shell are often much more important than those due to the ionic lattice.

Before proceeding to discuss the quantitative values of the shielding parameters one point which may easily give rise to confusion must be dealt with. This is that the nuclear quadrupole itself gives rise to a distortion of the electron cloud around it and may be thought of as inducing a concentric quadrupole moment of opposite sign in the spherically symmetric electron cloud. In fact the shielding produced by this induced quadrupole is exactly equal in magnitude to the shielding produced by the distortion due to the external field-gradient. This important statement will now be proved.

Let $\mathscr{H}_0$ be the Hamiltonian operator describing the undisturbed spherically symmetric electron distribution, and having the ground state eigenvalue E_0.

The perturbing potential due to the external field-gradient is represented by the symbol $\mathscr{H}_e$ and the perturbing potential due to the nuclear quadrupole moment $\mathscr{H}_Q$. Let Ψ_0 be the wave function of the unperturbed system and $\psi_0 + \psi_e$ be the *exact* wave function of the system perturbed *only* by the external field-gradient, while $\psi_0 + \psi_n$ is the exact wave function of the system perturbed *only* by the nuclear quadrupole moment. Then

$$(\mathscr{H}_0 + \mathscr{H}_e)(\psi_0 + \psi_e) = (E_0 + E_e)(\psi_0 + \psi_e) \tag{1}$$

and

$$(\mathscr{H}_0 - E_0)\psi_e = -(\mathscr{H}_e - E_e)\psi_0. \tag{2}$$

Similarly

$$(\mathscr{H}_0 + \mathscr{H}_n)(\psi_0 + \psi_n) = (E_0 + E_n)(\psi_0 + \psi_n) \tag{3}$$

$$(\mathscr{H}_0 - E_0)\psi_n = -(\mathscr{H}_n - E_n)\psi_0. \tag{4}$$

If both perturbations are allowed to occur simultaneously we have the Hamiltonian operator $(\mathscr{H}_0 + \mathscr{H}_e + \mathscr{H}_n)$ and the first order expression for the ground-state wave-function is given by $(\psi_0 + \psi_e + \psi_n)$. The energy corresponding to this wave function is thus

$$E_{\text{total}} = \frac{\int(\psi_0 + \psi_e + \psi_n)|\mathscr{H}_0 + \mathscr{H}_e + \mathscr{H}_n|(\psi_0^* + \psi_e^* + \psi_n^*)\,\mathrm{d}\tau}{\int(\psi_0 + \psi_e + \psi_n|\psi_0^* + \psi_e^* + \psi_n)\,\mathrm{d}\tau}. \tag{5}$$

We may rewrite the numerator of this expression (the denominator is of course simply the normalization factor to which we give the symbol N) as

$$\begin{aligned} E_{\text{total}} = \mathrm{N}\Big\{&(\psi_0 + \psi_e)|\mathscr{H}_0 + \mathscr{H}_e|(\psi_0 + \psi_e) + \int(\psi_0 + \psi_n)|\mathscr{H}_0 + \psi_n|(\psi_0 + \psi_n) \\ &+ \int\psi_0\mathscr{H}_0\psi_0 + 2\int\psi_e\mathscr{H}_n\psi_0 + 2\int\psi_n\mathscr{H}_e\psi_0 + 2\int\psi_n\mathscr{H}_0\psi_e \\ &+ \int\psi_e\mathscr{H}_n\psi_e + \int\psi_n\mathscr{H}_e\psi_n.\Big\} \end{aligned} \tag{6}$$

The last two terms are of third order in the perturbations and may be neglected, the sum of the first three terms gives, by the Schrödinger equations (2) and (4) simply $E_0 + E_e + E_n$. The fourth, fifth and sixth terms are those which interest us here. From the Schrödinger equations (2) and (4) it may easily be seen that these integrals have the following mutual relationship

$$\int\psi_0\mathscr{H}_n\psi_e = \int\psi_0\mathscr{H}_e\psi_n = -\int\psi_n\mathscr{H}_0\psi_e. \tag{7}$$

Thus multiplying (2) on both sides by ψ_n and integrating

$$\int \psi_n(\mathcal{H}_0 - E_0)\psi_e = -\int \psi_n(\mathcal{H}_e - E_e)\psi_0 \tag{8}$$

$$\int \psi_n\mathcal{H}_0\psi_e = -\int \psi_n\mathcal{H}_e\psi_0. \tag{9}$$

Similarly from equation (4)

$$\int \psi_n\mathcal{H}_0\psi_e = -\int \psi_e\mathcal{H}_n\psi_0. \tag{10}$$

It now remains to show that these integrals have the form of a quadrupole coupling constant. A more explicit form of the perturbation operators $\mathcal{H}_e$ and $\mathcal{H}_n$ is

$$\mathcal{H}_e = -r^2P_2(\cos\theta_e)eV_{zz} \tag{11}$$

$$\mathcal{H}_n = -r^{-3}P_2(\cos\theta_n)eQ. \tag{12}$$

$P_2(\cos\theta)$ is of course the second-order Legendre polynomial in the nuclear (θ_n) or electronic (θ_e) coordinate system. For the sake of simplicity it is assumed that the external field-gradient has axial symmetry. The symmetry properties of the Hamiltonian then allows us to write the wave-function ψ_e and ψ_n as

$$\psi_e = -r^2P_2(\cos\theta_e)eV_{zz}\,.\,\psi'_e \tag{13}$$

$$\psi_n = -r^{-3}P_2(\cos\theta_n)eQ\,.\,\psi'_n. \tag{14}$$

Thus the integrals in question become

$$\int \psi_n|H_e|\psi_e = e^2QV_{zz}\int r^{-3}P_2(\cos\theta_n)\psi'_n| - r^2P_2(\cos\theta_e)|\psi_0\,\mathrm{d}\tau \tag{15}$$

$$= \int \psi_e|\mathcal{H}_e|\psi_0 = e^2QV_{zz}\int - r^2P_2(\cos\theta_n)\psi'_e|r^{-3}P_2(\cos\theta_e)|\psi_0\,\mathrm{d}\tau \tag{16}$$

$$= \tfrac{1}{2}\gamma e^2Qq \quad \text{(say)}. \tag{17}$$

Thus, taking account of the relationships (7), the total induced quadrupole coupling constant is γe^2Qq.

This result could have been arrived at another way, simply by considering $\mathcal{H}_n$ as a perturbation on $\mathcal{H}_0 + \mathcal{H}_e$, eigenfunction $\psi_0 + \psi_e$

$$E_{\text{pert}} = \int (\psi_0 + \psi_e)|\mathcal{H}_n(\psi_0 + \psi_e) \tag{18}$$

$$= E_n + 2\int \psi_e\mathcal{H}_e\psi_0, \tag{19}$$

$$E_{\text{total}} = E_0 + E_e + E_n + 2\int \psi_e\mathcal{H}_n\psi_0. \tag{20}$$

Similarly by considering $\mathscr{H}_e$ as a perturbation on $(\mathscr{H}_0 + \mathscr{H}_n)$, eigenfunctions $\psi_0 + \psi_n$,

$$E_{\text{pert}} = \int (\psi_0 + \psi_n)\mathscr{H}_e(\psi_0 + \psi_n) \tag{21}$$

$$= E_n + 2\int \psi_n H_e \psi_0, \tag{22}$$

$$E_{\text{total}} = E_0 + E_e + E_n + 2\int \psi_n \mathscr{H}_e \psi_0. \tag{23}$$

We thus have the result, at first sight surprising, that the interaction of the induced quadrupole moment with the external field-gradient is exactly equal to the interaction of the induced field-gradient with the nuclear quadrupole moment. The total shielding is not however the sum of these two calculated separately, i.e. $2\gamma e^2Qq$, since one half of the interaction is cancelled by the cross term $\int \psi_n H_0 \psi_e$.

It should be noticed that the above relationships do not depend on any assumptions other than the validity of first order perturbation theory. Sternheimer has extended this proof to show that the theorem is also valid at the second-order level.[2] Their application in practice presupposes that ψ_0 is known and in general of course ψ_0 is only known approximately. As a consequence the integrals $\int \psi_e H_n \psi_0$ and $\int \psi_n H_e \psi_0$ may not in fact be exactly equal or equally convenient to evaluate. Usually one or other is preferable and according to the circumstance the shielding factor is usually evaluated in only one of the two possible ways.

Before continuing with the quantitative evaluation of Sternheimer shielding it is appropriate to first discuss qualitatively the ways in which a spherical cloud of electrons is deformed by an external field-gradient. Consider a uniform spherical shell of electrons (see Fig. 1a). If initially we confine

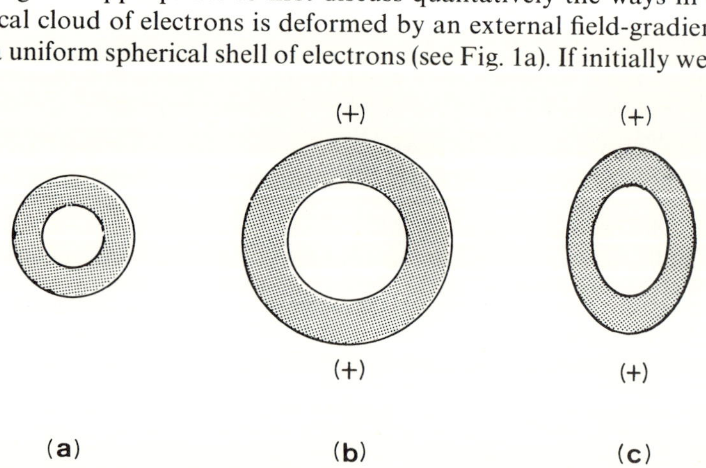

FIG. 5.1. Angular (b) and radial (c) distortions of an initially spherical shell (a) of electrons.

the electrons to the surface of the spherical shell then under the influence of an electrostatic field-gradient, illustrated by two symmetrically disposed positive charges, the electrons will concentrate in regions close to the positive charges. This is an *angular* distortion of the electron distribution (see Fig. 1b). If the shell is now allowed to change its radius it will take on an elliptical form (see Fig. 1c). This is a *radial* distortion.

We may illustrate these two distortions by considering the Sternheimer shielding of s and p functions. For the unperturbed system the Hamiltonian is

$$\mathscr{H}_0 = -\frac{\hbar^2}{2m}\nabla^2 + V_0 \tag{24}$$

where V_0 is the central-field atomic potential including interelectronic repulsion terms. The perturbation is given by

$$\mathscr{H}_1 = -\frac{Q(3\cos^2\theta - 1)}{2r^3} \tag{25}$$

where r is the length of the radius vector from the nucleus and θ the angle between the nuclear symmetry axis and this radius vector Q is the nuclear quadrupole moment.

As we have shown above it does not matter whether we consider the perturbation by the external field-gradient or by the nuclear quadrupole. The Schrödinger equation then becomes

$$(\mathscr{H}_0 + \mathscr{H}_1)(\psi_0 + \psi_1) = (E_0 + E_1)(\psi_0 + \psi_1) \tag{26}$$

where ψ_0 and ψ_1 are the perturbed and unperturbed parts of the wave function.

The zero-order wave function may be written

$$\psi_0 = \chi(r)Y^{lm}(\theta, \phi) \tag{27}$$

where $Y^{lm}(\theta, \phi)$ are the normalized spherical harmonics and $\chi(r)$ is a radial function normalized according to

$$\int_0^\infty \chi^2(r)\,\mathrm{d}r = 1. \tag{28}$$

Consider first an s-state

$$\psi_0 = \frac{\chi(r)}{\sqrt{4\pi}}. \tag{29}$$

From considerations of symmetry it may be easily seen that the first-order perturbation energy of an s-state by $\mathscr{H}_1$ is zero. The Schrödinger equation therefore becomes

$$(\mathscr{H}_0 + \mathscr{H}_1)(\psi_0 + \psi_1) = E_0(\psi_0 + \psi_1). \tag{30}$$

Expanding this to the first order in Q we obtain

$$(\mathscr{H}_0 - E_0)\psi_1 = -\mathscr{H}_1\psi_0 \tag{31}$$

$$(\mathscr{H}_0 - E_0)\psi_1 = \frac{Q}{2r^3}(3\cos^2\theta - 1).\frac{\chi(r)}{\sqrt{4\pi}}. \tag{32}$$

Since neither $\mathscr{H}_0$ nor E_0 are angularly dependent the whole angular dependence of the left-hand side of equation (32) (and this must be the same as the angular dependence of the right-hand side) must be contained in ψ_1. Thus ψ_1 may be written as

$$\psi_1 = \frac{Q\chi_1(r)(3\cos^2\theta - 1)}{4\sqrt{\pi}}. \tag{33}$$

The term Q has been introduced to emphasize that ψ_1 is linearly dependent on the magnitude of the perturbation by the nuclear quadrupole moment. The factor $1/4\sqrt{\pi}$ in introduced for convenience and facilitates comparison with Sternheimer's work. Note that $\chi_1(r)$ is a function dependent only on r and it is assumed to be normalized according to the equation

$$\int_0^\infty \chi_1^2(r)\,dr = 1. \tag{34}$$

It will be appreciated that ψ_1 is a "d" function and the perturbation corresponds to an angular distortion of the s-orbital.

The total quadrupole moment of the perturbed electron distribution is given by the expectation value of the operator $\sum_i r^2(3\cos^2\theta - 1)$ over the wave function $(\psi_0 + \psi_1)$. The summation is over the different i electrons in the system. The induced quadrupole moment is then given by

$$\Delta Q = 4\int_r \psi_0|r^2(3\cos^2\theta - 1)|\psi^0. \tag{35}$$

It has been assumed that the initial s-orbital contains two electrons. Thus

$$\begin{aligned}\Delta Q_s &= \frac{Q}{2\pi}\int_0^\infty \chi_0(r)\chi_1(r)r^2\,dr\int_0^\pi (3\cos^2\theta - 1)^2\sin\theta\,d\theta\int_0^{2\pi} d\phi\\ &= -\frac{5Q}{8}\int_0^\infty \chi_0(r)\chi_1(r)r^2\,dr.\end{aligned} \tag{36}$$

The evaluation of the induced quadrupole moment then depends upon the evaluation of the integral in equation (36) and hence of the perturbed radial function $\chi_1(r)$. From equation (32), $\chi_1(r)$ is given by

$$-\frac{d^2\chi_1(r)}{dr^2} + \frac{6\chi_1(r)}{r^2} + V_0(r)\chi_1(r) - E_0\chi_1(r) = \frac{\chi_1(r)}{r^3}. \tag{37}$$

The possibility of solving this equation depends on the form of $V_0(r)$. For hydrogen-like atoms in their ground states Dalgarno[3] has shown that the quadrupole shielding factor may be calculated exactly and is given by the formula

$$\frac{\Delta Q}{Q} = \gamma = \frac{1}{3Z} \tag{38}$$

where Z is the nuclear charge.

The calculation of the quadrupole shielding of a p-electron may be carried out in much the same way as for an s-electron above. Consider first a p-orbital with $m = 0$, i.e. directed along the axis of the perturbing field-gradient. Then the unperturbed wave-function may be written

$$\psi_0 = \sqrt{\frac{3}{4\pi}}\chi(r) \,.\, \cos\theta. \tag{39}$$

Unlike the case for s-electrons the first order perturbation energy, E_1, of p-electrons by a field-gradient or nuclear quadrupole moment does not vanish. It is given by

$$E_1 = -\frac{Q}{2}\int_0^{\pi}\int_0^{\infty} \psi_0^2 \frac{(3\cos^2\theta - 1)}{r^3}\sin\theta \, d\theta \tag{40}$$

$$= -\frac{2Q}{5}\left\langle\frac{1}{r^3}\right\rangle_{2p}. \tag{41}$$

As before the perturbation obeys the Schrödinger equation

$$(\mathscr{H}_0 - E_0)\psi_1 = -(\mathscr{H}_1 - E_1)\psi_0. \tag{42}$$

The above value of E_1 and the explicit form of $\mathscr{H}_1$ yields

$$-\mathscr{H}_0\psi_0 = \sqrt{\frac{3}{4\pi}}\frac{Q\chi_0(r)}{r^3}[(\tfrac{3}{2}\cos^3\theta - \tfrac{1}{2}\cos\theta)]. \tag{43}$$

The factor in brackets may be rewritten as the sum of an f and a p function

$$-\mathscr{H}_1\psi_0 = \sqrt{\frac{3}{4\pi}}\frac{Q}{r^3}\chi_0(r)[(\tfrac{3}{2}\cos^3\theta - \tfrac{9}{10}\cos\theta) + \tfrac{2}{5}\cos\theta]. \tag{44}$$

Substituting this in equation (42) yields

$$(\mathscr{H}_0 - E_0)\psi_1 = Q\cdot\sqrt{\frac{3}{4\pi}}\cdot(\tfrac{3}{2}\cos^2\theta - \tfrac{9}{10}\cos\theta)\frac{\chi_0(r)}{r^3}$$

$$+ Q\sqrt{\frac{3}{4\pi}}\cdot(\tfrac{2}{5}\cos\theta)\chi_0(r)\left[\frac{1}{r^3} - \left\langle\frac{1}{r^3}\right\rangle_{2p}\right]. \tag{45}$$

Thus ψ_1 is given by

$$\psi_1 = Q\sqrt{\frac{3}{4\pi}}(\tfrac{3}{2}\cos^3\theta - \tfrac{9}{10}\cos\theta).[\psi_1(r)]_f + Q\sqrt{\frac{3}{4\pi}}.\tfrac{2}{5}\cos\theta[\psi_1(r)]_p \tag{46}$$

where the radial functions $(\psi_1(r))_f$ and $(\psi_1(r))_p$ are obtained from the Schrödinger equations

$$\left(-\frac{\mathrm{d}}{\mathrm{d}r^2} + \frac{12}{r^2} + V_0 - E_0\right)(\psi_1(r))_f = \frac{\psi_0(r)}{r^3}; \tag{47}$$

$$\left(-\frac{\mathrm{d}}{\mathrm{d}r^3} + \frac{2}{r^2} + V_0 - E_0\right)(\psi_{.1}(r))_p = \psi_0(r)\left(\frac{1}{r^3} - \left\langle\frac{1}{r^3}\right\rangle\right)_p. \tag{48}$$

The quadrupole shielding of p-electrons is thus given by an angular distortion and a radial distortion. The induced moment due to the angular distortion of the $m = 0$ electrons, $\Delta Q_{f,0}$, is thus given by

$$\Delta Q_{f,0} = 6Q\int_0^\infty \chi_0(r)\chi_1(r)_f r^2\,\mathrm{d}r\int_0^\pi [\tfrac{3}{2}\cos^4\theta - \tfrac{9}{10}\cos^2\theta](3\cos^2\theta - 1)\sin\theta\,\mathrm{d}\theta$$

$$= \frac{216Q}{175}\int_0^\infty \chi_0(r)\chi_f(r)r^2\,\mathrm{d}r. \tag{49}$$

A similar treatment of the $m = \pm 1$ electrons yields for the angular distortion

$$\Delta Q_{f,\pm 1} = \frac{144Q}{175}\int_0^\infty \chi_0(r)\chi_f(r)r^2\,\mathrm{d}r. \tag{50}$$

Thus the total angular distortion for a complete p shell is

$$\Delta Q_f = \frac{504Q}{175}\int_0^\infty \chi_0(r)\chi_f(r)r_2\,\mathrm{d}r. \tag{51}$$

The induced moment due to the radial distortion of the $m = 0$ p-electrons, $\Delta Q_{p,0}$ is given by

$$\Delta Q_{p,0} = \frac{12Q}{5}\int_0^\infty \chi_0(r)\chi_p(r)r^2\,\mathrm{d}r \int_0^\pi \cos^2\theta\,.\,(3\cos^2\theta - 1)\sin\theta\,\mathrm{d}\theta$$

$$= -\frac{32Q}{25}\int_0^\infty \chi_0(r)\chi_p(r)r^2\,\mathrm{d}r. \qquad (52)$$

Similarly for the $m = \pm 1$ electrons

$$\Delta Q_{p,\pm 1} = \frac{8Q}{25}\int_0^\infty \chi_0(r)\chi_p(r)r^2\,\mathrm{d}r. \qquad (53)$$

The total induced moment due to radial distortion is thus

$$\Delta Q_p = \frac{48Q}{25}\int_0^\infty \chi_0(r)\chi_p(r)r^2\,\mathrm{d}r. \qquad (54)$$

Equations (52) and (53) bring out one interesting point concerning the different field-gradients induced in the sigma ($m = 0$) and pi ($m = \pm 1$) p-orbitals. Had we chosen to calculate the shielding in terms of the field-gradient induced in the p shell by a field-gradient along the z-axis we would have obtained an expression of just the same form, i.e.

$$(\Delta V_{zz})_{p,0} = \tfrac{32}{25}V_{zz}\int_0^\infty \chi(r)\chi_p(r)r^2\,\mathrm{d}r \qquad (55)$$

$$(\Delta V_{zz})_{p,\pm 1} = \tfrac{8}{25}V_{zz}\int_0^\infty \chi(r)\chi_p(r)r^2\,\mathrm{d}r. \qquad (56)$$

For the induced field-gradient of the pi orbitals along either the x or y axes we have, from Laplace's equation and the cylindrical symmetry of p-orbitals,

$$(\Delta V_{xx})_{p,\pm 1} = -\tfrac{16}{25}V_{zz}\int_0^\infty \chi(r)\chi_p(r)r^2\,\mathrm{d}r. \qquad (57)$$

Thus if the sigma orbital contracts the pi orbitals will expand by half as much. We will have occasion to return to this point in Chapter 6 where exact calculations of quadrupole couplings in molecules are discussed.

B. QUADRUPOLE SHIELDING IN CLOSED-SHELL IONS IN EXTERNAL FIELD-GRADIENTS

The most clear-cut type of shielding occurs when an ion having a closed shell electronic structure, and hence a rigorously spherically symmetric electron distribution, is located in a pure ionic crystal lattice at a point where the field-gradients are non-vanishing. It transpires that the experimental coupling constant can be as much as two orders of magnitude greater than that calculated by the product of the nuclear quadrupole moment and the electrostatic field-gradients calculated from the known positions of the various point charges in the lattice. The shielding in this case is usually expressed by the formal equation

$$e^2Qq_{\text{obs}} = (1 - \gamma_\infty)e^2Qq_{\text{ionic}} \tag{58}$$

where q_{obs} is the experimental field-gradient, q_{ionic} is the field-gradient calculated from the known distribution of ionic charges and γ_∞ the shielding parameter. The subscript emphasizes that the field-gradient arises from charges which are far removed from the electrons in the ion under study. Sternheimer's initial attempts at calculating quadrupole shielding factors were based on the Thomas-Fermi model of the atom. It turns out that this approach only yields shielding factors due to *angular* displacements of the electrons and is thus unsatisfactory for quantitative studies, but it was sufficient to indicate that the effect was appreciable. Most of Sternheimer's results have been based on the numerical integration of equations (47) and (48) using Hartree-Fock wave functions for the unperturbed ground state of the atom. This procedure is tedious and subject to cumulative errors in the numerical integration. A variational approach has been suggested by Das and Bersohn.[4] After consideration of the form of the perturbed wave-functions obtained by Sternheimer as well as the results of the exact calculations on hydrogen-like systems, they proposed that the perturbed radial wave-functions be written in the form

$$\chi'(r) = \frac{\chi_0(r)}{R^3}[(\alpha + \beta r + \gamma r^2)r^2 + \delta] \tag{59}$$

where α, β and γ are parameters whose values are altered to minimize the perturbation energy in the variation procedure. δ is a term introduced to maintain the orthogonality of $\chi'(r)$ and $\chi_0(r)$, it is thus given by

$$\delta = -\int_0^\infty (\chi_0'(r))^2(\alpha + \beta r + \gamma r^2)r^2\,\mathrm{d}r \tag{60}$$

$$= -[\alpha\langle r^2\rangle + \beta\langle r^3\rangle + \gamma\langle r^4\rangle]. \tag{61}$$

As usual of course the unperturbed wave-functions were taken to the Hartree–Fock functions for the atoms in question. In practice the effect of the parameter γ was very small and subsequent calculations by this method have often omitted this term from the trial wave-function. This method has a distinct advantage over that of numerical integration of equations (47) and (48) in that the shielding factors are obtained with much less computational effort.

Watson and Freeman[5] have proposed yet another method of calculating shielding factors. Hartree–Fock functions are in general written as products of atomic orbitals ϕ_{nlms} having the following form

$$\phi_{nlms} = \chi(r)_{nl} Y^{l,m}(\theta, \phi) P(s) \tag{62}$$

where $\chi(r)$ is a radial function depending only on the level index n and the orbital angular momentum quantum number l, $Y^{l,m}(\theta, \phi)$ is a spherical harmonic, dependent only on l and the projection quantum number m while $P(s)$ is one of the two spin functions and is independent of n, l and m. The validity of this separation of variables depends on the atoms having spherical symmetry. It is thus only strictly acceptable for closed-shell systems but up to the present it has been adopted for all systems in the interest of computational simplicity. We shall return to this point in the next section where coupling constants in atomic systems are discussed; for the moment we are concerned with ionic shielding constants and in this case the separation of variables is valid for the isolated ions. Watson and Freeman made the suggestion that the restriction of the independence of $\chi(r)$ on m be removed and the wave functions recalculated in the presence of a perturbing field-gradient. In this way the *radial* contribution to the shielding constant is obtained. To calculate the angular contribution would mean that the radial and angular terms could no longer be separated and the computational difficulties that this would involve induced Freeman and Watson to neglect this term. In the case of shielding in ions, however, it transpires as will be seen below that the radial perturbation contributes about 90 per cent of the shielding so that the angular shielding may be estimated approximately to give a satisfactory value for the total shielding.

Table 1 shows a comparison of the results of these three different methods of calculations for the Cu^+ and Cl^- ions. The results have been broken down to show the relative importance of the various distortions to the total shielding, thus the term $\gamma_\infty(2p \rightarrow p)$ is the shielding due to the radial distortion of the $2p$ shell. The angular terms are not however listed separately since they are so much smaller. It is most satisfactory that the results of calculations by Sternheimer's method (numerical integration) and by the variation approach are so similar since the basic approach of these methods is the same. The results obtained by Watson and Freeman are however very different

TABLE 5.1. Shielding factors (γ_∞) calculated for Cl^- and Cu^+ by three different methods.

	Cl^-			Cu^+		
	Sternheimer[7]	Das[8]	Watson[5]	Sternheimer[7]	Das[9]	Watson
$\gamma_\infty(2p \to p)$	−1·5	−0·992	−12·1	−0·62	−0·62	−1·2
$\gamma_\infty(3p \to p)$	−56·5	−56·246	−78·3	−7·9	−5·75	−7·8
$\gamma_\infty(3d \to d)$	—	—	—	−8·5	−8·59	−8·6
γ_∞(ang)	+1·4	+1·724	+4·5*	+2·0	+1·19	+0·6
γ_∞(total)	−56·6	−55·514	−85·9	−15·0	−13·77	−17·0

particularly for the Cl^- ion. Watson and Freeman point out that their method, unlike those of Sternheimer or of Das and Bersohn, allows for the mutual perturbation of the *distorted* $2p$ and $3p$ shells and ascribes their higher values to this cause, particularly the higher values of the inner shell contribution. They have pursued their calculations for a number of systems and in all cases obtain higher shielding factors than is yielded by Sternheimer's method. Sternheimer[2] has attempted, without success, to bring together the shielding constants calculated by the two methods by including second-order terms in his approach and for the moment there the matter rests. However,

TABLE 5.2. Quadrupole shielding factors (γ_∞) for s-state ions.

Configuration	Nucleus	Shielding factor γ_∞	Ref	Configuration	Nucleus	Shielding factor γ_∞	Ref
$1s^2$	H^-	1·051	10			0·117	11
		1·149	10		N^{5+}	0·1011	10
		1·131	2			0·0996	11
	He	0·4189	10		O^{6+}	0·0869	11
		0·4145	10		Fe^{7+}	0·0767	11
		0·4183	10		Ne^{8+}	0·0689	11
		0·424	2	$1s^22s$	Li	1·358	10
		0·396	11			1·083	10
		0·397	12			1·079	10
	Li^+	0·2570	10	$1s^22s$	Be^+	0·7088	10
		0·2567	10		B^{++}	0·5461	10
		0·263	2		C^{3+}	0·4508	10
		0·248	11		N^{4+}	0·3842	10
		0·249	12		O^{5+}	0·3340	10
	Be^{++}	0·1857	10	$1s^22s^2$	Li^-	2·642	10
		0·189	2		Be	1·356	10
		0·181	11, 12			1·349	10
	B^{3+}	0·1453	10			1·355	10
		0·142	11, 12			1·354	10
	C^{4+}	0·1193	10			0·786	11

TABLE 5.2.—*cont.*

Configuration	Nucleus	Shielding factor γ_∞	Ref
	B^{+}	0·9996	10
		0·768	2
		0·549	11
		0·72	12
	C^{++}	0·8021	10
		0·428	11
	N^{3+}	0·6721	10
		0·361	11
	O^{4+}	0·5798	10
		0·298	11
	F^{5+}	0·5107	10
		0·259	11
	Ne^{6+}	0·229	11
$1s^2 2s^2$	O^{2-}	−429·4	10
		−950·5	10
$2p^6$		−28·2*	13
$1s^2 2s^2 2p^6$	F^{-}	−23·03	10
		−22·15	10
		−22·12	10
		−22·00	10
		−22·53	2
		−23·22	13
	Ne	−9·262	10
		−7·810	10
		−7·956	10
		−7·937	10
		−7·872	10
	Na^{+}	−3·835	10
		−3·657	10
		−3·663	10
		−4·56	2
	Mg^{++}	−3·038	10
	Al^{3+}	−2·236	10
		−2·36	2
	Si^{4+}	−1·743	10
	P^{5+}	−1·412	10
$1s^2 2s^2 2p^6 3s$	Na	−3·835	10
		−3·657	10
		−3·663	10
	Mg^{+}	−2·472	10
	Al^{++}	−1·787	10
	Si^{3+}	1·353	10
	P^{4+}	−1·059	10
	S^{5+}	−0·8502	10
$Ne3s^2$	Na^{-}	−2·131	10

Configuration	Nucleus	Shielding factor γ_∞	Ref
	Mg	−1·776	10
	Al^{+}	−1·270	10
	Si^{++}	−0·8995	10
	P^{3+}	−0·6502	10
	S^{4+}	−0·4783	10
	Cl^{5+}	−0·3567	10
Ar	Cl^{-}	−66·56	10
		−53·91	10
		−56·6	2
		−50·07	13
	Ar	−5·146	10
		−19·85	10
		−25·22	10
		−25·58	10
		−25·83	10
		−30·19	10
	K^{+}	−12·17	10
		−17·32	2
	Ca^{++}	−12·12	10
	Sc^{3+}	−9·461	10
	Ti^{4+}	−7·721	10
	V^{5+}	−6·503	10
$Ar4s$	K	−15·70	10
	Ca^{+}	−11·91	10
$Ar4s^2$	K^{-}	−14·33	10
	Ca	−11·26	10
$Ar3d^5 4s^2$	Mn^{2+}	−11·37	2
		−6·81	13
	Fe^{3+}	9·14	2
		−6·17	13
$Ar3d^{10} 4s^2$	Cu^{+}	−15·0	2
		−17·0	5
	Ga^{3+}	−9·50	2
		−6·94	13
Kr	Br^{-}	−123·0	2
		−100	14
	Kr	−40·96	13
	Rb^{+}	−47·2	2
		−50	14
$Kr4d^{10}$	Ag^{+}	−34·9	2
		−22·11	13
Xe	I^{-}	−138·4	2
		−175	14
	Cs^{+}	−102·5	2
		−110	14

* For contracted O^{2-} in MgO lattice.

as Sternjeimer points out the question is one of mainly academic interest since in any case the wave-function for an ion in a crystal will be significantly different from that for the isolated ion. The important thing is that the order of magnitude of the shielding is the same whatever the method used to calculate it.

Table 2 shows the results of calculations of γ_∞ for closed-shell ions. Many of these are hypothetical entities and their calculated shieldings are of a somewhat limited interest.

TABLE 5.3. Quadrupole shielding factors (γ_∞) for rare-earth ions

	Shielding factor γ_∞	Reference
La^{3+}	−68·0	14
Ce^{3+}	−71·0	14
Pr^{3+}	−80·9	2
	−80·487	15
Tm^{3+}	−75·3	2
	−74·199	15
Yb^{3+}	−79·0	14

Table 3 shows shielding factors calculated for a number of rare-earth ions for which quadrupole coupling constants are becoming available from Mössbauer spectroscopy. These are not of course in general ions having a closed-shell structure and the *f*-electrons give rise to a field-gradient. However *f*-electrons are quite diffuse and their gradients are often less than those produced by distant charges in the lattice when shielding is taken into account.

C. STERNHEIMER SHIELDING IN ATOMS OR IONS HAVING OPEN-SHELL STRUCTURES

In many ways the question of Sternheimer shielding in open-shell structures is an artefact introduced by our inability to cope with calculations on complicated electronic structures. As pointed out in the previous section the form of the Hartree–Fock wave-functions usually adopted (equation (62)) is only strictly valid for closed-shell spherically symmetric structures. This assumption for open-shell systems implies that the field-gradient produced by all the inner completed shells is zero and that the field-gradient in the isolated atom is due entirely to the valence electrons. The fact that this is not the case is the origin of Sternheimer shielding in open-shells systems and it is clear from the foregoing that this is really a "non-phenomenon". However, we may think of building up the open-shell atom or ion by starting with the nearest closed-shell structure, for which now the usual Hartree–Fock assumption is valid, and adding the various electrons in the appropriate atomic orbitals. The field-gradients that these give rise to may then distort the inner spherical shell of electrons and thus produce a Sternheimer shielding. Note that the distortion produced by the nuclear quadrupole moment itself is still present and the theorem that the two sources of shielding produce the same contribution to the total is still true. Nevertheless almost all calculations have been made from the point of view of the perturbation produced by the valence electrons. The rather paradoxical situation holds that the total distortion of the spherical inner shell is the same whether or not the nucleus has a quadrupole moment or not, despite the fact that the nucleus does indeed distort the electron cloud. Thus we introduce our probe —an aspheric nucleus—without at all disturbing the system.

Conventionally the shielding in open-shell systems is denoted by the equation

$$e^2Qq_{\text{obs}} = e^2Qq_{\text{atomic}}(1 - R). \tag{63}$$

If in addition there is a field-gradient from external sources this term is treated separately by the methods of the previous section and the total coupling is formally expressed by the equation

$$e^2Qq_{\text{obs}} = e^2Qq_{\text{atomic}}(1 - R) + e^2Qq_{\text{ionic}}(1 - \gamma_\infty). \tag{64}$$

The methods used for calculating R values are essentially the same as those discussed in the previous section although few if any calculations seem to have been made using the variation approach of Das and Bersohn.[4] Intellectually the most satisfying is the method of Watson and Freeman[5] where one is essentially just going back and doing the Hartree–Fock calculations properly; however the results of Sternheimer's method seem to produce results very similar to those of Watson and Freeman.

It is fortunate that in general R is much smaller than γ_∞, generally lying in the range $0 < R < 1$. The reason for this may be seen very easily. Consider the radial distortions produced in a thin shell of negative charge by two point-negative charges symmetrically disposed about the centre of the shell. If the point charges are outside the shell the latter will distort to take on the shape illustrated in Fig. 2(a) and produce a field-gradient at the nucleus which reinforces that of the point charges. On the other hand if the point charges are inside the shell distorts to an elliptical form (see Fig. 2b) and the induced field-gradient opposes that produced by the point charges. Thus as a point charge is brought up to a closed-shell atom or ion from infinity, where the shielding is γ_∞, to a point where it begins to penetrate the electron cloud the total shielding will gradually diminish and can even change sign. This is illustrated by Fig. 3, taken from a paper by Sternheimer.[6]

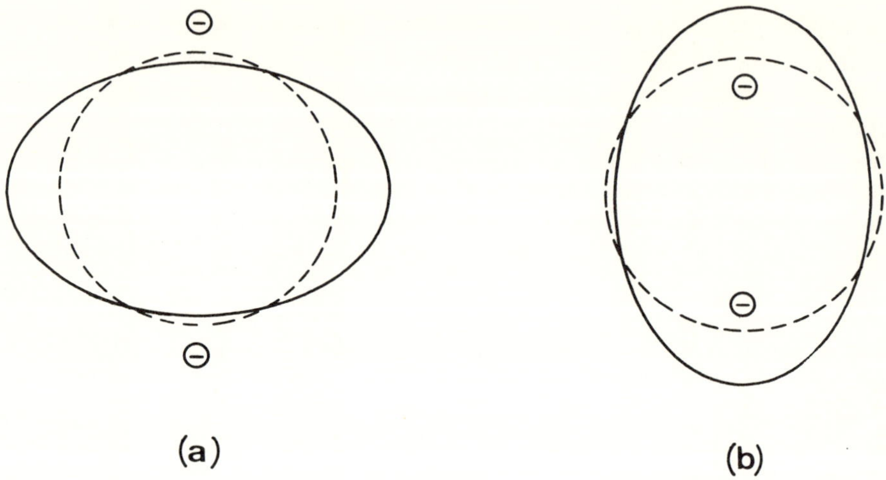

FIG. 5.2. Radial deformation of a spherical shell of negative charge (dotted line) by point negative charges outside (a) and inside (b) the shell.

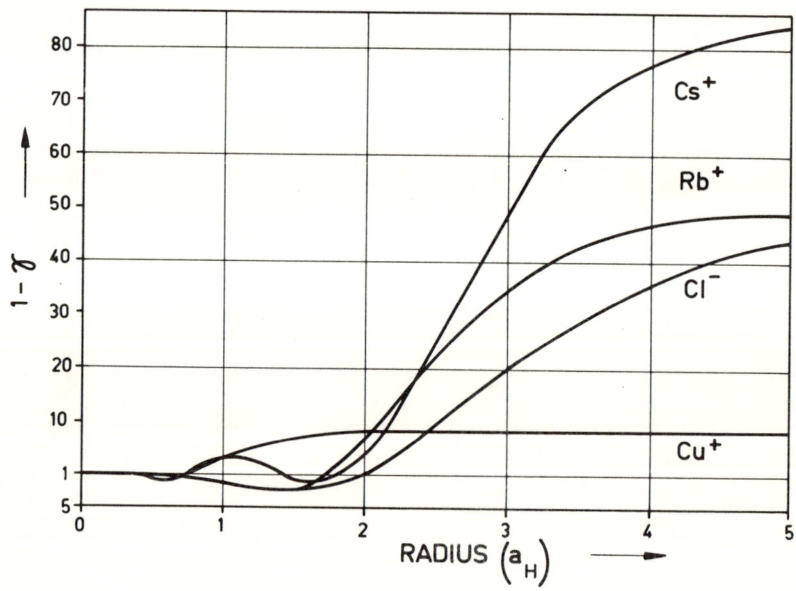

FIG. 5.3. $1 - \gamma(r)$ for Cl^-, Cu^+, Rb^+ and Cs^+.

Table 4 shows the results of Sternheimer's systematic calculations of shielding constants in atoms in various spectroscopic states. For the Cl^+ atom Freeman and Watson have obtained similar values for R although their actual calculated field-gradients are somewhat different. This is

TABLE 5.4.

Element	State	R	$1/(1 - R)$
B	$2p$	0·143	1·17
Na	$3p$	−0·243	0·805
Cl	$3p^5$	0·112	1·13
K	$4p^a$	−0·188	0·842
K	$5p^a$	−0·142	0·876
K	$4p^b$	−0·009	0·991
K	$5p^b$	−0·003	0·997
Rb	$5p^a$	−0·271	0·787
Rb	$6p^a$	−0·209	0·827
Rb	$5p^b$	−0·141	0·876
Rb	$6p^b$	−0·120	0·893
Rb	$7p^b$	−0·110	0·901
Cs	$6p^a$	−0·332	0·751
Cs	$7p^a$	−0·275	0·784
Cs	$6p^b$	−0·204	0·831
Cs	$7p^b$	−0·178	0·849

probably mainly due to the fact that whereas they used Hartree–Fock wave-functions for neutral chlorine as their starting point, Sternheimer was forced to use chloride ion functions since these were the only ones available at the time. A number of other shielding constants have been calculated for atomic systems in individual cases as they have become necessary to derive nuclear quadrupole moments from atomic spectral data. These are in general in the same range as those shown in Table 4.

D. STERNHEIMER SHIELDING IN MOLECULES

This is even more of an artefact than Sternheimer shielding in open-shell atoms. In order to achieve satisfactory results in *a priori* calculations of field-gradients in molecules it is essential to allow for distortion of the inner shell, i.e. to describe the bonding in terms of all the atomic wave-functions of the component atoms. This question is however more conveniently discussed in Chapter 6.

REFERENCES

1. R. M. Sternheimer, *Phys. Rev.* **80**, 102 (1950); also *Phys. Rev.* **84**, 244 (1951); ibid. **86**, 316 (1952); ibid. **95**, 736 (1954); ibid. **105**, 158 (1957); ibid. **130**, 1423 (1963); ibid. **132**, 1637 (1963); ibid. **146**, 140 (1966).
R. M. Sternheimer and H. M. Foley, *Phys. Rev.* **92**, 1460 (1953); ibid. **102**, 731 (1956).
H. M. Foley, R. M. Sternheimer and D. Tycko, *Phys. Rev.* **93**, 734 (1954).

2. R. M. Sternheimer, *Phys. Rev.* **146**, 140 (1966).
3. A. Dalgarno and J. T. Lewis, *Proc. R. Soc.*, A, **233**, 70 (1956); ibid. **240**, 284 (1957); *Proc. Phys. Soc.*, A, **69**, 628 (1957).
4. T. P. Das and R. Bersohn, *Phys. Rev.* **102**, 733 (1956).
5. R. E. Watson and A. J. Freeman, *Phys. Rev.* **131**, 250 (1963).
6. H. M. Foley, R. M. Sternheimer and D. Tycko, *Phys. Rev.* **93**, 734 (1954).
7. R. M. Sternheimer and H. M. Foley, *Phys. Rev.* **102**, 731 (1956).
8. P. G. Khubchandani, R. R. Sharma and T. P. Das, *Phys. Rev.* **126**, 594 (1962).
9. E. G. Wikner and T. P. Das, *Phys. Rev.* **109**, 360 (1958).
10. F. W. Langhoff and R. P. Hurst, *Phys. Rev.* **139**A, 1415 (1965).
11. J. Lahiri and A. Mukerhji, *Phys. Rev.* **141**, 428 (1966).
12. A. Dalgarno, *Adv. Phys.* **11**, 281 (1962).
13. G. Burns and E. G. Wikner, *Phys. Rev.* **121**, 155 (1961).
14. R. E. Watson and A. J. Freeman, *Phys. Rev.* **135**A, 1209 (1964).
15. M. N. Ghatikar, A. K. Raychaudhuri and D. K. Ray, *Proc. Phys. Soc.* **86**, 1235 (1965).

CHAPTER 6

EXACT CALCULATIONS OF FIELD-GRADIENTS

A. GENERAL

In this chapter we will mainly be concerned with the rigorous evaluation of the field-gradients at the nuclei of small covalent molecules using molecular orbitals which are usually the eigenfunctions of self-consistent field (S.C.F.) Hamiltonian operators. It is convenient at this stage to discuss briefly the way in which these orbitals have been obtained.

The S.C.F. wave function is an antisymmetrized product of one-electron orbitals $\phi_{i\lambda\alpha}$ where λ labels the symmetry species, α the sub-species and i labels orbitals not distinguishable by symmetry. These orbitals $\phi_{i\lambda\alpha}$ are expressed as linear combinations of basis functions $\chi_{p\lambda\alpha}$

$$\phi_{i\lambda\alpha} = \sum_p \chi_{p\lambda\alpha} c_{i\lambda p}. \tag{1}$$

The basis functions are usually but not always normalized Slater-type functions

$$\chi_{p\lambda\alpha}(N) = R_{\lambda p}(r_N) Y^{lm}(0_N, \phi_N) \tag{2}$$

for the various nuclei, N, in the molecule. $Y^{lm}(\theta_N, \phi_N)$ are the familiar spherical harmonics while the radial functions $R_{\lambda p}(r_N)$ are given by the Slater function

$$R_{\lambda p}(r_N) = [(2n_{\lambda p})!]^{-\frac{1}{2}}(2\zeta_{\lambda p})^{n_{\lambda p}+\frac{1}{2}} r_N^{n_{\lambda p}-1} \exp(-\zeta_{\lambda p} r_N). \tag{3}$$

The expansion coefficients $C_{i\lambda p}$ are obtained usually by the iterative self-consistent Roothaan procedure.[1] The radial functions $R_{\lambda p}(r_N)$ have been the point at which most variety of choice has been exercised. The earliest calculations used radial functions calculated by Slater's rules (Slater Atom Molecular Orbitals, SAMO). The next improvement used values of the exponent, ζ, calculated to produce the best agreement with atomic properties (Best Atomic Molecular Orbitals, BAMO), while yet another treated ζ as a parameter in the molecular calculation (Best Limited Molecular Orbital). All these three basis sets have just one function for each atomic orbital of the nucleus N up to and including the orbitals in the valence shell, i.e. are minimal sets. The next step is to have two (or more) basis functions for each atomic orbital, each atomic orbital being represented by two different values of the exponent ζ (double zeta orbitals). Finally of course excited atomic orbitals can also be included.

One other set of basis functions has recently been employed. These are the Hartree–Fock–Roothaan S.C.F. orbitals for the various atoms, themselves expressed as sums of Slater-type functions. Not unsurprisingly it would appear that these yield better results more quickly than the basis sets described above. Very clear accounts of these methods can be found in papers by Ransil,[2] Huo[3] and Cade *et al.*[4]

Two more points remain before we can proceed to a review of these calculations. As the following discussion will show the major contribution to the electrostatic field-gradient at a nucleus arises from the valence p-electrons of the atom in question. It is consequently no use starting a calculation at all if the basis functions, when applied to the central-field atomic case, do not produce something like the correct field-gradient. At distances relatively far from the nucleus Slater orbitals are a fair approximation to the Hartree–Fock atomic orbitals and give reasonably good results when used in calculations of molecular binding energies since the main effects of molecule formation is to perturb the atomic wave functions at such distances. It is well-known however that at distances close to the nucleus they are very poor approximations to the Hartree–Fock orbitals and since the field-gradients at the nucleus depend on $1/r^3$, it is precisely the electron distribution close to the nucleus which dominate in this case.

In Chapter 4 it was seen that the values of $\langle 1/r^3 \rangle$ calculated from Slater orbitals were quite unsatisfactory for all atoms except those belonging to the first short period. Consequently, all calculations using minimal basis sets of Slater orbitals on such atoms are bound to give ridiculous results for the field-gradients. Even for the members of the first short period this can be a very serious source of error and we will illustrate this point below.

The objections to Slater orbitals are considerably attenuated if more than a minimal basis set is employed since a linear combination of two or more Slater orbitals begins to look much more like a Hartree–Fock orbital. Figure 1 shows a comparison of the Hartree–Fock wave functions for the $3p$ orbital of chlorine, the conventional Slater orbital and a linear combination of three Slater orbitals. In practice, however, it is not possible to increase the size of the basis set indefinitely since subsidiary minima in the energy as a function of the parameters begin to appear, thus giving rise to the possibility that the iterative computation required for a solution of the Roothaan equations converges on a false solution. In most cases so far the use of double-zeta basis functions including excited atomic orbitals has represented the limit of complexity and even here much care has to be exercised to avoid these subsidiary minima. As we will see below the field-gradients calculated from such functions are reasonably satisfactory. When the basis set employed is the Hartree–Fock–Roothaan atomic orbitals obviously on this score at least there is no objection to their suitability.

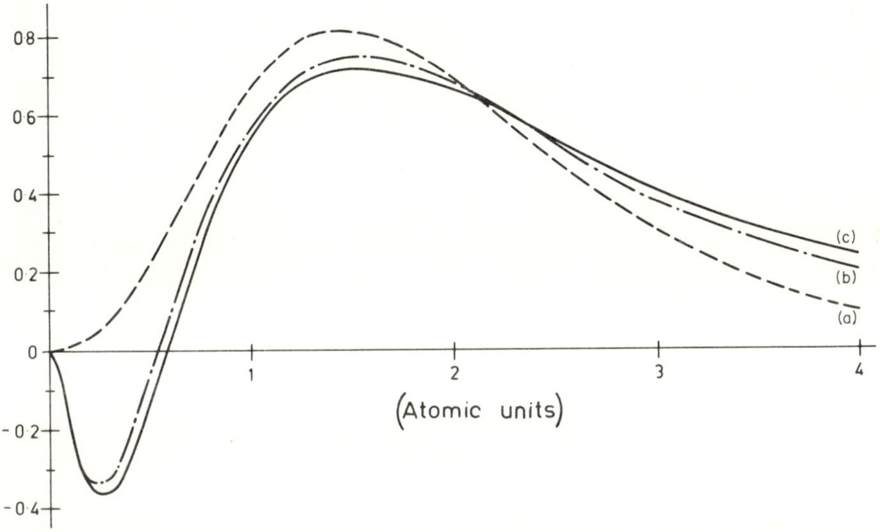

FIG. 6.1. 3*p* radial functions for a free chlorine atom. (a) one Salter orbital, (b) a linear combination of three Slater-type orbitals, (c) Hartree–Fock function.

The second point concerns the question of the so-called Sternheimer shielding. This shielding is really an artefact of the approximate methods used in calculations and provided we are talking about S.C.F. calculations a need to consider it disappears as the size of the basis set is increased. If calculations are only carried out on the valence orbitals obviously the inner shell retains its spherical symmetry and a Sternheimer shielding correction may be properly applied. If a minimal basis set is used, i.e. all occupied orbitals specifically included, this naturally already allows for some distortion of the valence shells while the use of double zeta functions and the inclusion of excited atomic orbitals of appropriate symmetry gives all the latitude necessary for a correct inclusion of inner shell polarization.

As in the case of atomic field-gradients the field-gradient at a nucleus k is a one particle operator and is thus a sum over single particle terms

$$(\mathscr{H}_q)_k = \sum_p (\mathscr{H}_q)_{kp} + \sum_n (\mathscr{H}_q)_{kn} \tag{4}$$

where the subscripts p and n denote electrons and nuclei respectively. The nuclear terms may be calculated by straightforward electrostatics since the nuclei may be considered in fixed positions. For extremely refined calculations averages may be taken over the vibrational motion of the molecule.

The electronic term is given by the expectation value of the operator over the molecular electronic wave function, Ψ_M,

$$q_{\text{electronic}} = \sum_p \int \Psi_M (\mathscr{H}_q)_k \Psi_M^* \, \mathrm{d}\tau \tag{5}$$

$$(\mathscr{H}_q)_{kp} = \frac{3\cos^2\theta_{kp} - 1}{r_{kp}^3}. \tag{6}$$

As we have seen above Ψ_M is usually expressed as a product of sums over basis functions. The actual integrals which will have to be evaluated therefore fall into four classes,

$$\int \chi_k (\mathscr{H}_q)_k \chi_k \, \mathrm{d}\tau, \qquad \int \chi_k (\mathscr{H}_q)_k \chi_j \, \mathrm{d}\tau,$$

$$\int \chi_j (\mathscr{H}_q)_k \chi_j \, \mathrm{d}\tau, \qquad \int \chi_i (\mathscr{H}_q)_k \chi_j \, \mathrm{d}\tau$$

where the subscript i, j, k labels the nucleus at the origin of the basis functions χ. For almost any basis function one would think of using the resolution of the first integral is usually trivial, but the remaining three can be more difficult. If the basis set is composed of Slater orbitals all the integrals are calculable although such calculation for the three-centre integrals is lengthy and in some early attempts at calculating field-gradients they were neglected. However, as we will see their numerical values are often comparable with those of the two-centre terms.

B. EXAMPLES OF CALCULATIONS IN SIMPLE COVALENT MOLECULES

1. Carbon Monoxide $C^{17}O$

Carbon monoxide represents one of the most convenient systems for illustration of the foregoing points. The coupling constant of the ^{17}O nucleus in this molecule is well-known[5] and the recent experimental measurements on atomic oxygen[6] combined with the Hartree–Fock–Roothaan atomic orbitals have yielded a reliable value of the nuclear quadrupole moment.[7] The experimental coupling constant can thus be translated into an electrostatic field-gradient.

After an initial, and surprisingly successful, attempt by Ransil[8] to obtain the electronic field-gradient by a simple population analysis of his S.C.F.

wave-functions,[2] Richardson[9] analysed the same wave-function rigorously to obtain a much less satisfactory value. His results are summarized in Table 1.

TABLE 6.1. Field-gradients in CO using S.C.F. LCAO molecular orbitals with Slater orbitals as bases.

$q_{nuclear}$	$q_{electronic}$	q_{total}	$q_{experimental}$
1·238 a.u.	−1·389 a.u.	−0·151 a.u.	−0·785 a.u.*

* This value differs from that in Richardson[9] since it is based on a more recent estimate of the quadrupole moment of ^{17}O.

As indeed remarked by Richardson himself and as discussed above part of this lack of success is undoubtedly due to the use of Slater orbitals in calculating the field-gradient and this is particularly important for gradients due to electrons in orbitals of the quadrupolar nucleus itself. If that part of the field-gradient arising from the electrons in the oxygen $2p$ orbitals is replaced by the empirical value for the gradient, the electronic term is increased to −1·518 a.u. and the calculated total gradient to −0·28 a.u., a distinct improvement although still far from satisfactory.

The figures in Table 1 illustrate clearly one of the great difficulties inherent in many calculations of field-gradients and certainly of most diatomic molecules. That is that the total field-gradient is the difference of two large terms, one due to the nuclei and the other to the electrons. The experimental value of the electronic field-gradient is 2·025 a.u. and the electronic field-gradient calculated by Richardson represents some 70 per cent of this. Owing to the presence of the large constant nuclear term, however, the calculated total field-gradient is only about 20 per cent of the experimental total field-gradient. To get a calculated total field-gradient within 10 per cent of the experimental value than represents a very considerable achievement and the corresponding wave function must be considered to represent very well the actual electron distribution.

More recently there have been two rather more exhaustive studies of carbon monoxide. The first by Lefebre-Brion *et al.*[10] used double zeta Slater orbitals as a basic set. The results of a simple S.C.F. calculation were supplemented by two further calculations introducing configuration interaction and yielded the much improved values of the field-gradient in the range −0·621 to −0·639 a.u. Huo[3] has studied CO, using both the double zeta Slater orbitals and Hartree–Fock–Roothaan atomic orbitals as bases. The former set of course yield values essentially those of Lefebre-Brion *et al.* while the molecular orbitals having Hartree–Fock–Roothaan atomic orbitals as a basis give a somewhat higher value for the field-gradient at the oxygen atom ($-0{\cdot}679$ a.u.).

In connection with the calculations on carbon monoxide the interesting point arises as to the internuclear distance which should be used in calculating the molecular properties. Both Lefebre–Brion *et al.*'s calculations and those of Huo were carried out for various internuclear distances, yielding the potential energy curve and it transpires that in all cases the equilibrium internuclear distance thus obtained is slightly shorter than that observed experimentally (e.g. R_e (expt) $= 2{\cdot}132$ bohr, R_e (calc., Huo) $= 2{\cdot}081$ bohr). Furthermore in all cases if the field-gradient is calculated for the theoretical internuclear distance there is obtained much better agreement with experiment than if it is calculated for the experimental internuclear distance (e.g. Lefebre-Brion *et al.*, $-0{\cdot}693$ a.u.; Huo, (HFR), $-0{\cdot}797$ a.u.).

2. Hydrogen Chloride

Before proceeding in what might appear to be a logical manner to a discussion of calculations on the molecule isoelectronic with carbon monoxide, namely N_2, we will discuss the results of a calculation on hydrogen chloride where the field-gradient at the chlorine atom is available. This calculation, by Scrocco and Tomasi,[11] is particularly interesting for the care which the authors have taken to lay out the results in such a way that the sources of the various contributions to the field-gradient are clearly visible. The wave function uses a double zeta set for the chlorine orbitals up to and including the valence levels but not including any excited configurations. For the hydrogen atom a single term was used for the 1*s* orbital although the orbital exponent was increased above the value of unity appropriate to an isolated hydrogen atom and an excited $2p_z$ orbital was also introduced. These orbitals are set out in Table 2. The following Table 3 shows the

TABLE 6.2. Basis functions for HCl.

	Orbital exponent		Orbital exponent
$1s_{H} = h$	1·27	$2p_{xCl} = 2x$	8·35
$2p_{\sigma H} = p$	2·20	$2p'_{xCl} = 2\xi$	4·50
$1s_{Cl} = 1s$	16·7	$3p_{xCl} = 3x$	2·64
$2s_{Cl} = 2s$	8·35	$3p'_{xCl} = 3\xi$	1·42
$2s'_{Cl} = 2\sigma$	4·50	$2p_{yCl} = 2y$	8·35
$3s_{Cl} = 3s$	2·64	$2p'_{yCl} = 2\eta$	4·50
$3s'_{Cl} = 3\sigma$	1·42	$3p_{yCl} = 3y$	2·64
$2p_{zCl} = 2z$	8·35	$3p'_{yCl} = 3\eta$	1·42
$2p'_{zCl} = 2\zeta$	4·50		
$3p_{zCl} = 3z$	2·64		
$3p'_{zCl} = 3\xi$	1·42		

TABLE 6.3. Molecular orbitals for HCl

	$\phi_{1\sigma}$	$\phi_{2\sigma}$	$\phi_{3\sigma}$	ϕ^{b}	ϕ^{nb}	$\phi_{1\pi}, \phi_{1\bar{\pi}}$	$\phi_{2\pi}, \phi_{2\bar{\pi}}$
h	0·00011	−0·00430	−0·00273	−0·52557	0·00065		
p	0·00031	−0·01744	−0·00696	−0·04993	−0·00684		
$1s$	−0·98923	0·32065	−0·00470	−0·01254	−0·09150		
$2s$	−0·02604	−0·26299	0·00512	0·00825	0·04627		
2σ	0·00996	−0·91454	0·01101	0·07009	0·40699		
$3s$	−0·00361	0·22769	−0·00088	−0·09986	−0·98661		
3σ	0·00131	−0·08721	0·00213	0·08842	−0·17552		
$2z$	−0·00024	−0·00261	−0·50080	0·09804	−0·03102		
2ζ	0·00025	−0·01502	−0·58176	0·15235	−0·03588		
$3z$	−0·00026	0·01142	0·07092	−0·49223	0·12745		
3ζ	0·00004	−0·00158	−0·02434	−0·24368	0·14082		
$2x$						0·50349	−0·12807
2ζ						0·57855	−0·17719
$3x$						−0·06830	0·60161
3ζ						0·02495	0·53581
eq_{zz}^{HCl}	0·0	−0·025	−158·601	−6·589	−0·545	159·268	10·282

calculated molecular orbitals for HCl and the last row of this table shows the contribution of the various molecular orbitals to the electronic field-gradient. The functions ϕ^b and ϕ^{nb} are linear combinations of the orbitals $\phi_{4\sigma}$ and $\phi_{5\sigma}$

$$\phi^b = 0{\cdot}473218\,\phi_{4\sigma} + 0{\cdot}880945\,\phi_{5\sigma} \tag{7}$$

$$\phi^{nb} = 0{\cdot}880945\,\phi_{5\sigma} - 0{\cdot}473218\,\phi_{4\sigma} \tag{8}$$

chosen so as to minimize the contributions of hydrogenic orbitals in ϕ^{nb} so that ϕ^b and ϕ^{nb} closely approximate the bonding orbital and the chlorine lone-pair orbital respectively of conventional hybridization theory.

The total electronic field-gradient from this wave-function is 3·789 a.u. which together with the contribution from the hydrogen nucleus of 0·143 a.u. yields a total field-gradient of 3·932 a.u. corresponding to a coupling constant for the ^{35}Cl nucleus of 72·9 Mc/s in very good agreement with experiment (68·0 ± 2·8 Mc/s). The inner shell electrons essentially occupy the orbitals $\phi_{1\sigma}, \phi_{2\sigma}, \phi_{3\sigma}, \phi_{1\pi}, \phi_{1\bar{\pi}}$, and the contribution of these to the field-gradient is 0·642 a.u. or 17 per cent of the total. This term of course corresponds approximately to the Sternheimer shielding. Written in this way, therefore, the main source of the field-gradient comes from the bonding orbital ϕ^b and the non-bonding orbitals ϕ^{nb}, $\phi_{2\pi}, \phi_{2\bar{\pi}}$.

A further point is brought out if we rewrite these orbitals in another way,

$$\phi^b = 0{\cdot}07084\,\mathrm{S}^b + 0{\cdot}63967\,\mathrm{P}_Z^b + 0{\cdot}52794\,\mathrm{H}^b \tag{9}$$

$$\phi^{nb} = 0{\cdot}96974\,\mathrm{S}^{nb} - 0{\cdot}23809\,\mathrm{P}_Z^{nb} - 0{\cdot}00686\,\mathrm{H}^{nb} \tag{10}$$

where the terms S^b, P_Z^b, S^{nb}, P_Z^{nb} collect together terms having the same magnetic quantum number, l, and centred on the chlorine atom while the last term collects together the hydrogenic orbitals. The explicit forms for these are

$$\left\{\begin{aligned} \mathrm{S}^b &= 0{\cdot}17705(1s) - 0{\cdot}11641(2s) - 0{\cdot}98942(2\sigma) + 1{\cdot}40975(3s) \\ &\quad - 1{\cdot}24819(3\sigma) \\ \mathrm{P}_Z^b &= -0{\cdot}15327(2z) - 0{\cdot}23816(2\zeta) + 0{\cdot}76951(3z) + 0{\cdot}38095(3\zeta) \\ \mathrm{H}^b &= 0{\cdot}99552(h) + 0{\cdot}09458(p) \end{aligned}\right.$$

$$\left\{\begin{aligned} \mathrm{S}^{nb} &= 0{\cdot}09435(1s) - 0{\cdot}04771(2s) - 0{\cdot}41969(2\sigma) + 1{\cdot}01740(3s) \\ &\quad + 0{\cdot}18099(3\sigma) \\ \mathrm{P}_Z^{nb} &= -0{\cdot}13028(2z) - 0{\cdot}15072(2\zeta) + 0{\cdot}53533(3z) + 0{\cdot}59147(3\zeta) \\ \mathrm{H}^{nb} &= 0{\cdot}09468(h) - 0{\cdot}99855(p) \\ \mathrm{P}_\pi &= -0{\cdot}12807(2x) - 0{\cdot}17719(2\xi) + 0{\cdot}60161(3x) + 0{\cdot}53581(3\xi). \end{aligned}\right. \tag{11}$$

In Fig. 2 the radial functions of the two orbitals P_Z^b, P_Z^{nb} are plotted together with the radial function of the corresponding Hartree–Fock orbital. The correspondence between these three orbitals is remarkable.

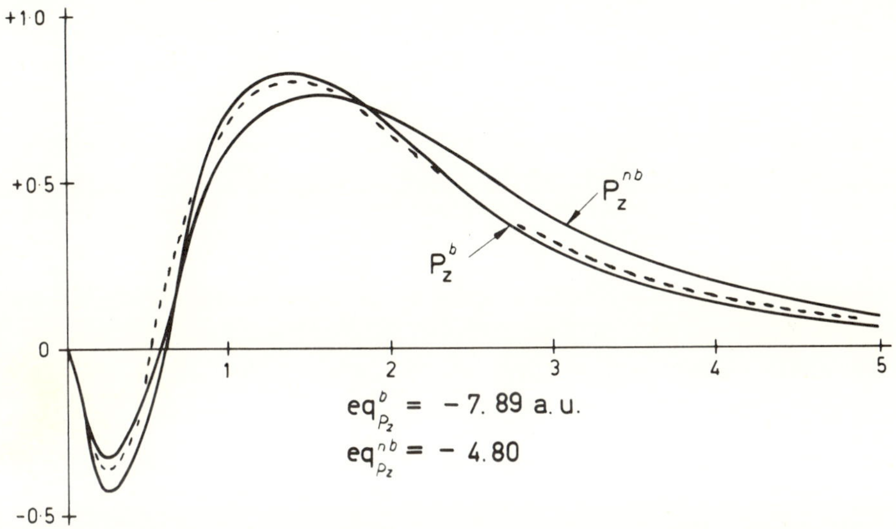

FIG. 6.2. P_Z^b and P_Z^{nb} functions (see text) for HCl [$P = rR(r)$]. The dotted line is the Hartree-Fock function for the 3p orbital of the neutral chlorine atom.

Yet another interesting break-down of the contribution of the orbitals ϕ^b and ϕ^{nb} to the field-gradient is afforded by Tables 4 and 5 where the diagonal and non-diagonal contributions of the orbitals S^b, P_Z^b, etc. are shown. Clearly the overlap terms are almost negligible being much less than 1 per cent

TABLE 6.4.

ϕ^b	S^b	P_z^b	H^b
	0·00502	0·04531	0·03740
S^b	0·0	0·0	0·02128
	0·0	0·0	0·00159
	0·04531	0·40917	0·33771
P_z^b	0·0	−7·89044	−0·04196
	0·0	−6·45683	−0·02834
	0·03740	0·33771	0·27872
H^b	0·02128	−0·04196	−0·14070
	0·00159	−0·02834	−0·07843

TABLE 6.5.

ϕ^{nb}	S^{nb}	P_z^{nb}	H^{nb}
	0·94039	−0·23088	−0·00665
S^{nb}	0·0	0·0	0·05972
	0·0	0·0	−0·00079
	−0·23088	0·05668	0·00163
P_z^{nb}	0·0	−4·79659	0·04596
	0·0	−0·54379	0·00015
	−0·00665	0·00163	0·00005
H^{nb}	0·05972	0·04596	−0·21277
	−0·00079	0·00015	−0·00002

of the total while although the contribution of the H^b orbitals are a little greater than this they go a long way towards compensating for the field-gradient produced by the hydrogen nucleus. The most important terms are those due to P_Z orbitals and the contribution of P_Z^b is more than ten times that of P_Z^{nb}. Despite the superficial similarity of the orbitals P^b and P^{nb} the field-gradients along Z-axis produced by an electron in them are respectively −7·89 a.u. and −4·80 a.u., while the P^π orbitals have a corresponding field-gradient of −5·14 a.u. We will return to this point below when discussing the approximate Townes and Dailey method of analysing quadrupole coupling constants.

Cade[12] has calculated the coupling constant of the ^{35}Cl nucleus in HCl as −67·9 Mc/s, in excellent agreement with experiment, but details of these calculations are not at present available.

3. Nitrogen, N_2

An appreciation of the significance of the results of a calculation of a field-gradient in molecular nitrogen and indeed in any nitrogen-containing molecule is complicated by the absence of a reliable value for the nuclear quadrupole moment of ^{14}N. This is due to the lack of experimental information about the quadrupole coupling constant of any atomic state of ^{14}N, the ground state of atomic nitrogen being the spherically symmetric ^4S so that conventional atomic beam studies yield no information on this point while the coupling constant is so small (not more than 10–20 Mc/s), that its effect on the excited levels of atomic nitrogen would be undetectable in atomic spectroscopy. The only hope for such a measurement would appear to be the atomic beam resonance spectroscopy of an excited state of atomic nitrogen. However one consequence of this state of affairs is that a value for the quadrupole moment must be derived from coupling constants of molecules and the number of calculations of the field-gradients in small nitrogen-containing molecules is correspondingly large.

Because a value for the coupling constant of molecular nitrogen (4·65 Mc/s) has only become available relatively recently,[13] molecular nitrogen was far from being one of the first molecules to be investigated in this way. It is however likely to give more accurate values of the nuclear field-gradient than the isoelectronic carbon monoxide, which as we have seen above already seems to have yielded a satisfactory value for the field-gradient at the oxygen nucleus. As for this last molecule the first calculations on N_2 were due to Richardson[9] and Tables 6 and 7 show a summary of his results. They clearly illustrate the considerable sensitivity of the field-gradient to the basis set and the results for the BLMO and double zeta basis sets illustrate very well the point that the exponents of the p_σ and p_π orbitals may be very different.

The nitrogen molecule has been the subject of a recent very searching investigation by Cade *et al.*[4] which probably has resolved the question of the nuclear quadrupole moment of ^{14}N once and for all. Two separate types of

TABLE 6.6. Orbital exponents for the N_2 wave functions.

	Slater-zeta	Wave function Best limited	Double-zeta*
$1s$	6·7	6·6675	6·6645
$2s$	1·95	1·917	1·93
$2p\sigma$	1·95	2·252	2·30
$2p\pi$	1·95	1·909	1·89

* Listed for $\zeta(2s)$ and $\zeta(2p\sigma)$ are what would be the best over-all ζ's obtained by an approximate averaging over each of the MO's.

TABLE 6.7 Contributions to q_A^a in N_2

	Slater AO's	Wave function Best limited	Double-zeta
$q_{A,AA}$	−0·397	−1·323	−0·992
$q_{A,AB}$	−0·459	−0·532	−0·522
$q_{A,BB}$	−1·038	−1·071	−1·138
Electronic contributions	−1·894	−2·926	−2·652
$q_{A,B}$ nucleus	+1·584	+1·584	+1·584
Net molecular q_A	−0·310	−1·341	−1·068
Calculated Q (units of 10^{-24} cm^2) for N^{14}	+0·064	+0·015	+0·019

* Expressed in a.u. A refers to one nucleus; B to the other.

basis functions were employed. In the first, starting with Ransil's BLMO functions,[2] the size of a Slater-type orbital basis set was gradually increased including both double zeta functions and excited atomic orbitals, and the various molecular properties calculated at each stage. The results for the field-gradient at the Nitrogen nucleus are illustrated in Fig. 3; the details of each computational stage will be found in the original paper. It is amusing to note that this field-gradient after 20 stages has settled down to almost the original value! The rather violent oscillations in the early stages of the computation were paralleled by the behaviour of other computed molecular properties. Also shown in Fig. 3 are the results of a shorter investigation using as a basis set the Hartree–Fock–Roothaan atomic orbitals for atomic nitrogen, and clearly computations based on this set attain stability quite

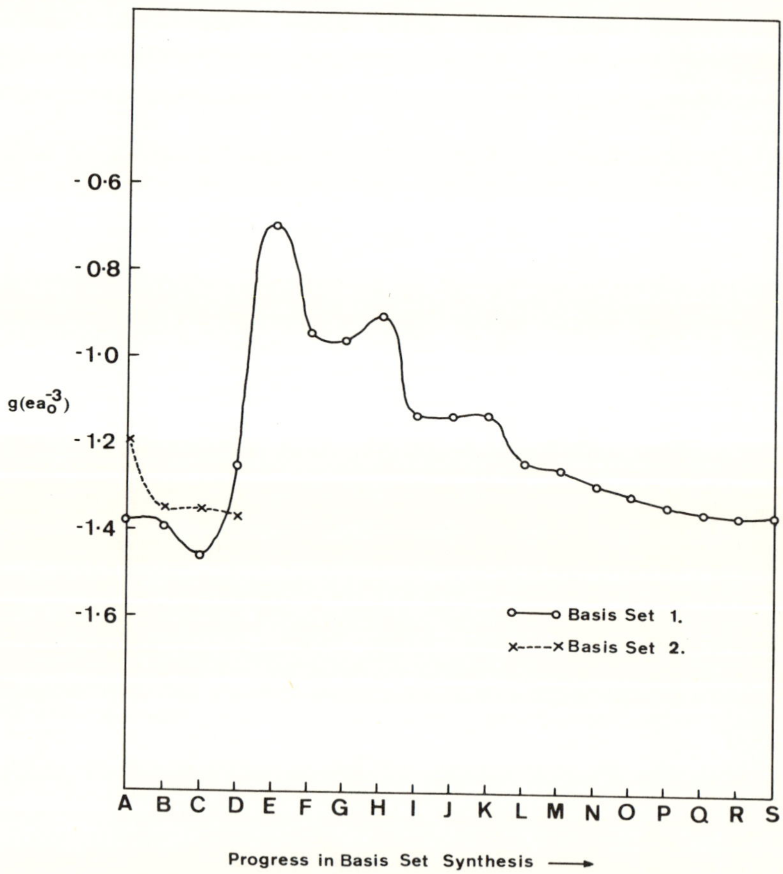

FIG. 6.3. The field-gradient at the nitrogen atom in the N_2 molecule as a function of the size and type of the basis set. 0—0 Basis set 1; x--x basis set 2.

rapidly. It seems fairly certain that the field-gradient in N_2 cannot be far-removed from $-1{\cdot}35$ a.u. implying a quadrupole moment for ^{14}N of $+0{\cdot}015$ barn.

The results of a number of other calculations on nitrogen-containing molecules of greater or less sophistication are summarized in Table 8. An interesting estimate of the field-gradient in ammonia has been made by Bader and Jones[14] from an electron distribution obtained from the simple requirement that in the equilibrium configuration the forces exerted by the nuclei on the electrons are exactly balanced by the internuclear repulsions. The basis functions used in obtaining this distribution were essentially conventional Slater orbitals although the hydrogen atom exponent was increased from unity to 1·29, so that the inadequacies of Slater orbitals discussed above naturally still apply although they are not too serious for nitrogen. Nevertheless the results obtained by a method which involves no quantum mechanical computation in the conventional sense are striking and the introduction of such restraints into complex S.C.F.–M.O. calculations will surely rapidly become standard practice.[15]

An interesting attempt at obtaining a direct measurement of the ^{14}N nuclear quadrupole moment has been made by Lin[16] in a study of the paramagnetic NO molecule. A consideration of the hyperfine levels of this molecule shows that the quadrupole Hamiltonian connects magnetic levels in which the electronic magnetic moment is parallel or anti-parallel to the molecular axis and introduces a term proportional to the field-gradient produced by the unpaired electron. In a manner analogous to that employed in the evaluation of the field-gradient in atoms, this field-gradient (or a quantity related to it) can be obtained from the experimentally observed magnetic fine structure so that the quadrupole moment of ^{14}N may be determined from the ratio of two experimental quantities. Lin obtains $Q_{^{14}N} = +0{\cdot}016 \pm 0{\cdot}007 \times 10^{-24}\,\text{cm}^2$. The rather large uncertainty arises from difficulties in the experimental determination of the parameters involved rather than in the theory behind it. It is gratifying that this range embraces the majority of the values given in Table 8, and the central value is very close to that obtained by Cade *et al.* from their calculations on N_2.

4. Lithium and Lithium Hydride

Because the quadrupole moment of ^{7}Li, like ^{14}N, is not available from atomic measurements a rather considerable amount of effort has gone into the calculation of the field-gradient at the lithium atom in these two gaseous diatomic molecules. For Li_2, which for a long time was the only molecule for which experimental results were available, the coupling constant[17] is small (60 kc/s) and early considerations[18] of the best wave-functions then

TABLE 6.8. Calculations of field-gradient in nitrogen-containing molecules

Molecule	Wave-function	Electronic gradient	Nuclear gradient	Total gradient	Experimental coupling constant	^{14}N Quadrupole moment (barns)	Ref.
HCN	a	−3·903	1·174	−2·729	−4·58	+0·007	40
HCN	b	−1·655	1·174	−0·481	−4·58	+0·0406	28
NH_3	c	−1·39467	−0·224115	−1·61879	−4·10	+0·0108	41
NH_3	c	−1·39467	−0·246876	−1·64155	−4·10	+0·0109	41
NH_3	c	−1·39467	−0·23133	−1·62600	−4·10	0·0107	41
NH_3	c	−1·61045	−0·23133	−1·84178	−4·10	0·0094	41
NH_3	c	−1·60973	−0·23133	−1·84106	−4·10	0·0094	41
NH_3	c	−1·25567	−0·23133	−1·48700	−4·10	0·0117	41
NH_3	c	−1·08839	−0·23133	−1·31972	−4·10	0·0132	41
NH_3	c	−0·94909	−0·23133	−1·15536	−4·10	0·0150	41
NH_3	c	−1·889	−0·23133	−2·120	−4·10	0·0082	13
NH_3	c	−1·824	−0·23133	2·055	−4·10	0·0085	13
$[N^*–N–N]^-$	d	−0·580	1·661	1·081	±1·7	±0·0067	28
$[N^*–N–N]^-$	e	−1·287	1·661	0·374	±1·7	±0·0194	28
$[N–N^*–N]^-$	d	−2·568	2·953	0·385	±1·0	±1·0111	28
$[N–N^*–N]^-$	e	−4·294	2·953	−1·341	±1·0	±0·0032	28

[a] L.C.A.O. Slater orbitals.
[b] L.C.A.O.–S.C.F. best atom Slater orbitals.
[c] L.C.A.O.–S.C.F. Slater orbitals.
[d] L.C.A.O.–S.C.F. best atom Slater orbital.
[e] L.C.A.O.–S.C.F.

available showed that the electronic and nuclear terms almost exactly cancelled and the sign of the theoretical field-gradient (and hence of the ^{7}Li nuclear quadrupole moment) changed with a slight change in wave-function. This behaviour of course is due to the fact that the valence electron in lithium is in the 2*s* shell although the energy difference between the 2*s* and 2*p* levels (1·85 e.V.) does allow appreciable participation of the 2*p* orbitals. Even when better wave-functions were used it was still only possible to show that the quadrupole moment of ^{7}Li was negative and of the order of 10^{-25} cm^2 [19]

The ^{7}Li coupling in LiH is fortunately much larger[20] (346 ± 1 kc/s) and the wave-function can be checked by the coupling constant of the deuteron in LiD (Section **B5**). Both Kahalas and Nesbet[21] and Browne and Matsen[22] have made calculations of this sort and the latter's probably most accurate value for the quadrupole moment of ^{7}Li is $-4{\cdot}44 \times 10^{-26}$ cm^2. The value obtained by Kahalas and Nesbet differs only slightly from this. Chong and Brown[15] have extended the calculations of Browne and Matsen to obtain a result differing insignificantly from them.

The calculations of Browne and Matsen are remarkable for the care which they have attempted to take in calculating the field-gradient. Thus the nuclear term in the field-gradient was obtained by averaging the nuclear field-gradient over the vibrational motion to obtain* 0·0696 ea_0^{-3} whereas $2/R_e^3$ gives the value 0·0730 ae_0^{-3}.

5. Deuterium Coupling Constants

Many of the deuteron coupling constants that are available have given rise to fairly sophisticated attempts at calculating the appropriate electrostatic field-gradients and since in any case the number of such coupling constants is small it is convenient to review all the results at this point.

Because of the small value of the deuteron quadrupole moment and even more because the valency electron of hydrogen is in a 1*s* orbital, deuterium quadrupole coupling constants are small, the highest value so far recorded being 340 kc/s for DF. Consequently pure quadrupole resonance is out of the question and almost all the values result from the observation of hyperfine structure in pure-rotation spectra, where special techniques are necessary in order to improve spectrometer resolving power, or in the deuteron magnetic resonance spectra of solids. For a few compounds values have been estimated by observation of relaxation times in their deuteron magnetic resonance spectra in solutions and splittings due to quadrupole coupling have also been seen in the deuteron magnetic resonance spectra of compounds forming nematic phases. Table 9 shows a list of all the deuteron coupling constants known to the author at the date of writing.

* This figure is twice that shown in Browne and Matsen's paper, where the gradients are divided by twice the electronic or nuclear charge.

TABLE 6.9

Molecule	Coupling constants (kc/s)	Method	Reference
HD	227 ± 4	Mol. Beam. Res.	J. M. B. Kellog, I. J. Rabi, N. F. Ramsey and J. R. Zacharias, *Phys. Rev.* **57**, 677 (1940)
D_2	225 ± 0·1	Mol. Beam Res.	N. J. Harrick, R. G. Barnes, P. J. Bray and N. F. Ramsey, *Phys. Rev.* **90**, 260 (1953)
LiD	33 ± 1	Mol. Beam. Res.	L. Wharton, L. P. Gold and W. Klemperer, *J. Chem. Phys.* **37**, 2149 (1962)
DF	340 ± 40	Mol. Beam Res.	H. M. Nelson, J. A. Leavitt, M. R. Barker and N. F. Ramsey, *Phys. Rev.* **122**, 856 (1961)
HDO	318·6 ± 2·4	Mol. Beam Maser	P. Thaddeus, L. C. Krisher and J. M. H. Loubser, *J. Chem. Phys.* **40**, 257 (1964)
	310·3 ± 3·4		
	314·3 ± 1·5		E. B. Treacy and Y. Beers, *J. Chem. Phys.* **36**, 1473 (1962)
	315 ± 7	Microwave	D. W. Posener, *Aust. J. Phys.* **10**, 276 (1957)
HDS	154·7 ± 1·6	Mol. Beam Maser	P. Thaddeus, L. C. Krisher } *J. Chem. Phys.*, **40**, 257 (1964)
HDCO	170 ± 2·5	Mol. Beam Maser	P. Thaddeus, L. C. Krisher } *J. Chem. Phys.*, **40**, 257 (1964)
H_2DN	282 ± 12	Mol. Beam Maser	P. Thaddeus, L. C. Krisher, P. Cahill, *J. Chem. Phys.* **41**, 1542 (1964)
D—C≡C.F	212 ± 10	Mol. Beam	V. Weiss and W. H. Flygare, *J. Chem. Phys.* **45**, 8 (1966)
D—C≡C.Cl	175 ± 20	Microwave	R. L. White, *J. Chem. Phys.* **23**, 253 (1955)
	225 ± 18	Mol. Beam	V. Weiss and W. H. Flygare, *J. Chem. Phys.* **45**, 8 (1966)

TABLE 6.9—*continued*

Molecule	Coupling constants (kc/s)	Method	Reference
D—C≡N	290 ± 120	Microwave	R. L. White, *J. Chem. Phys.* **23**, 253 (1955)
D.C≡C.D	200 ± 10	Mol. Beam	J. N. Pinkerton, Thesis Harvard 1961
D.C≡C.CH_3	208 ± 10	Mol. Beam	V. Weiss and W. H. Flygare, *J. Chem. Phys.* **45**, 8 (1966)
D_2O (ice)	213·2 ± 0·8 216·4 ± 1·0	NMR (solid)	P. Waldstein, S. R. Rabideau and J. A. Jackson, *J. Chem. Phys.* **41**, 3407 (1964)
$D_2O[Ba(COO_3)_2D_2O]$	243·5 ± 1·5	NMR (solid)	T. Chiba, *J. Chem. Phys.* **39**, 947 (1963)
$D_2O[Li_2SO_4.D_2O]$	237 ± 10	NMR (solid)	S. Ketudat and R. V. Pound, *J. Chem. Phys.* **26**, 708 (1957)
$(ND_4)^+(ND_4D_2PO_4)$	~130	NMR (solid)	T. Chiba, *Bull. Soc. Chem. Japan*, **38**, 490 (1965)
C_6D_6	193	NMR (solid)	W. D. Phillips, J. C. Rowell and L. R. Melby, *J. Chem. Phys.* **41**, 2551 (1964)
C_6D_6	147	NMR (solution)	G. Bonera and A. Rigamonti, *J. Chem. Phys.* **42**, 175 (1965)
$(CD_3)_2CO$	135	NMR (solution)	G. Bonera and A. Rigamonti, *J. Chem. Phys.* **42**, 175 (1965)
$D_2C{=}C{=}O$ (D \ C=C=O / D)		Microwave	W. H. Flygare and V. W. Weiss, *J. Am. Chem. Soc.* **87**, 5317 (1965)
C_6D_6	200 ± 10	NMR (solution)	J. G. Powles and M. Rhodes, *Proc. XIV Colloque Ampère*, Ljubljana 1966
C_6D_5Br	175 ± 10	NMR (solutions)	J. G. Powles and M. Rhodes, *Proc. XIV Colloque Ampère*, Ljubljana 1966

TABLE 6.10.

Molecule	Electronic field g dt (ea_0^{-3})	Nuclear field g dt (ea_0^{-3})	Total field g dt (ea_0^{-3})	Deuteron coupling constant (kc/s)	Reference
D—C≡C—Cl	−0·9364	1·5014		368	R. L. White, *J. Chem. Phys.* **23**, 253 (1955)
	−0·9435	1·5270	0·5835	380	R. Bersohn, *J. Chem. Phys.* **32**, 85 (1960)
HDO	−1·7441 (zz) 0·9308 (yy) −0·2942 (yz)	2·7313 (zz) −1·3438 (yy) 0·0571 (yz)	0·9572 (zz) −0·4130 (yy) −0·2371 (yz)	649	R. Bersohn, *J. Chem. Phys.* **32**, 85 (1960)
LiD	−0·1506	0·2189	0·0683	45·3	SAMO H. J. Kolber and M. Karplus, *J. Chem. Phys.* **36**, 960 (1962)
	−0·1495	0·2189	0·0694	46·1	BAMO H. J. Kolber and M. Karplus, *J. Chem. Phys.* **36**, 960 (1962)
	−0·1615	0·2189	0·0574	38·1	BLMO H. J. Kolber and M. Karplus, *J. Chem. Phys.* **36**, 960 (1962)
DF*	−2·8306	3·4584	0·6278	416·2	SAMO H. J. Kolber and M. Karplus, *J. Chem. Phys.* **36**, 960 (1062)
	−2·8310	3·4584	0·6274	415·9	BAMO H. J. Kolber and M. Karplus, *J. Chem. Phys.* **36**, 960 (1962)
	−2·8964	3·4584	0·5620	372·5	BLMO H. J. Kolber and M. Karplus, *J. Chem. Phys.* **36**, 960 (1962)
HDCO	−1·44106 (zz) −0·21646 (yz) 0·63415 (yy)	1·70813 (zz) 0·17818 (yz) −0·77817 (yy)	0·26707 (zz) −0·03828 (yz) −0·14402 (yy)	176, $\eta = 0·09$	J. T. Lowe and W. H. Flygare, *J. Chem. Phys.* **41**, 2153 (1964)
D.C≡N	−1·219	1·691	0·472	310	C. W. Kern and M. Karplus, *J. Chem. Phys.* **42**, 1062 (1965)

TABLE 6.10—*cont.*

Molecule	Electronic field $g\,\mathrm{d}t\ (ea_0^{-3})$	Nuclear field $g\,\mathrm{d}t\ (ea_0^{-3})$	Total field $g\,\mathrm{d}t\ (ea_0^{-3})$	Deuteron coupling constant (kc/s)		Reference
D.C≡C.D	−1·183	1·656	0·473	311	BAMO	C. W. Kern and M. Karplus, *J. Chem. Phys.* **42**, 1065 (1965)
	−1·233	1·656	0·423	278	BI	
	−1·245	1·656	0·411	270	BII	
	−1·026	1·656	0·630	414	BIII	
	−1·219	1·656	0·437	287	BIV	
C_6H_6	−1·1714	1·4085	0·2371	154		G. Bonera and A. Rigamonti, *J. Chem. Phys.* **42**, 175 (1965)
CD_3COCD_3	−1·1708	1·3568	0·1860	121		G. Bonera and A. Rigamonti, *J. Chem. Phys.* **42**, 175 (1965)

* P. E. Cade (personal communication) and *Bull. Am. Chem. Soc.* **11**, 11 (1966), has calculated the deuteron coupling in DF as +358 kc/s in excellent agreement with experiment.

The fact that hydrogen has only one electron, the 1*s* valency electron, on the one hand makes deuteron coupling constants capable of quite exact interpretation and on the other hand makes them untypical of other quadrupole coupling constants. The absence of inner closed-shell electrons removes any worries about Sternheimer shielding and the fact that accurate wave-functions are available for molecular hydrogen makes possible an accurate evaluation of the quadrupole moment of the deuteron so that real electrostatic field-gradients may be derived from the measured constants of other molecules. However, the spherical symmetry of the hydrogen valency electron has the consequence that the main sources of field-gradients are the electrons and nuclei of neighbouring atoms. Since electrons are diffuse their field-gradient is usually outweighed by that due to the nuclei so that deuteron coupling constants (the deuteron has a positive quadrupole moment) are expected to be positive and in fact for all molecules where the sign has been determined this has been found to be the case.

In Table 10 are summarized the results of calculations of deuteron coupling constants. An early attempt by White[23] to calculate the field-gradient at the deuteron in the fragment D—C≡, using conventional Slater orbitals, led to a value twice that which he had observed in D—C≡C—Cl and the subject was not taken up again until Bersohn[24] attempted a more systematic study of the coupling constants then available, concentrating his attention on the value for HDO and using the "accurate" wave-functions for the water-molecule calculated by Ellison and Shull.[25] Here again the calculated maximum field-gradient and asymmetry parameter were twice those observed, while the direction of the maximum field-gradient was 10° away from the OD bond direction compared with an experimental value of $>20°$. Because of the computational difficulties involved Bersohn, however, had neglected most of the three-centre integrals involved in the evaluation of the coupling constant from the wave-function and it is probable that their inclusion would have decreased the gap between theory and experiment. More success attended the efforts of Kolker and Karplus[26] who studied the simpler diatomic molecules LiD and DF—where no three-centre integrals arise—using wave-functions calculated by Ransil.[2] They were fortunate in having available in fact three wave-functions for each of these molecules using different atomic orbitals as their bases (SAMO, BAMO, BLMO). As expected the last set gives best agreement with the experimental binding energy and with the observed coupling constants but the results for all three calculations are of interest in that they show how a modification in wave function which brings only a small improvement in binding energy brings a very significant improvement in field-gradient. (The binding energy given by BAMO is 0·1032 atomic units less than the experimentally observed energy while BLMO gives 0·1004.) They also served to support Bersohn's suggestion that the reason for the poor field-gradient of Ellison and Shull's wave-function lay in the choice of neutral atom Slater orbitals as a basis set for the construction of molecular orbitals. Guided by these results Lowe and Flygare[27] used Foster and Boys[28] self-consistent field wave-functions for formaldehyde to calculate the field-gradient at the deuteron in HDCO, taking care to derive exact values for the three-centre integrals. Considerable agreement with experiment was quite satisfactory.

A similar investigation of C_2D_2 and DCN has been made by Kern and Karplus[29] using a variety of wave-functions and here they showed that approximately 15 per cent of the calculated field-gradient has its origin in the terms involving three-centre integrals. The agreement with experiment was not, however, as good as that achieved by Lowe and Flygare and it is perhaps significant that the best agreement with experiment (Lowe and Flygare[27]; Kolker and Karplus[26] BLMO) was obtained when the exponent of the 1*s* orbital of the hydrogen atom was increased above the isolated atom value of unity. The apparent success of Bonera and Rigamonti[30] in their calculations on Benzene and Acetone using ordinary Slater orbitals must be viewed with some suspicion since they only considered the D—C≡ fragments and thus side-stepped any problems involving three-centre integrals. In any case their experimental value for benzene is not in agreement with that obtained by other workers (see Table 9) and Powles[31] has shown that this is due to an error in their analysis of their data. Their value for acetone is therefore presumably likewise incorrect.

It is probably true to say that the deuteron quadrupole coupling constant of simple molecules is one of the most important criteria available for the testing of molecular wave-functions and we are certain to hear more about them in the future, both for their experimental determination and their calculation. Nevertheless they occupy a peculiar and private corner of the field of nuclear quadrupole coupling in that as much as anything else they reflect the molecular geometry and internuclear distance. Thus the greatest part of the difference between the coupling for HDO and HDS is due to the difference in the D—O (0·19572 Å) and the D—S (1·3452 Å) bond-lengths rather than to the undoubted electronic differences which exist between these two bonds. Similarly the deuteron coupling constants in the series FC≡CD, ClC≡CD, DC≡CD and CH_3C≡CD are experimentally indistinguishable.

C. THE CALCULATION OF FIELD-GRADIENTS AT LATTICE SITES IN IONIC CRYSTALS

The field-gradients at lattice sites in ionic crystals are in many ways only of subsidiary interest to the main theme of this book. Furthermore, the considerable but uncertain magnitude of the Sternheimer shielding coefficient for ions makes it difficult to achieve any great accuracy in such calculations. Nevertheless in some cases they make possible a distinction between an ionic structure and one in which the bonds have considerable covalent character even though the component atoms have high coordination numbers. It is not however intended to deal with this subject in any great depth.

If the question of Sternheimer shielding is left aside the problem becomes essentially one of classical electrostatics. The component ions are replaced by point charges and the field-gradients calculated by a lattice summation. However these lattice sums only converge quite slowly since, even aside from the fact that the crystal will contain both positive and negatively charged ions, the summation contains both positive and negative terms. Much care must thus be taken in deciding the point at which the summation is broken off. However by breaking up the lattice into a series of parallel planes and calculating the field-gradient along the normal to the plane produced by each plane and then summing the contributions from the individual planes, the series can often be made to converge more rapidly.[32,33] Other transformations for producing more rapid convergence have also been devised.[34,35]

Since ions are not point charges but polarizable spheres of charge they will in the presence of an electric field give rise to an induced electrostatic dipole moment. Several authors[35,36,37,38,39] have shown that these induced dipoles contribute significantly to the total field-gradient. In order to calculate them a knowledge of the dipole polarizability of the various ions is required and of course the lattice sums are now even more complex. A good general discussion of the problems involved has been given by Sholl.[40]

REFERENCES

1. C. C. J. Roothaan, *Rev. Mod. Phys.* **23**, 69 (1951); *Rev. Mod. Phys.* **32**, 179 (1960).
2. B. J. Ransil, *Rev. Mod. Phys.* **32**, 245 (1960).
3. W. Huo, *J. Chem. Phys.* **43**, 624 (1965).
4. P. E. Cade, K. D. Sales ad A. C. Wahl, *J. Chem. Phys.* **44**, 1973 (1966).
5. B. Rosenblum and A. H. Nethercot Jr., *J. Chem. Phys.* **27**, 828 (1957).
6. S. Harvey, to be published.
7. N. Bessis, H. Lefebre-Brion and C. Moser, *Phys. Rev.* **128**, 213 (1963).
8. B. J. Ransil, *J. Chem. Phys.* **30**, 1113 (1959).
9. J. W. Richardson, *Rev. Mod. Phys.* **32**, 461 (1960).
10. H. Lefebre-Brion, C. Moser, R. K. Nesbet and M. Yamazaki, *J. Chem. Phys.* **38**, 2311 (1963).
11. E. Scrocco and J. Tomasi, *Th. Chim. Acta* **2**, 386 (1964).
12. P. E. Cade, Personal communication and *Bull. Am. Phys. Soc.* **11**, 11 (1966).
13. T. A. Scott, *J. Chem. Phys.* **36**, 1459 (1962).
14. R. F. W. Bader and G. A. Jones, *J. Chem. Phys.* **38**, 2791 (1963).
15. D. P. Chong and W. Byers Brown, *J. Chem. Phys.* **45**, 392 (1966).
16. C. C. Lin, *Phys. Rev.* **119**, 1027 (1960).
17. R. A. Logan, R. E. Coté and P. Kusch, *Phys. Rev.* **86**, 280 (1952).

18. E. G. Harris and M. A. Melkanoff, *Phys. Rev.* **90**, 585 (1953).
19. R. M. Sternheimer and H. M. Foley, *Phys. Rev.* **92**, 1460 (1953); I. Mannari and T. Arai, *J. Chem. Phys.* **28**, 28 (1958).
20. L. Wharton, L. P. Gold and W. Klemperer, *J. Chem. Phys.* **37**, 2149 (1962).
21. S. L. Kahalas and R. K. Nesbet, *J. Chem. Phys.* **39**, 529 (1963).
22. J. C. Browne and F. A. Matsen, *Phys. Rev.* **135**A, 1227 (1964).
23. R. L. White, *J. Chem. Phys.* **23**, 253 (1955).
24. R. Bersohn, *J. Chem. Phys.* **32**, 85 (1960).
25. F. O. Ellison and H. Shull, *J. Chem. Phys.* **23**, 2348, 2358, (1955).
26. H. J. Kolker and M. Karplus, *J. Chem. Phys.* **36**, 960 (1962).
27. J. T. Lowe and W. H. Flygare, *J. Chem. Phys.* **41**, 2153 (1964).
28. J. M. Foster and S. F. Boys, *Rev. Mod. Phys.* **32**, 303 (1960).
29. C. W. Kern and M. Karplus, *J. Chem. Phys.* **42**, 1062 (1965).
30. G. Bonera and A. Rigamonti, *J. Chem. Phys.* **42**, 175 (1965).
31. J. G. Powles, private communication.
32. F. W. de Wette, *Phys. Rev.* **123**, 103 (1961).
33. F. W. de Wette and G. E. Schacher, *Phys. Rev.* **137**A, 78, 92 (1964).
34. R. R. Hewitt and T. T. Taylor, *Phys. Rev.* **125**, 524 (1962).
35. T. T. Taylor and T. P. Das, *Phys. Rev.* **133**A, 1327 (1964).
36. D. K. Ray, *Proc. Phys. Soc.* **82**, 47 (1963).
37. R. R. Sharma and T. P. Das, *J. Chem. Phys.* **41**, 3581 (1964).
38. H. C. Bolton and C. A. Sholl, *J. Nucl. Mat.* **14**, 265 (1964).
39. G. Burns, *J. Chem. Phys.* **42**, 377 (1965).
40. C. A. Sholl, *Proc. Phys. Soc.* **87**, 897 (1966).
41. A. Bassompierre, *Disc. Faraday Soc.* **19**, 260 (1955).
42. Y. Kato, *J. Chem. Phys.* **34**, 619 (1961).

CHAPTER 7

APPROXIMATE METHODS

As we have seen in the previous chapter the calculation of electrostatic field-gradients in molecules is not something which can be undertaken lightly. We are, however, more usually interested in the reverse process of using quadrupole coupling constants to find out something about electron distribution in complex molecules without having recourse to computations which at this stage would be prohibitively time-consuming. It is consequently necessary to devise a simple method of analysis of the quadrupole coupling data which at the same time yields results which have some significance. Before we can do this, however, we must make up our minds very clearly about what we want.

It is a salutory reflection that as soon as they are no longer concerned with an essentially phenomenological description of nature very few branches of the sciences can be classified as exact. This is nowhere more true than in chemistry and the chemist at an early age becomes familiar with such extremely useful but often rather loose concepts such as "electronegativity", "electron configuration", "aromaticity", "hybridization", and so on. These limitations are often not clearly appreciated by the mathematical physicist or spectroscopist. Most of the ideas so useful to the chemist are in fact hierarchical concepts, to be associated with the symbols "less than" or "greater than" rather than with "equals" and they are useful as long as they remain guidelines for thought and present a stimulus to the imagination. As soon as they show a tendency to become more important than the chemistry they attempt to describe they must be hastily put in a pigeon-hole for the attention of the historian.

To the chemist then all things are relative and we must take a lesson from him when developing approximate methods for the analysis of quadrupole coupling constants. The only way to treat the coupling constant of one molecule in isolation from all others is to have recourse to the exact calculations discussed in the previous section. Conversely the only way to use approximate analyses is to consider groups of molecules and to use the coupling constant data to analyse the trends in bonding with molecular type. The chemist is not very much further advanced if he is presented with a detailed contour map of the electron distribution in say a diatomic molecule for he knows that as soon as he wants to do anything with it he must put it in solution, bring another molecule close to it or electronically excite it and the contour map immediately becomes blurred.

Our ambition then in an approximate analysis of nuclear quadrupole coupling constants is to translate the data into terms with which the chemist is familiar and which will provide for him a meaningful description of molecular ground states. We want to provide an at least semiquantitative evaluation of the effects which he has so often invoked to explain, or rather rationalize, chemical reactivity, for example "ionic character of covalent bonds", "partial double-bond character of single bonds", "valence shell expansion" and so on, but we must never forget that any definitions of these terms depend themselves on certain basic assumptions as to, for example, the nature of a single pure covalent bond. In other words, all these terms only have significance with reference to a model and tend to disappear in any complete *a priori* calculation of molecular properties. Quantum chemists in fact devote considerable time and ingenuity to devising methods of transforming where possible their molecular wave-functions into forms exhibiting a recognizable relationship to the familiar localized two-electron bond structures. The significance which we attach therefore to the numbers which our analysis of the coupling data will yield must not be greater than the significance of the model upon which these numbers are based.

As seen in Chapter 6, the field-gradient at a nucleus α in a molecule is given by a sum over electronic and nuclear terms

$$q_z^\alpha = q_{z,\text{electronic}}^\alpha + q_{z,\text{nuclear}}^\alpha. \tag{1}$$

The term q_{nuclear} can be calculated from the known molecular geometry while the electronic contribution is given by the expectation value of the field-gradient operator (a one-electron operator) summed over all the electrons in the molecule.

$$q_{z,\text{electronic}}^\alpha = \sum_p \int \Psi_p \left(\frac{3\cos^2\theta_{p\alpha} - 1}{r_{\alpha p}^3}\right) \Psi_p^* \, d\tau \tag{2}$$

where Ψ_p is the molecular wave-function of electron p. In general such molecular wave-functions are expressed as linear combinations of atomic wave-functions, ϕ_i

$$\Psi_p = \sum_i C_{ip}\phi_i. \tag{3}$$

The electronic contribution to the field-gradient then becomes

$$q_{z,\text{electronic}}^\alpha = \sum_p \sum_i \sum_j C_{ip} C_{jp} q_{ij}^\alpha \tag{4}$$

where

$$q_{ij}^\alpha = \int \phi_i \left(\frac{3\cos^2\theta_{p\alpha} - 1}{r_{p\alpha}^3}\right) \phi_j^* \, d\tau.$$

As discussed in Chapter 6 the various integrals q_{ij}^{α} can be one centre, two centre or three centre terms and we may bring this out by expanding equation (4).

$$q_{z,\text{electronic}}^{\alpha} = \sum_{p} \left\{ \sum_{i} (C_{ip})^2 q_{ii}^{\alpha} + \sum_{j} (C_{jp})^2 q_{jj}^{\alpha} + \sum_{i} \sum_{j} C_{ip} C_{jp} q_{ij}^{\alpha} + \sum_{j} \sum_{k} C_{jp} C_{kp} q_{jk}^{\alpha} \right\}. \tag{5}$$

The subscript i indicates an atomic orbital centred on the nucleus α while the subscripts j and k indicate orbitals centred on other nuclei.

Equation (5) was drastically simplified by Townes and Dailey[1] in a pioneering paper which has formed the basis of almost all approximate treatments of coupling constants. Their simplifications are best discussed by reference to a particular example and since on the one hand coupling constants from the halogens, and ^{35}Cl in particular, are more abundant than for all other nuclei put together and on the other hand the detailed S.C.F. calculations of Scrocco and Tomasi are available for this molecule, hydrogen chloride is an obvious choice. Moreover as it is a linear molecule the field-gradient tensor has axial symmetry and the principal axis is defined by the molecular framework.

The simple picture of the bonding in hydrogen chloride, so familiar to all chemists, is drawn only in terms of the valence orbitals. The symmetry of a diatomic molecule enables these to be split up into sigma and pi states and it is easy to show that four valence electrons are in sigma states and four in pi states. Since no excited orbitals of hydrogen are considered the pi-electrons are supposed to be completely localized on two of the chlorine $3p$ orbitals while the four sigma electrons are shared between the hydrogen $1s$ orbital, the chlorine $3s$ orbital and the remaining chlorine $3p$ orbital. The chemist, moreover, has the habit of thinking of a single bond as involving just two electrons and he therefore prefers to divide up the sigma electrons between a bonding and a non-bonding orbital. To do this he has recourse to the device of hybridization. The chlorine $2s$ and $2p$ orbitals are thought of as mixing together in such a way as to yield two equivalent orbitals, one of which overlaps strongly with the hydrogen $1s$ orbital while the other overlaps so weakly that its contribution to the bonding can be neglected (see Fig. 1).

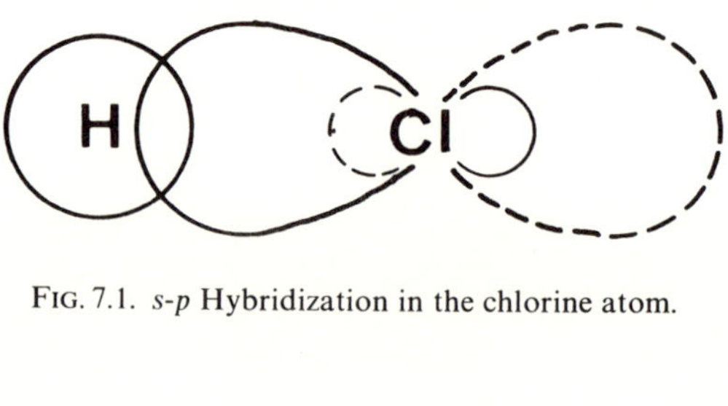

FIG. 7.1. s-p Hybridization in the chlorine atom.

The reader will recognize the similarity of this picture to the ϕ^b and ϕ^{nb} orbitals of Tomasi and Scrocco[2] (Chapter 6). If this is now put in the quantitative terms we have the orbitals shown below.

$$\sigma_b = \phi_{1s} + b\{(1 - \alpha^2)^{\frac{1}{2}}\phi_{3p_z} + \phi_{3s}\},$$
$$\sigma_{nb} = \phi_{3p_z} - (1 - \alpha^2)^{\frac{1}{2}}\phi_{3s}, \qquad \pi = \phi_{3p_x}, \qquad \pi = \phi_{3p_y}. \tag{6}$$

Since this wave function does not include the inner closed-shell electrons it necessarily implies that they are unperturbed by bond-formation and retain their spherical symmetry thus contributing nothing to the field-gradient. As far as the orbitals centred on the nucleus α are concerned, we may therefore restrict the summation in equation (5) to the valence orbitals and electrons. This is of course where Sternheimer shielding comes in and we have seen from Tomasi and Scrocco's calculations[2] that HCl perturbation of the inner $1s^2\ 2s^2\ 2p^6$ shell accounts for 17 per cent of the field-gradient at the chlorine nucleus. This then is an important source of error in the approximate treatment.

If the wave function for hydrogen chloride is assumed to be a simple product of one-electron orbitals

$$\Psi = (\sigma_{nb}\alpha)^1(\sigma_{nb}\beta)^2(\pi\alpha)^3(\pi\beta)^4(\pi\alpha)^5(\pi\beta)^6(\sigma_b\alpha)^7(\sigma_b\beta)^8 \tag{7}$$

then it is easily seen that the expectation value of the quadrupole Hamiltonian operator becomes

$$\begin{aligned}
(q_{\mathrm{Cl}})_{z,\mathrm{electronic}} = {} & 2(\alpha^2 - \alpha^2 b^2 + b^2)\int \phi_{3pz}\,.\,\mathscr{H}_{q_z}\,.\,\phi_{3p_z}\,\mathrm{d}\tau \\
& + 2\int \phi_{3p_x}\,.\,\mathscr{H}_{q_z}\phi_{3p_x}\,\mathrm{d}\tau + 2\int \phi_{3p_y}\,.\,\mathscr{H}_{q_z}\,.\,3p_y\,.\,\mathrm{d}\tau \\
& + 2(1 - \alpha^2 + a^2\alpha^2)\int \phi_{3s}\,.\,\mathscr{H}_{q_z}\,.\,\phi_{3s}\,\mathrm{d}\tau \\
& + 2\alpha(1 - \alpha^2)(b^2 - 1)\int \phi_{3s}\mathscr{H}q_z\phi_{3p_z}\,\mathrm{d}\tau \\
& + 2a^2\int \phi_{1s}\mathscr{H}q_z\phi_{1s}\,\mathrm{d}\tau \\
& + 2ab(1 - \alpha^2)^{\frac{1}{2}}\int \phi_{1s}\mathscr{H}q_z\phi_{3p_z}\,\mathrm{d}\tau \\
& + 2ab\alpha\int \phi_{1s}\mathscr{H}q_z\phi_{3s}\,\mathrm{d}\tau.
\end{aligned} \tag{8}$$

Because the quadrupole operator is a one-electron operator no additional terms would have been introduced if the correct antisymmetrized product form of the wave-function had been used.

The first five integrals in equation (8) are one-centre integrals of which the fourth and fifth are identically zero. In general integrals involving spherical harmonics and the quadrupole Hamiltonian $\mathscr{H}_q$ of the type $\int Y_{l_1m_1} . \mathscr{H}_q . Y_{l_2m_2} . d\tau$ are identically zero unless $|l_1 - l_2| = 2$ or $l_1 = l_2$. When $l_1 = l_2 = 0$ the integral is zero because of the spherical symmetry of the spherical harmonics. Since only s and p electrons are considered in the present example no integrals fulfilling the condition $|l_1 - l_2| = 2$ occur here. Since they are in general much smaller than the integrals where $l_1 = l_2$, when they do occur they are neglected in the Townes and Dailey approximation.[1] Since HCl is a diatomic molecule no three-centre integrals q^{α}_{jk} occur: as seen in Chapter 6 they are small but by no means inconsiderable, however they are again neglected in the Townes and Dailey analysis.

The next step in the Townes and Dailey approximation is to set off the nuclear field-gradient $q^{\alpha}_{z,\text{nuclear}}$ against the integrals $\int \phi_{1s}\mathscr{H}_q\phi_{1s}\, d\tau$ and $\int \phi_{1s}\mathscr{H}_q\phi_{3p_z}$, etc. We have seen that for HCl, Tomasi and Scrocco's calculations show that indeed these terms do cancel to a great extent. Table 1 which gives a break-down of the electronic contributions[3] for N_2 shows that a BLMO SCF function which appears to yield something like the correct field-gradient indicates that here again the cancellation is almost complete. Table 1 however also shows a similar break-down for an SAMO-SCF function for CO where the cancellation is far less exact, however this function does not give good agreement with the experimental coupling for reasons which we have discussed earlier.

TABLE 7.1. Contributions to the field-gradient at the nitrogen atom in molecular N_2 and the oxygen atom in carbon monoxide. q_{AA} are the sum of terms from orbitals centred on the nucleus under study, q_{BB} from orbitals centred on the neighbouring nucleus while q_{AB} are overlap terms (Ref. 3).

	N_2	CO
q_{AA}	−1·323	−0·308
q_{AB}	−0·532	−0·373
q_{BB}	−1·071	−0·709
q_{nuclear}	+1·584	+1·238

At this point we must remember that in our approximate analysis we are concerned not so much with determining absolutely the electron distribution in molecules from the quadrupole coupling data, but more with the difference in electron distribution between two covalent molecules of not too-different bond-type. Thus we are not very interested in the difference between sodium chloride and carbon tetrachloride; the chemist knows perfectly well what the difference is. We are however interested in the differences between carbon tetrachloride and silicon tetrachloride. In these circumstances the above approximation represents only a second-order effect and while the Sternheimer shielding is certainly more important it is likely that the changes in Sternheimer shielding will parallel the changes in bond character. However we will return to this again below.

These simplifications then reduce the expression for the total field-gradient to

$$(q_{\mathrm{Cl}})_z = 2(\alpha^2 - \alpha^2 b^2 + b^2)\int \phi_{3p_z}\mathscr{H}q_z\phi_{3p_z}\,\mathrm{d}\tau + 2\int \phi_{3p_x}\mathscr{H}q_z\phi_{3p_x}\,\mathrm{d}\tau + 2\int \phi_{3p_y}\mathscr{H}q_z 3p_y\,\mathrm{d}\tau. \tag{9}$$

It is next assumed that the $3p_x$, $3p_y$ and $3p_z$ orbitals all have the same radial functions. The reader will recall that this was certainly not the case either for HCl or for N_2. However introducing this assumption means that the integrals in equation (9) have the simple relationship

$$\int 3p_x H_{q_z} 3p_x\,\mathrm{d}\tau = \int 3p_y H_{q_z} 3p_y\,\mathrm{d}\tau = -\tfrac{1}{2}\int 3p_z H_{q_z} 3p_z\,\mathrm{d}\tau \equiv -\tfrac{1}{2}q_0 \tag{10}$$

and the expression for $(q_z)_{\mathrm{total}}$ reduces to

$$(q_{\mathrm{Cl}})_z = -2(1-\alpha^2)(1-b^2)\,q_0. \tag{11}$$

It will be recognized that equation (11) simply amounts to adding up the electronic contribution of the population of each valence p-orbital. This amounts to neglecting all terms in equation (5) except the first. The final step is to equate the integral in equation (10) with the experimentally determined field-gradient of a p-electron in a neutral chlorine atom. The corresponding simple analysis for atomic chlorine in the state $1s^2 2s^2 2p^6 3s^2 3p^5$ shows that here the field-gradient is given by

$$(q_{\mathrm{Cl}})_{\mathrm{atomic}} = -q_0. \tag{12}$$

Hence

$$(e^2Qq_{\mathrm{Cl}})_z = 2(1-\alpha^2)(1-b^2)(e^2Qq_{\mathrm{Cl}})_{\mathrm{atomic}}. \tag{13}$$

This is the essence of the Townes and Dailey approximate analysis. It is easily extended to much more complicated molecules and examples of this will follow in this chapter and those dealing with experimental results.

Let us now consider from the view point of this simple method the chlorine molecule Cl_2. If we write the electronic structure in terms of only the valence orbitals we have the sigma molecular orbitals and corresponding occupation numbers set out in Table 2.

TABLE 7.2. Configuration of valency orbitals in Cl_2. S is the total overlap integral between the two $3s$-$3p$ hybrid sigma orbitals.

$$(\sigma_1) = \sqrt{\frac{1}{2(1+S)}}\{\alpha^2(\phi_{3s})_1 + (1-\alpha^2)^{\frac{1}{2}}(\phi_{3p_z})_1 + \alpha^2(\phi_{3s})_2 + (1-\alpha^2)(\phi_{3p_z})_2\}$$

$$(\sigma_{nb})_1 = \alpha^2(\phi_{3p_z})_1 - (1-\alpha^2)^{\frac{1}{2}}(\phi_{3s})_1$$

$$(\sigma_{nb})_2 = \alpha^2(\phi_{3p_z})_2 - (1-\alpha^2)^{\frac{1}{2}}(\phi_{3s})_2$$

$$\pi_1 = (\phi_{3p_x})_1$$

$$\bar{\pi}_1 = (\phi_{3p_y})_1$$

$$\pi_2 = (\phi_{3p_x})_2$$

$$\bar{\pi}_2 = (\phi_{3p_y})_2$$

A population analysis of the electronic field-gradient at one of the chlorine atoms then yields

$$q_z = -q_0(1-\alpha^2)\left(2 - \frac{1}{1+S}\right). \tag{14}$$

The coupling constant of atomic chlorine[4] is $-109{\cdot}6$ Mc/s while that of solid molecular chlorine[5] is 108·9 Mc/s. The coupling of the free chlorine molecule is likely to be higher than this (see Chapter 8) and Julg,[6] by interpolating from the gas-phase values for FCl, BrCl and ICI estimates that it is around 116 Mc/s. If this latter value is assumed and values of S are taken from tables of overlap integrals for Slater orbitals[7] then equation (14) may be solved graphically for α^2 to yield $\alpha^2 = 0{\cdot}21$.

It will be readily appreciated that we have been able to solve the above equation only because we have been discussing a homonuclear diatomic molecule. For a heteronuclear system, R—Cl, the sigma orbital would have taken on the more general form

$$\sigma_i = \frac{1}{\sqrt{1+2Sa(1-a^2)^{\frac{1}{2}}}}\{a\psi_R + (1-a^2)^{\frac{1}{2}}[\alpha^2(3s) + (1-\alpha^2)(3p)]\} \tag{15}$$

where the field-gradient is given by

$$q_{Cl} = 2(1-\alpha^2)\left\{\frac{1-a^2}{(1+2Sa(1-a^2)^{\frac{1}{2}}} - 1\right\} q_0. \tag{16}$$

This cannot be solved without making an assumption as to the value of either α^2 or a.

If instead of a molecular orbital description of the bond in Cl_2 the Heitler–London method had been applied the equation for the coupling constant would have been

$$q_z = -q_0(1 - \alpha^2)\left(2 - \frac{1}{1 + S^2}\right). \tag{17}$$

In these circumstances the amount of *s*-character required by the experimental coupling constant is 0·09. This disagreement between the two simplest methods of describing a chemical bond combined with the complications introduced by the overlap term together with the fact that the then currently accepted coupling constant of chlorine (−108·9 Mc/s; solid) led to the rather high values for α^2 of 0·26 and 0·19 for the LCAO and Heitler-London methods respectively induced Townes and Dailey[1,9] and Gordy[8] to neglect the overlap term altogether, i.e. to use a Hückel molecular orbital description of the chemical bond. In these circumstances the expression for the coupling constant reduces to

$$e^2Qq_z = -e^2Qq_0(1 - \alpha^2). \tag{18}$$

If, as was originally done, the coupling constant for solid chlorine (108·9 Mc/s) is substituted in this equation, then the very satisfactory result is achieved that $\alpha^2 = 0$, i.e. a bond is formed between two pure $3p$ orbitals of chlorine. Of course if the higher value of 116 Mc/s is employed no value of α in equation (18) can achieve agreement with the experimental coupling constant and this led Julg[6] to suggest that this fact shows clearly that *d*-orbitals participate in the bonding in Cl_2. It is quite clear from the foregoing discussion that it shows nothing of the sort.

Before continuing the critical examination of the Townes–Dailey approximate analysis it is now an appropriate moment to develop some of the formulae relating bond parameters and coupling constants to which it gives rise. Originally the analysis was developed in terms of resonance theory which does indeed for the present purpose have something to recommend it. We continue for the moment to restrict our attention to a halogen atom such as chlorine in one of its singly-coordinated compounds, R—Cl. If only singly bonded forms are considered we have the three following resonance structures,

$$R^+Cl^- \leftrightarrow R - Cl \leftrightarrow R^-Cl^+.$$

The quadrupole coupling constant of the molecule RCl may then be calculated as the weighted mean of the coupling constants of the three resonance forms. The coupling of the ionic form R^+Cl^- is zero, that of the covalent form R—Cl depends on the state of hybridization of the halogen sigma orbital. If only *s-p* hybridization is considered the above analysis shows

$$(e^2Qq)_{\text{covalent}} = -(1 - s)e^2Qq_0 \tag{19}$$

where s ($= \alpha^2$ in equation 18) is the fractional s-character of the sigma orbital. If there is also the possibility of d-hybridization and it is assumed that the contribution of a d-electron to the field-gradient can be neglected (cf. Chapter 4) then the above expression becomes replaced by

$$(e^2Qq)_{\text{covalent}} = -(1 - s + d)e^2Qq_0. \tag{20}$$

The field-gradient for the Cl^+ ion presents a little more of a problem. The most self-consistent way of looking at the problem is to consider that the configuration of the ion is that arising from the removal of the electron from the sp hybrid bonding orbital. In that case the total p-electron defect is $2(1 - s)$. However this view point is not always taken and the configuration of the Cl^+ ion is considered to be $3s^2 3p_x^2 3p_y^2$ having a p-electron defect of just 2.

Before proceeding to write down the field-gradient due to either of these configurations we must however consider the effect of charge on the p-orbitals and the field-gradients to which the electrons in them give rise. It is obvious that a positive ion has a greater effective negative charge than a neutral atom; this causes the orbitals to contract and hence to have a greater value of $\langle r^{-3} \rangle$ and a greater value of the field-gradient at the nucleus. It is found empirically from atomic spectroscopy that for a given atom the expression is

$$\langle r^{-3} \rangle_n = \langle r^{-3} \rangle_0 (1 + \varepsilon)^n \tag{21}$$

where n is the degree of ionization and ε is a constant for a particular atom. Townes and Dailey[9] have listed values for ε (Table 3) obtained from magnetic hyperfine splitting in atomic emission spectra. In much of their early work they used the average value 0·25. Although it does not arise in the present case since the Cl^- ion has spherical symmetry and no field-gradient they have extended this equation to negative ions where n takes on negative values.

TABLE 7.3. Values of ε for various elements. The quadrupole coupling per electron is proportional to $(1 + \varepsilon)^n eQq$, where eQq is the coupling constant for the neutral atom and n is the degree of ionization (+1 for single positive ionic charge, −1 for single negative ionic charge) (Ref. 9).

Be 0·9	B 0·54	C 0·47	N 0·32	O 0·20	F 0·17
Mg 0·5	Al 0·27	Si 0·28	P 0·18	S 0·19	Cl 0·14
Sr 0·7	Sc 0·37	Ge 0·26	As 0·16	Se 0·20	Br 0·13
				Te 0·20	I 0·12

With this result we may now write the field-gradient for a Cl^+ ion as either $-2(1 - s)(1 + \varepsilon)e^2Qq_0$ or $-2(1 + \varepsilon)e^2Qq_0$ according to the configuration we have chosen. The final assumption is that only one of the ionic structures is important for any particular molecule, depending on the relative electronegativities of R and Cl. The fractional ionic character of the molecule is then i and the covalent character $(1 - i)$. With the first, more logical choice for the configuration of Cl^+ we then have the simple formula

$$e^2Qq = -(1 - s)(1 - i)e^2Qq_0 \tag{22}$$

if the halogen atom is the more electronegative partner and

$$e^2Qq = -(1 - s)[1 + i(1 + 2\varepsilon)]e^2Qq_0 \tag{23}$$

if it is the less electronegative partner.

We must now consider the possibility that the lone-pair $3p_x$ and $3p_y$ electrons of the chlorine atom participate in the bonding so that the R–Cl bond takes on a partial multiple character. We must distinguish between two cases. When R is an atom or a group having a three-fold or greater symmetry axis along the RCl bond both the $3p_x$ and $3p_y$ enter equally into the bonding and the field-gradient tensor retains its axial symmetry. When, however, R only has a two-fold symmetry axis or none at all then the $3p_x$ and $3p_y$ orbitals will not in principle interact to the same extent and the field-gradient tensor loses its axial symmetry, i.e. must be specified by both its component in the z-direction and the asymmetry parameter. A typical example of this kind which we will be discussing in more detail later on is chlorobenzene where only one of the $3p$ orbitals can conjugate with the pi-system of the aromatic ring.

TABLE 7.4. Resonance structures and field-gradients in RCl where the R—Cl bond has a partial triple character.

Resonance form	Weight	Field-gradient along z-axis
R^+Cl^-	i	0
$R^-{=}Cl^+$	π	$-q_0(0{\cdot}5 - s)(1 + \varepsilon)$
$R^-{=}Cl^+$	π	$-q_0(0{\cdot}5 - s)(1 + \varepsilon)$
R—Cl	$(1 - i - 2\pi)$	$-q_0(1 - s)$

To continue our development in terms of resonance theory we must add to the resonance structures a form $R{=}Cl^+$. In the first case however when both $3p_x$ and $3p_y$ orbitals contribute equally and the field retains its cylindrical symmetry there are two such forms each having identical weights, which we will represent by the symbol π. We then have for a bond where the chlorine atom is the negative partner the data shown in Table 4. These contributions may be summed to yield

$$e^2Qq = -e^2Qq_0[(1 - i)(1 - s) - \pi(1 - \varepsilon - 2\varepsilon s)]. \tag{24}$$

Since both ε and s are of the order of 0·1–0·2 this is usually approximated by the formula

$$e^2Qq = -e^2Qq_0[(1 - i)(1 - s) - \pi(1 - \varepsilon)]. \tag{25}$$

If the chlorine atom is at the positive end of the bond a similar procedure yields

$$e^2Qq = -e^2Qq_0\{(1 - s)[1 + i(1 + 2\varepsilon)] - \pi(1 - \varepsilon)\}. \tag{26}$$

It is evident that these equations contain three unknowns so that two of them must take on assumed values if we are to make use of the quadrupole coupling constants to obtain values for the remaining parameter.

When only one of the $3p$ lone-pair orbitals can interact to form a pi-bond the principal value of the field-gradient tensor is still given by the above formula with the exception that $\pi/2$ replaces π. However the field-gradient tensor now is asymmetric and we can easily work out the partial double bond character in terms of the asymmetry parameter. Thus carrying out the same analysis as given above but this time for the x and y components of the field-gradients we have, where the $3p_x$ orbital is the one participating in double bond formation, the data shown in Table 5. Hence:

$$q_{zz} \cdot \eta = (q_{xx} - q_{yy}) = \tfrac{3}{2}\pi(1 + \varepsilon)|q_0| \tag{27}$$

or

$$\pi = \frac{2}{3}\left|\frac{e^2Qq_{zz}}{e^2Qq_0}\right| \cdot \eta(1 + \varepsilon). \tag{28}$$

This formula was originally given by Bersohn[10] without the factor $(1 + \varepsilon)$.

TABLE 7.5. Resonance forms and field-gradients in RCl where the R—Cl bond has partial double character

Resonance form	Weight	q_{xx}	q_{yy}
R^+Cl^-	i	0	0
$R^-{=}Cl^+$	π	$-\frac{1}{2}(1 + s)(1 + \varepsilon)\lvert q_0\rvert$	$-\left(1 + \frac{s}{2}\right)(1 + \varepsilon)\lvert q_0\rvert$
R—Cl	$1 - i - \pi$	$\frac{1}{2}(1 - s)\lvert q_0\rvert$	$\frac{1}{2}(1 - s)\lvert q_0\rvert$

We will leave any extensive discussion of the results from it until Chapter 10 where the coupling constants of the conjugated halobenzenes are presented.

The above formulae have been derived for a singly-coordinated nucleus with ^{35}Cl as an example and may be used directly for any nucleus in this situation. The analysis of the coupling constant data for a multi-coordinated nucleus proceeds in general in a very similar fashion with the exception that there is one additional complication. For a singly-coordinated nucleus even when the bond axis is not a molecular symmetry axis, as for example the C—Cl bond in CH_2Cl_2, the field-gradient z axis has been found to almost coincide with the bond axis. Thus the directions of the field-gradient axes in the molecular framework are essentially known. For an unsymmetrically substituted multi-coordinated nucleus, for example the ^{14}N nucleus in dimethylamine, $(CH_3)_2NH$, no such assumption can be made about the directions of the field-gradient axes and they must in fact be determined experimentally. If they are not known it is impossible to analyse the quadrupole coupling data, or more precisely, the direction cosines of the field-gradient axes in a co-ordinate system appropriate to the molecular framework will appear as additional parameters in the Townes–Dailey equation. When they are known, however (and in general they are known because the directions are determined by the molecular symmetry, e.g. trimethylamine $(CH_3)_3N$) and when the molecular geometry is also known then usually the equations relating the coupling constants to the bond parameters contain fewer unknown parameters than the corresponding formulae for singly-coordinated nuclei. This is of course because from the formal point of view at least, the extent of s-p hybridization of the central atom is known from the bond angles. Thus for example in trimethylamine where there is little possibility of pi-bonding, if it is assumed that the lone-pair orbital holds just two-electrons and the hybridization is determined entirely by the known molecular geometry, then the quadrupole coupling constant can be analysed uniquely in terms of the ionicity of the C—N bonds. Since the precise form of the corresponding Townes–Dailey formulae varies with the coordination number and geometry there is no point in setting out here a list of formulae and they will be derived in the later section as they are required. We must, however, mention in passing that the formalism of resonance theory gets much too cumbersome for multi-coordinate nuclei and a molecular orbital population analysis is to be preferred—certainly at least by the present author!—and will be adopted henceforth.

To return to singly-coordinated nuclei we have seen above how even the simplifying assumptions of the Townes–Dailey method yield formulae for the coupling constant in terms of three parameters: ionic character (i), hybridization of the sigma orbital (s) and multiple-bond character (π). Even when multiple-bond character is unlikely we are left with the question of hybridization and this point was the subject of considerable and, in view of the extreme approximations involved in the theory, misplaced dispute when Townes and Dailey first developed their approximate analysis. While Gordy[11] preferred to put s always equal to zero, i.e. considered the bonds as involving only the $3p$ orbitals, Townes and Dailey assigned a value to s by a rule whose only virtue is its simplicity, namely that when the halogen under study is at the negative end of the bond and its electronegativity is 0·25 units more than that of its partner then the sigma orbital is considered to have 15 per cent s-character, i.e. $s = 0{\cdot}15$. In all other circumstances s is put equal to zero.

It is clear that neither of these procedures is entirely satisfactory although in the author's opinion that due to Gordy has more to recommend it than the sudden discontinuity resulting from the use of Townes and Dailey's assumptions. In general there are three factors which lead to a mixing of s and p orbitals in a diatomic molecule. The first is purely electrostatic arising from the simple polarization of s and p orbitals in an electrostatic potential gradient which in principle can be calculated straightforwardly by first order perturbation theory. Thus if the two valency orbitals are given by the functions

$$\psi_{ns} = \frac{1}{\sqrt{4\pi}} f(r) \tag{29}$$

$$\psi_{np_z} = \frac{1}{\sqrt{2\pi}}\sqrt{\frac{3}{2}} = \sqrt{\frac{3}{4\pi}} f(r)\cos\theta \tag{30}$$

where $f(r)$ is a normalized radial function of some sort, and if the energy difference, ΔE, between these orbitals is much less than between either of them and any other orbitals, then in a potential field $V_0 + V_1 r\cos\theta$, i.e. a field with a potential gradient along the z axis, the two orbitals mix to give

$$\psi_1 = \frac{1}{\sqrt{1+\lambda^2}}\{\psi_{ns} + \lambda\psi_{np_z}\} \tag{31}$$

$$\psi_2 = \frac{1}{1+\lambda^2}\{\psi_{np_z} - \lambda\psi_{ns}\} \tag{32}$$

and λ is given by

$$\lambda = \int_T \frac{\psi_{ns} V_1 r\cos\theta\,\psi_{np_z} r^2 \sin\theta\,\mathrm{d}r\,\mathrm{d}\theta\,\mathrm{d}\phi}{\Delta E} \tag{33}$$

which may be evaluated from the form of ψ_{ns} and ψ_{np_z} as

$$\lambda = \frac{\mathrm{e.V.}}{\Delta E}\,\frac{\langle r\rangle_{\mathrm{av}}}{2\sqrt{3}}. \tag{34}$$

If the field-gradient is produced by a neighbouring fractional electronic charge, xe, which normally would be roughly a distance $2\langle r\rangle$ away we have

$$V_1 = \frac{xe}{4\langle r^2\rangle} \tag{35}$$

and

$$\lambda = \frac{1}{2\sqrt{3}}\,\frac{xe^2}{4\langle r\rangle\Delta E} \tag{36}$$

As an order of magnitude $\langle r \rangle$ may be taken as $2a_0$ where a_0 is the Bohr radius then,

$$\lambda = \frac{0{\cdot}98x}{\Delta E} \tag{37}$$

where ΔE is measured in electron volts.

It is obvious that as long as ΔE is not too large *s-p* hybridization of up to 10 per cent can easily arise from this source. The second and third sources are more intimately concerned with the chemical bonding. Thus in general the overlap integral of an *s-p* hybrid σ-orbital with the orbital of the neighbouring atom is usually greater than that of either *s* or *p* orbitals alone. It was considerations of this sort which originally led Pauling to propose the whole idea of *s-p* hybridization. Figure 2 shows the variation in overlap integral for a *s-p* hybrid Slater orbital of chlorine with (a) a hydrogen 1(*s*) orbital and (b) another identical chlorine *s-p* hybrid which illustrates that on the one hand "a little *s*-character goes a long way" and on the other hand that there is not

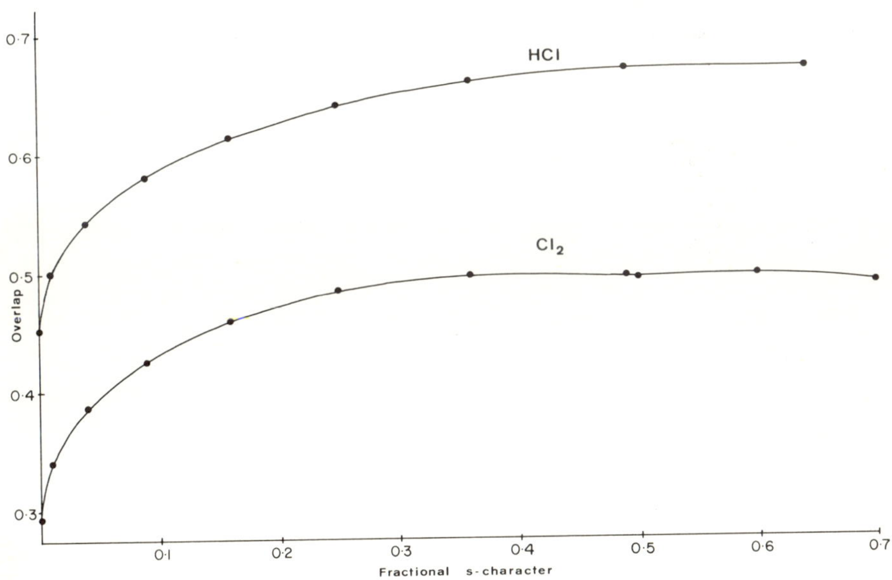

FIG. 7.2. Overlap and hybridization in HCl and Cl_2 using Slater orbitals for the chlorine atom.

much to be gained by increasing the *s*-character above 20–25 per cent. Finally *s-p* hybridization can arise purely from energetic considerations. Figure 3 illustrates in a purely general way the formation of a single sigma bond between a chlorine atom and an atom having just one orbital. When the partner

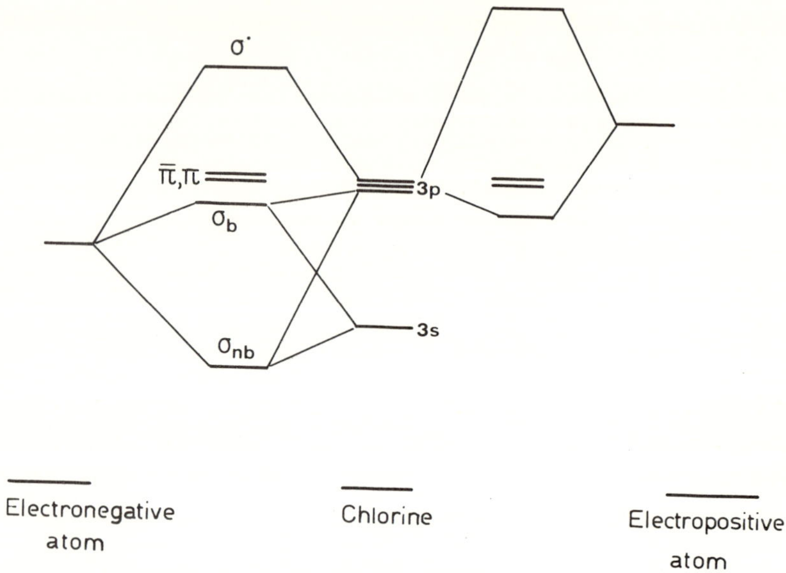

FIG. 7.3. Correlation diagram for bonds formed by chlorine with more electropositive and more electronegative elements.

is an electronegative element its orbital energy will be intermediate between that of the chlorine s and p orbitals, and the contributions of the s orbitals to σ_b and σ_{nb} will be considerable. When it is a more electropositive element its orbital energy will however be greater than that of the chlorine p-orbital and hence even greater than that of the chlorine s-orbital. All other things being equal the contribution of the latter to the bonding will now be much less than when the partner is a more electronegative element than chlorine. It will be recognized that this is just the opposite to the trend implied by Townes and Dailey's "rule".

Whitehead and Jaffé[12] have attempted a more systematic treatment of s-p hybridization in the halogens which we may illustrate simply by reference to HCl. The ionic character of a bond is presumed to be given by a simple-relationship first proposed by Gordy[11]

$$i = \tfrac{1}{2}|\chi_A - \chi_B| \tag{38}$$

where χ_A and χ_B are the electronegativities of the atoms A and B respectively. When a s-p hybrid orbital is used for bonding Whitehead and Jaffé introduce the concept of orbital electronegativity given by

$$\chi_{\text{hyb}} = s\chi_s + (1 - s)\chi_p \tag{39}$$

where $s\,(= \alpha^2)$ is the fractional s-character of the hybrid orbital.

Then if χ_H is the electronegativity of the hydrogen atom, we have for HCl

$$i = \tfrac{1}{2}\{s\chi_s^{\mathrm{Cl}} + (1 - s)\chi_p^{\mathrm{Cl}} - \chi_{\mathrm{H}}\} \tag{40}$$

and substitution of this into equation (22) yields

$$\frac{e^2Qq}{e^2Qq_0} = \left(\frac{\chi_s^{\mathrm{Cl}} - \chi_p^{\mathrm{Cl}}}{2}\right)s^2 - \left(\frac{\chi_s^{\mathrm{Cl}}}{2} - \chi_p^{\mathrm{Cl}} + 1 + \frac{\chi^{\mathrm{H}}}{2}\right)s + \left(1 + \frac{\chi^{\mathrm{H}}}{2} - \frac{\chi_p^{\mathrm{Cl}}}{2}\right). \tag{41}$$

Substitution of the observed value of ρ and the values of $\chi_{\mathrm{H}}, \chi_s^{\mathrm{Cl}}, \chi_p^{\mathrm{Cl}}$ previously calculated by Hinze and Jaffé[13] from the spectroscopic data, then yields the value of s.

This method was applied to the tetrahalides of Group IV where the hybridization of the central atom is defined by the tetrahedral symmetry and to the trihalides of Group III where the s-p hybridization of the central atom was obtained from the experimentally determined bond-angles. In this manner they obtained values for s shown in Table 6. In making this analysis Whitehead and Jaffé assumed that the lone-pair p-π electrons of the halogens played no part in the bonding. This is almost certainly not the case for the tetrahalides of Silicon, Germanium and Tin and in the same paper Whitehead and Jaffé made an attempt to take this factor into account. It is not now however an appropriate place to discuss this point and we will return to it in Chapter 11.

TABLE 7.6. Fractional s-character in halogen sigma orbital for various compounds (Ref. 12, *Theor. Chim. Acta* **1**, 209 1963)

Compound X	F—X	H—X	CX_4	SiX_4	GeX_4	SnX_4
Cl	0·048	0·069	0·097	0·371	0·323	0·338
Br	0·067	0·107	0·069	0·315	0·286	0·307
I	—	0·050	0·004	0·262	0·293	0·278

Although there does not exist any general experimental method to determine unambiguously the relative amount of s and p character in a bond formed by a singly coordinated atom, the nuclear electron spin–spin hyperfine coupling tensor of free radicals will in favourable cases yield values of this ratio for the molecular orbital containing the unpaired electron. Thus this hyperfine coupling tensor can be split up into two parts: the first an isotropic term, and the second a traceless tensor. In solution where the paramagnetic species are undergoing a rapid trumbling motion only the isotropic splitting is apparent, measurement of the anisotropic part requires a study on an oriented paramagnetic molecule. It is a standard result of electron spin resonance theory that the isotropic coupling is proporticnal to the value of the square of the unpaired electron's wave function at the nucleus in question, $\psi^2_{(0)}$, while the anisotropic coupling in any particular orientation is proportional to the quantity

$$a_{\text{anisotropic}} = \left\langle \frac{3\cos^2\theta - 1}{r^3} \right\rangle . \tag{42}$$

Here θ is the angle between the direction of the magnetic field and the vector joining the electron and nuclear coordinates and r is the length of this vector. The brackets $\langle\ \rangle$ denote averaging over the electronic motion. If the unpaired electron's molecular orbital is expressed as

$$\psi_0 = \sum_i C_{0_i}\phi_{A_i} + \sum_j C_{0_j}\phi_{B_j} \tag{43}$$

where ϕ_{A_i} are atomic orbitals of the atom A giving rise to the hyperfine coupling and ϕ_{B_j} are atomic orbitals of other atoms in the molecule, then it can be seen from the above definitions that only s-orbitals of A contribute to the isotropic coupling while, contributions from the p, d, f . . . orbitals of A will normally be dominant in the anisotropic coupling owing to the dependence on $\langle r^{-3}\rangle$. The values of such terms can be calculated for any assumed atomic wave-function and indeed often be determined experimentally from atomic beam resonance spectra. If now an unpaired electron is moving in an essentially s-p hybrid sigma orbital when it is in the neighbourhood of the nucleus in question, then it is a straightforward matter to compare the isotropic and anisotropic couplings with those theoretically and/or experimentally determined for the pure atomic s and p orbitals and hence obtain the relative weights of the s and p contributions.

Relevant to the present discussion are the electron spin resonance results for interhalogen diatomic radical-anions which have been studied as species trapped in alkali-halide matrices. Table 7 summarizes the data available for these compounds.[14] The unpaired electron in these radicals is moving in the antibonding sigma orbital and in Table 7 the coupling data have been analysed to yield the coefficients of this orbital given by

$$\psi_0 = \{\alpha_x\phi_s^x + \beta_x\phi_p^x\} - \{\alpha_y\phi_s^y + \beta_y\phi_p^y\} \tag{44}$$

where (ϕ_s^x) and (ϕ_p^x) are the s- and p-atomic valence orbitals of atom x.

For the neutral interhalogen molecules we are of course more interested in the bonding sigma orbital, however if we rewrite the antibonding orbital as

$$\psi_0 = \frac{a}{\sqrt{a^2 + b^2 - 2abs}}\{\alpha'_x\phi_s^x + \beta'_x\phi_p^x\} - \frac{b}{\sqrt{a^2 + b^2 - 2abs}}\{\alpha'_y\phi_s^y + \beta'_y\phi_p^y\} \tag{45}$$

TABLE 7.7. $S(\alpha^2) - p(\beta^2)$ hybridization for the radical anions F_2^-, FX^- and X_2^- trapped in alkali-halide matrices (Ref. 14).

	α_F^2	β_F^2	$\alpha_F^2 + \beta_F^2$	$\frac{\alpha_F^2}{\alpha_F^2 + \beta_F^2}$	α_x^2	β_x^2	$\alpha_x^2 + \beta_x^2$	$\frac{\alpha_x^2}{\alpha_x^2 + \beta_x^2}$	$\sum_{x,F} (\alpha_i^2 + \beta_i^2)$	e^2Qq_x (Mc/s)
F_2^-	0·016	0·587	0·603	0·026	—	—	—	—	1·206	—
FCl^-	0·011	0·556	0·567	0·019	0·031	0·707	0·738	0·042	1·305	28 ± 28
FBi^-	0·009	0·526	0·535	0·017	0·029	0·722	0·751	0·039	1·286	308 ± 56
FI^-	0·005	0·497	0·502	0·001	0·036	0·693	0·729	0·049	1·231	(~1300)
Cl_2^-	—	—	—	—	0·024	0·585	0·609	0·039	1·218	50 ± 6
Br_2^-	—	—	—	—	0·019	0·587	0·606	0·031	1·212	224 ± 22
I_2^-	—	—	—	—	0·020	0·563	0·583	0·034	1·166	(~930)

where S is the overlap integral between the hybrid orbitals then the bonding orbital is given by

$$\psi_{\text{bonding}} = \frac{b}{\sqrt{a^2 + b^2 + 2abS}}\{\alpha'_x\phi_s^x + \beta'_x\phi_p^x\}$$
$$+ \frac{a}{\sqrt{a^2 + b^2 + 2abS}}\{\alpha'_y\phi_s^y + \beta'_y\phi_p^y\}. \quad (46)$$

Thus the relative amounts of s- and p-character in the antibonding and bonding orbitals are the same.

These results show a number of very distinct trends. In the first place whether the atom in question is at the positive or the negative end of the dipole the s-hybridization is very small, never exceeding 5 per cent (it would, however, be very interesting to have the results for ICl^-, IBr^- and $BrCl^-$). Furthermore, in complete disagreement with the trend suggested by Townes and Dailey's "rule" for assessing hybridization the percent s-character is greater when an atom is at the positive end of the dipole than it is when it is at the negative end. This is of course in agreement with the qualitative discussion of the expected trend given above.

Rather imprecise values of the nuclear quadrupole coupling constants are also available from these data. It is easily shown that the Townes and Dailey analysis would indicate that the coupling constants of the radical anions should be roughly one half that of the corresponding neutral molecule. In general they appear to be rather less than this.

The discussion given above concerning the factors which influence the extent of s-p hybridization in a halogen sigma orbital of a singly-coordinated halogen atom, together with the experimental demonstration that the extent of hybridization is small, suggests that it is unreasonable to attempt to separate the effects of ionicity and hybridization. We will follow this practice in the rest of this book.

This is an appropriate point to show that the Townes and Dailey analysis does indeed give results which appear reasonable. Table 8 shows values of e^2Qq_0 for valence shell p-electrons for various nuclei obtained from atomic spectral or atomic beam data. They are all for atoms in which the field-gradient arises from a single electron or hole. Figure 4 shows a plot of the ionic character of diatomic halides as yielded by equation (22) and the corresponding nuclear quadrupole coupling constants against the electronegativity differences of the two atoms. It is assumed that $s = 0$. Particularly interesting are the results for ICl and BrCl where estimates of the ionic character are available from results at both ends of the molecule. The straight line represents the function

$$i = \tfrac{1}{2}(\chi_A - \chi_B) \quad (47)$$

TABLE 7.8. Values of e^2Qq_0 from atomic measurements.

Nucleus	Electron	e^2Qq_0(Mc/s)	Ref.
Na^{23}	$3p$	$-5{\cdot}16$	19
K^{39}	$4p$	$+5{\cdot}6$	20
Rb^{85}	$5p$	$-48{\cdot}8$	21
Rb^{87}	$5p$	$-23{\cdot}6$	21
Cs^{133}	$6p$	$-0{\cdot}92 \pm 1{\cdot}0$	20
B^{11}	$2p$	$-5{\cdot}390$	22
Al^{27}	$3p$	$-37{\cdot}52$	23
Ga^{69}	$4p$	$-125{\cdot}044940$	24
Ga^{71}	$4p$	$-78{\cdot}79808$	24
In^{115}	$5p$	$-899{\cdot}1048$	25
Cl^{35}	$3p$	$+109{\cdot}746$	26
Cl^{37}	$3p$	$+86{\cdot}510$	26
Br^{79}	$4p$	$-769{\cdot}756$	27
Br^{81}	$4p$	$-643{\cdot}032$	27
I^{127}	$5p$	$+2292{\cdot}712$	28

suggested by Gordy[11] for the relationship between ionic character and electronegativity difference while the dotted line corresponds to a smooth curve through the points. We will of course be returning to the data given in this figure in later chapters and will not discuss the details any more at this point, it nevertheless should serve to convince the reader that the Townes and Dailey analysis has something to recommend it. We must now look into the reasons for this success.

Cotton and Harris[15] have suggested an approximate method of analysis of nuclear quadrupole coupling constants which is somewhat different to that given by Townes and Dailey. Starting from equation (5) they make the following assumptions:

(1) The contribution to the field-gradient of the inner closed-shell electrons on nucleus α—the Sternheimer shielding term—is given the symbol P.

(2) The sum of $q^{\alpha}_{\text{nuclear}}$ and the two-centre integrals q^{α}_{jj} where orbital j is an inner shell or core orbital is approximated by the term q^{α}_{core}, the field-gradient produced by the other atoms in the molecule less their valence electrons. Thus, for example, in HCl $q^{\text{Cl}}_{\text{core}}$ is the field-gradient produced by the H^+ nucleus. This assumption is quite justified if the inner shell electrons are, as will be usually the case, close to the nucleus.

(3) The sum of all *non-core* two-centre integrals a^{α}_{jj} plus one half the sum $\sum_i \sum_j C_{i_p} C_{j_p} q^{\alpha}_{jj}$ is set off against q^{α}_{core} defined in (2) above. The justification for this lies in a consideration of actual numerical results for a number of halogen-containing molecules and Cotton and Harris estimate that this introduces errors of only "a few per cent".

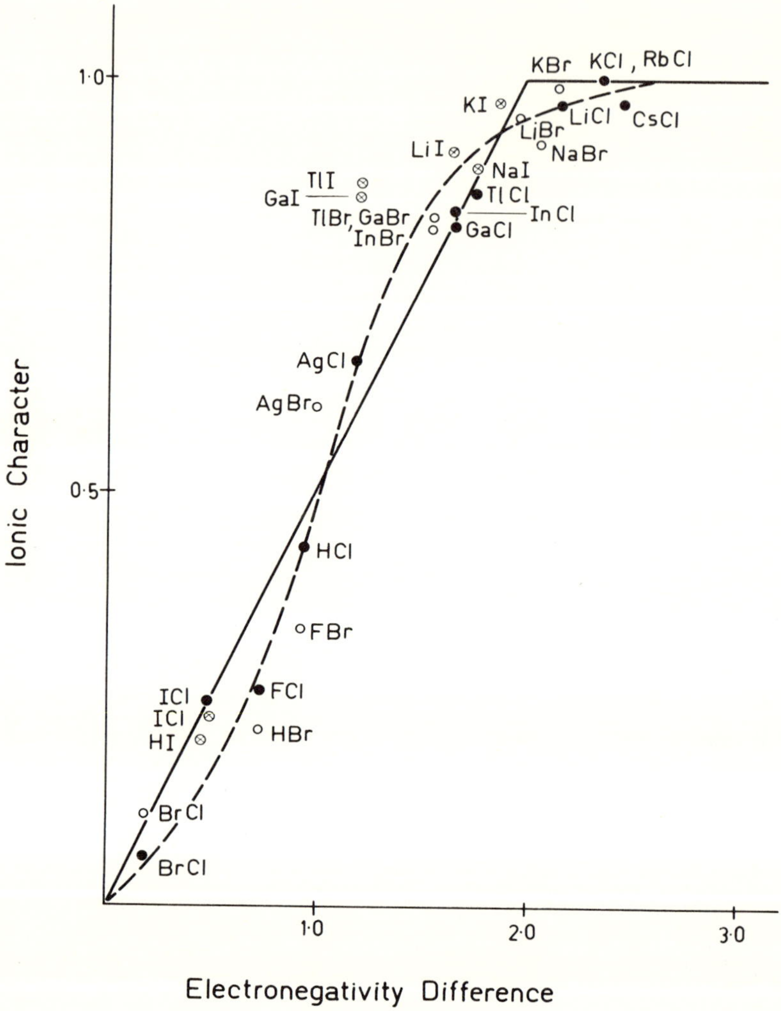

FIG. 7.4. The relationship between ionic character, calculated from the halogen coupling constants assuming no *s-p* hybridization of the chlorine sigma orbital, and the electronegativity difference for gaseous diatomic halides.

(4) As in Townes and Dailey's approximation all three-centre integrals are neglected. Thus equation (5) can be rewritten as

$$q_{\mathrm{a}} = \sum_{p} \left\{ \sum_{i} C_{i_p}^2 \overline{q_{ii}^{\alpha}} + \sum_{i \neq j}\sum C_{i_p} C_{j_p} \overline{q_{jj}^{\alpha}} \right\} + P \tag{48}$$

The bars over the integrals q_{ii}^{α} indicate that summation is now restricted to the core orbitals.

Cotton's next step was to suggest that the two-centre integrals q_{ij}^{α} may be approximated by the relationship

$$q_{ij}^{\alpha} = S_{ij} q_{ii}^{\alpha} \tag{49}$$

where S_{ij} is the overlap integrals between orbitals ϕ_i and ϕ_j. Equation (38) then becomes

$$q^{\alpha} = \sum_i f_i \bar{q}_{ii}^{\alpha} + P \tag{50}$$

where

$$f_i = \sum_p (C_{i_p})^2 + C_{i_p} \sum_{j>i} C_{j_p} S_{ij} \,. \tag{51}$$

The coupling constant is then given by

$$e^2 Q^{\alpha} q^{\alpha} = \sum_i f_i \left\{ e^2 Q^{\alpha} \bar{q}_{ii}^{\alpha} + \frac{e^2 Q^{\alpha} P}{\sum_i f_i} \right\}. \tag{52}$$

It is then assumed that the term in P, although perhaps large, does not vary much from molecule to molecule or indeed from atom to molecule. The term in brackets may then be replaced by the free-atom value for the coupling constant,

$$e^2 Q^{\alpha} q^{\alpha} = \sum_i f_i \,.\, e^2 Q^{\alpha} q_{\text{atomic}}^{\alpha} \,. \tag{53}$$

Clearly equation (53) reduces to Townes and Dailey's expression if S_{ij} is put equal to zero.

A closer consideration of equation (53) reveals that the quantity f_i defined in equation (51) has the same definition as the quantity $N(rk)$ called by Mulliken[16] the total gross population of the atomic orbital ϕ_{rk}. Chirgwin and Coulson[17] and Löwdin[18] had previously shown that for certain molecules this quantity is identical with the population, ρ_{rk} of atomic orbital ϕ_{rk} obtained in the L.C.A.O.–M.O. method with neglect of overlap

$$\rho_{rk} = \sum_i^m (C_{rki}^{\circ})^2. \tag{54}$$

The necessary and sufficient conditions for the equality of ρ_{rk} and $N(rk)$ is discussed in references 17 and 18; in practice $N(rk)$ and ρ_{rk} are identical for homonuclear molecules. For heteronuclear molecules the two quantities are not identical but it would appear that they are often not too different.

Inasmuch as Cotton and Harris's approximations are reasonable the above discussion combined with equation (54) provides a justification for Townes and Dailey's neglect of overlap in equation (17). The reader can easily verify for example that for the chlorine molecule with the molecular orbitals given in

* Equation (49) almost certainly overestimates the value of q_{ij}^{α}.

equation (6), equation (53) yields for the coupling constant of the chlorine atom

$$(e^2Qq)_{\text{Cl}_2} = (1 - \alpha^2)e^2Qq_{\text{atomic}} \tag{55}$$

identical with equation (18).

Cotton and Harris originally developed equation (53) in order to obtain the halogen coupling constants from the molecular orbitals they had calculated for transition-metal complexes. In other words, they wished to go from a wave-function to a coupling constant. As we have just shown this equation in fact provides some justification for going from nuclear quadrupole coupling constants to orbital population via the Townes and Dailey equations. We now, however, interpret the thus derived orbital populations as Mulliken's total gross atomic orbital populations rather than Hückel charge densities, ρ_r. Mulliken[16] has discussed the significance of the terms $N(rk)$ and shown that they correspond closely to real charge distributions.

The above discussion seems therefore to go some way towards justifying the use of the Townes and Dailey method for the analysis of nuclear quadrupole coupling data. There remains, however, one point in which it is deficient. In the step linking equations (52) and (53) it is assumed that the term in brackets in equation (52) is a universal constant for a particular type of orbital of a particular atom. For example, for the chlorine atom it has a particular value for the $3p$ orbitals irrespective of whether these orbitals are $3p_x$, $3p_y$ or $3p_z$. Table 9 shows a break-down of the various contributions to shielding for atomic chlorine. Two things are very striking, firstly that the angular distortion terms are smaller than the radial distortion terms and secondly that the sum of angular distortion and radial distortion for the inner shell, i.e. the total inner-shell shielding, is approximately zero. Almost the whole of the atomic shielding in Cl˙ is due to the $3p \rightarrow p$ radial distortion.

TABLE 7.9. Shielding in atomic chlorine. Listed are the field-gradients (in units of a_0^{-3}) produced by the various distortions of the Hartree–Fock wave-functions for chlorine (Ref. R. M. Sternheimer, *Phys. Rev.* **105**, 158 (1957).

Angular distortions:	$1s \rightarrow d$	0·130
	$2s \rightarrow d$	0·050
	$2p \rightarrow f$	0·086
	$3s \rightarrow d$	0·036
	$3p \rightarrow f$	0·042
Radial distortions:	$2p \rightarrow p$	−0·29
	$3p \rightarrow p$	0·59
Total:	—	0·644

It is quite easy to show (Chapter 5, equations 55, 56 and 57) that the shielding due to radial distortion of the p functions is composed of two terms: that due to the contraction (or expansion) of the orbital, p_z, having the unpaired electron and that due to the expansion (or contraction) of the remaining p_x and p_y orbitals. Furthermore the shielding produced by contraction (or expansion) of the p_z orbital is *exactly* twice that produced by the expansion (or contraction) of the p_x or p_y orbitals. Note that the distortion of the p_x and p_y orbitals produces a shielding of the *same* sign as that due to the distortion of the p_z orbitals despite the fact that one is contracted and the other expanded; this is of course because the field-gradient of the electrons in the p_x and p_y orbitals along the z direction is of opposite sign to that due to an electron in the p_z orbital. In formal terms therefore we have

$$(q_{n,l,m=0})_z = q_0(1 + 2\delta) \quad (56)$$

$$(q_{n,l,m=\pm 1})_z = -\tfrac{1}{2}q_0(1 - \delta). \quad (57)$$

The field-gradient of atomic chlorine is thus

$$(q_{nlm=0})_z + 4(q_{nlm=\pm 1})_z = -q_0(1 - 4\delta). \quad (58)$$

The similarity between the behaviour of the atomic orbitals when the restriction of independence of the radial function on magnetic quantum number is relaxed and the behaviour of the same orbitals when the atom takes part in molecule formation is too close to be coincidental. It is quite clear that a good part of the contraction of sigma orbitals and the smaller expansion of pi-orbitals noted by Tomasi and Scrocco[2] has nothing at all to do with the bonding but is simply a part of the Sternheimer shielding of the free atom.

A consequence of the Sternheimer shielding of atomic chlorine is that if we wish to improve the Townes and Dailey method it is not permissible to equate the coupling due to an electron in a p-orbital with the experimentally observed coupling in atomic chlorine. For singly-coordinated chlorine for example it is clear that two separate p-orbital gradients should be used, q_σ and q_π. For HCl, therefore, using the orbitals in equation (6) and separate values for q_σ and q_π we may therefore derive from equation (9) the expression

$$q_{\mathrm{Cl}} = 2\{(\alpha^2 - \alpha^2 b^2 + b^2)q_\sigma - q_\pi\} \quad (59)$$

$$\equiv \{N_{\mathrm{Cl},p_\sigma} \cdot q_\sigma - 2q_\pi\}. \quad (60)$$

Equation (60) has been written as above to avoid having to distinguish between the effects of ionicity and hybridization. N_{Cl,p_σ} is, as the discussion above has shown, the total gross population of the valence $3p$-σ orbital.

As it stands above equation (60) does not have the correct behaviour as N_{Cl,p_σ} approaches the value (2-a) for the chloride ion where the field-gradient is identically zero. Nor does it have the correct behaviour for a changing pi-population. We may overcome this if we assume that relationships such as (56) and (57) hold good for chlorine atoms within a molecule. If we write

$$q_\sigma = q_0\{1 + 2\delta(N_{\text{Cl},p_\pi} - N_{\text{Cl},p_\sigma})\} \tag{61}$$

$$q_\pi = q_0\{1 - \delta(N_{\text{Cl},p_\pi} - N_{\text{Cl},p_\sigma})\} \tag{62}$$

where N_{Cl,p_π} is the total gross population of one of the pi orbitals of the chlorine atom so that the correction terms in δ disappear as the field-gradient disappears, and substitute these in equation (60) we obtain

$$q = q_0(N_{\text{Cl},p_\sigma} - N_{\text{Cl},p_\pi})\{1 - \delta(2N_{\text{Cl},p_\sigma} + N_{\text{Cl},p_\pi})\}. \tag{63}$$

In this equation we have allowed the $3p_x, 3p_y$ ($\equiv 3p_\pi$) orbitals of chlorine to participate in bonding while retaining the cylindrical symmetry of the H—Cl bond.

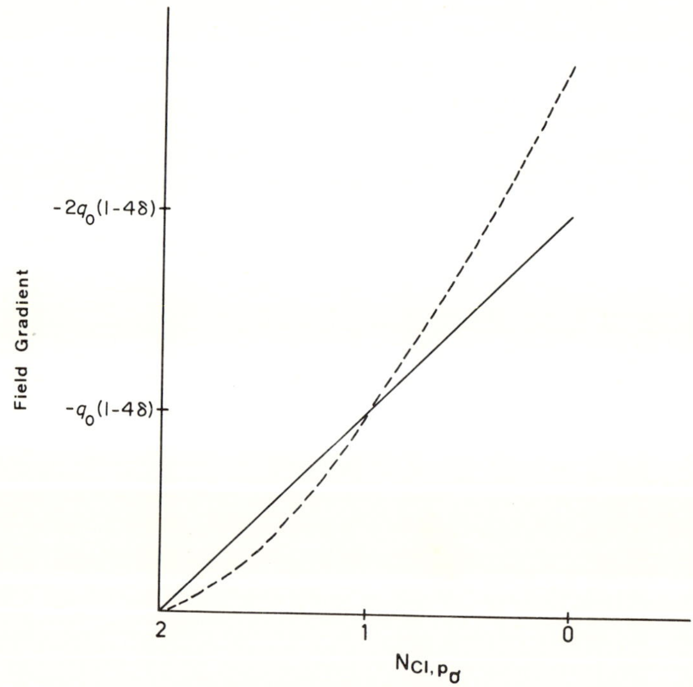

FIG. 7.5. Modified relationship between halogen atom sigma orbital population and quadrupole coupling constant (equation 63).

Figure 5 compares the predictions of equation (63) with the predictions of the Townes and Dailey equation using a unique value of the field-gradient produced by a chlorine $3p$-electron. (Note that according to equation (54) this is given by $q_0(1 - 4\delta)$). This figure has been drawn assuming that the pi-electrons of the chlorine atom do not participate in the bonding, and

N_{Cl,p_π} has the value 2·0. Two interesting observations may be made concerning this figure. In the region between $2 > N_{\text{Cl},p_\sigma} > 1$ the Townes and Dailey equation predicts a higher value for the field-gradient given by a particular value of N_{Cl,p_σ} than does equation (63). Putting it the other way round the value of N_{Cl,p_σ} deduced from the observed coupling constant is higher according to the Townes and Dailey equation than according to equation (63). In this connection it is interesting to note that one of the principle reasons which induced Townes and Dailey[9] to postulate such large amounts of s-character in R—Cl bonds precisely in this region was that they considered that otherwise the ionic characters estimated by their equation were much too high. Secondly in the region $1 > N_{\text{Cl},p_\sigma} > 0$ the simple Townes and Dailey equation underestimates the field-gradient for a particular value of N_{Cl,p_π}. Townes and Dailey overcame this by introducing the factor $(1 + \varepsilon)$ equation (21) due to the contraction of the p-orbitals under the influence of the increased effective nuclear charge. Figure 5 suggests that perhaps much of this arises from the differential Sternheimer shielding effect.

This analysis may be extended to Bersohn's equation (equation 28) for partial double-bonding in R—Cl bonds. The reader may easily verify that equation (28) becomes simply

$$\pi = \frac{2}{3}\left|\frac{e^2Qq_{zz}}{e^2Qq_\pi}\right|\eta(1 + \varepsilon). \qquad (64)$$

Since for a pure covalent bond $q_\pi = q_0(1 - \delta)$ whereas $q_{\text{atomic}} = q_0(1 - 4\delta)$ equation (28) overestimates the double bond character. Bersohn's original equation did not include the term $(1 + \varepsilon)$ and it seems likely that the compensation produced by the omission of this term to give the equation originally proposed by Bersohn

$$\pi = \frac{2}{3}\left|\frac{e^2Qq_{zz}}{e^2Qq_{\text{atomic}}}\right| \qquad (65)$$

yields reasonably satisfactory values of partial double bond character. In what follows, equation (65) will be used in this form.

The analysis given above can be extended without difficulty to multi-coordinate systems. The problem is that it introduces another parameter, δ, whose value must be estimated from calculations on atomic systems or from comparison of the results for a series[9] of molecules. For this reason we have contented ourselves in the survey of experimental results which follows with analysing the results according to the original Townes and Dailey equation although interpreting the results as yielding the total gross orbital populations. The qualitative effect of differences between the various orbital field-gradients is then indicated when this is appropriate. For singly-bonded compounds such as the halides no attempt has been made to separate effects of hybridization and ionicity: for multiply-bonded systems the hybridization is assumed to be given by the molecular geometry.

In conclusion, we remind the reader that the above approximate analyses should mainly be used to compare the results for different compounds and that not too much stress should be given to the values of bond parameters thus obtained for isolated molecules. As long as this is kept in mind then the method is extremely useful but we should not expect too much from it nor overlead with significance an apparently anomalous or paradoxical result.

REFERENCES

1. C. H. Townes and B. P. Dailey, *J. Chem. Phys.* **17**, 782 (1949).
2. E. Scrocco and J. Tomasi, *Th. Chim. Acta.* **2**, 386 (1964).
3. J. W. Richardson, *Rev. Mod. Phys.* **32**, 461 (1960).
4. L. Davis, B. T. Feld, C. W. Zabel and J. R. Zacharias, *Phys. Rev.* **76**, 1076 (1949). V. Jaccarino and J. G. King, *Phys. Rev.* **83**, 471 (1951). J. G. King and V. Jaccarino, *Phys. Rev.* **91**, 209 (1953); also *Phys. Rev.* **94**, 1610 (1954).
5. R. Livingston, *J. Phys. Chem.* **57**, 496 (1953).
6. A. Julg, *Tetrahedron* **19**, Supplement 2, 139 (1963).
7. R. S. Mulliken, C. A. Rieke, D. Orloff and H. Orloff, *J. Chem. Phys.* **17**, 1248 (1949).
8. W. Gordy, *Dis. Faraday Soc.* **19**, 14 (1955).
9. B. P. Dailey and C. H. Townes, *J. Chem. Phys.* **23**, 118 (1955).
10. R. Bersohn, *J. Chem. Phys.* **22**, 2078 (1954).
11. W. Gordy, "Microwave Spectroscopy", John Wiley & Sons, New York (1953).
12. M. A. Whitehead and H. H. Jaffé, *Trans. Faraday Soc.* **57**, 1854 (1961); *J. Chem. Phys.* **34**, 2204 (1961); *Theor. Chim. Acta* **1**, 209 (1963).
13. J. Hinze and H. H. Jaffé, *J. Am. Chem. Soc.* **84**, 540 (1962).
14. D. Schoemaker, *Phys. Rev.* **149**, 693 (1966).
15. F. A. Cotton, and C. B. Harris, *Proc. Nat. Acad. Sci.* **56**, 12 (1966).
16. R. S. Mulliken, *J. Chem. Phys.* **23**, 1833; 1841; 2338 and 2343 (1955).
17. B. H. Chirgwin and C. A. Coulson, *Proc. R. Soc.* **201**A, 196 (1950).
18. P. O. Löwdin, *J. Chem. Phys.* **18**, 365 (1950).
19. M. Perl, I. I. Rabi and B. Senitzky, *Phys. Rev.* **98**, 611 (1955).
20. P. Buck, I. I. Rabi and B. Senitzky, *Phys. Rev.* **104**, 553 (1956).
21. B. Senitzky and I. I. Rabi, *Phys. Rev.* **103**, 315 (1956).
22. G. Wessel, *Phys. Rev.* **92**, 1582 (1953).
23. H. Lew and G. Wessel, *Phys. Rev.* **90**, 1 (1953).
24. R. T. Daly Jr. and J. H. Holloway, *Phys. Rev.* **96**, 539 (1954).
25. P. Kusch and T. G. Eck, *Phys. Rev.* **94**, 1799 (1954).
26. V. Jaccarino and J. G. King, *Phys. Rev.* **63**, 471 (1951).
27. J. G. King and V. Jaccarino, *Phys. Rev.* **94**, 1610 (1954).
28. V. Jaccarino, J. G. King, and R. H. Satten and H. H. Stroke, *Phys. Rev.* **94**, 1798 (1954).

CHAPTER 8

ENVIRONMENTAL EFFECTS ON NUCLEAR QUADRUPOLE COUPLING CONSTANTS

A. INTRODUCTION

In this chapter we will be concerned with several topics which are loosely related by their being about effects of molecular environment on nuclear quadrupole coupling constants. For the most part they are specifically concerned with quadrupole coupling constants measured on nuclei in molecules forming part of a solid phase. A discussion of them is necessary before we can go on to a survey of experimental data.

B. THE DIFFERENCE BETWEEN COUPLING CONSTANTS IN THE SOLID AND GAS PHASES

If the quadrupole coupling constant of a molecule is measured both in the gas phase and in the solid phase it is almost always observed that the absolute magnitude of the coupling constant in the gas phase is about 5 per cent higher than that in the solid phase. Table 1 illustrates this phenomenon. This statement must of course be qualified by the remark that the molecule under study should form a molecular crystal in the solid phase—it is obviously ridiculous to compare sodium chloride vapour with crystalline sodium chloride—and furthermore that the intermolecular forces present in the molecule crystal should be non-specific. Thus the presence of intermolecular hydrogen bonds could mean that the systems in the gas phase and the solid phase are quite different. This is particularly true of a nucleus such as nitrogen which is well-known to form strong hydrogen bonds, and Table 1 shows the results for ammonia in order to illustrate the proportionately much greater changes which take place in this case. For the present purposes such changes in quadrupole coupling constants are trivial in the sense that the two systems are now quite dissimilar although of course their

TABLE 8.1. Pure quadrupole coupling constants of various compounds in the gas and solid phases. Where quoted the estimated errors in the gas phase coupling constants are those quoted by the authors of the measurements, solid state coupling constants are known to at least ±0·01 Mc/s.

Molecule	Nucleus	Gas-phase data e^2Qq_{zz}(Mc/s)	η	Ref.	Solid state data at 77°K e^2Qq_{zz}(Mc/s)	η	Ref.
CH_3Cl	Cl^{35}	−74·77 ± 0·10	0·0	11	68·4	(0)	24
CF_3Cl	Cl^{35}	−77·93 ± 0·1	0·0	12	−77·58	(0)	24
$(CH_3)_3CCl$	Cl^{35}	−66·9 ± 1·5	0·0	13	62·12	(0)	25
C_6H_5Cl	Cl^{35}	−71·10		14	69·14	(0)	25
CH_3Br	Br^{79}	577·15 ± 0·10	0·0	11	528·9	(0)	26
CF_3Br	Br^{79}	619	0·0	15	604·0	(0)	27
$(CH_3)_3CBr$	Br^{79}	511·6 ± 5·0	0·0	16	467·0	(0)	28
C_6H_5Br	Br^{79}	567·0 ± 4·0	0·03	17	537·71	(0)	29
CH_3I	I^{127}	−1929	0·0	18	1767·3	0·0008	30
CF_3I	I^{127}	2143·8 ± 2·0	0·0	19	2069·2	(0)	30
NF_3	N^{14}	−7·07	0·0	20	7·068*	0·006*	31
$(CH_3)_3N$	N^{14}	−5·47	0·0	21	5·1939	0·0	32
NH_3	N^{14}	−4·084	0·0	22	3·1607	0·0	33
C_5H_5N	N^{14}	−4·88	0·414	23	4·60	0·4	34

* Extrapolated to 0°K.

intrinsic interest is often quite great. It will also often be quite difficult to draw the line between specific and non-specific intermolecular association and indeed we will not attempt to do so.

This effect of the crystalline state on quadrupole coupling constants also manifests itself in two other ways. In the first place coupling constants measured in the solid state are temperature dependent, increasing with decreasing temperature. For a typical covalent chlorine compound the resonance frequency increases by about 0·5 Mc/s on going from room temperature to 77 °K. The theory of this phenomenon will be discussed below. Secondly nuclei which in the isolated molecule are chemically identical but which in the crystalline state are not in crystallographically equivalent positions usually have slightly different frequencies. Table 2 illustrates this phenomenon. Allied to it is the fact that different crystalline forms of the same molecule have different coupling constants. Before we discuss the theoretical reasons for this we must stress the practical point which is very relevant to the main theme of this book, namely that solid state coupling constants do not represent entirely accurately the coupling constants in the isolated molecule so that small differences between the coupling constant of a particular nucleus in two different compounds should not be endowed with extreme significance. The example of chloral hydrate shown in Table 2 is probably the most extreme case of frequency shifts due

TABLE 8.2. Multiple quadrupole resonance frequencies for chemically equivalent nuclei

Molecule	Nucleus	Pure quadrupole resonance frequencies (Mc/s at 77°K)	Reference
$CCl_3CH(OH)_2$	Cl^{35}	37·513, 38·699, 38·781	35
CCl_3COOH	Cl^{35}	39·967, 40·165, 40·240	35
o-$Cl_2C_6H_4$	Cl^{35}	35·580, 35·755, 35·824	36
C_5H_5N*	N^{14}	2·956, 2·987, 2·992, 3·001, 3·856, 3·893, 3·904, 3·914	34

* Note that for this compound the two sets of four closely spaced resonances are due to the presence of a non-zero asymmetry parameter since the nitrogen nucleus has a unit nuclear spin.

to crystallographically inequivalent nuclei and it has been suggested[1] that the difference is due to hydrogen-bonding in the solid state. It is certainly true that one of the C—Cl bond lengths is markedly shorter than the other two[2] (1·79, 1·78 and 1·72 Å). This view has, however, recently been contested.[3] It is worth noting that even in the isolated molecule if the CCl_3 group is not rapidly rotating with respect to the molecular framework then the three chlorine atoms are not strictly chemically equivalent and the different frequencies may well reflect real intramolecular differences in electronic environment. In general, however, when the nuclei in question are strictly equivalent for the rigid isolated molecule then the spread in frequencies is small. For Cl^{35}, for example, a difference of 0·5 Mc/s in resonant frequency between two different molecules can usually be regarded as significant.

The reasons for the solid-state frequency shifts is straightforward at a qualitative level. In the first place most of the compounds studied are polar so that in the solid state each molecule lies in the field of a number of neighbouring electrostatic dipoles. As will be discussed below pure quadrupole resonant frequencies exhibit a shift in the presence of an electrostatic field due to the polarization of the electrons in the neighbourhood of the nucleus under study, and quantitative studies of the shifts produced by external fields together with estimates of the magnitude of the crystalline electrostatic field show that this effect is quite large enough to account for the observed shifts. In addition the dipoles will also give rise to field gradients at this nucleus, but this effect is likely to be less important.

It remains to put this qualitative theory on a more quantitative basis but of course this can only be done for each compound in the light of the knowledge of its crystal structure. Bersohn[4] has attempted with fair success to do so for methyl chloride and 1,3,5-trichlorobenzene. An absolute calculation of the solid state shift involves a number of assumptions which really preclude anything but a result correct in sign and order of magnitude. This was achieved with no difficulty for methyl chloride but 1,3,5-trichlorobenzene presents the interest that it exhibits three resonance frequencies—corresponding to chemically equivalent halogen atoms in crystallographically inequivalent positions—for which the theory predicts the relative shift. Bersohn's results are shown in Fig. 1 where the observed frequencies are plotted against the calculated internal field (or to be more precise the component of this field lying along the C—Cl bond vector, c.f. the discussion on externally applied electric fields given below). This field was calculated from the known crystal structure of 1,3,5-trichlorobenzene by placing a point dipole, μ, at a point midway between the carbon and chlorine nuclei and directed along the C—Cl bond.

It is interesting to note that Bersohn's calculation for this molecule imply that, contrary to what he finds for chloromethane, the internal field *increases* the coupling constant over that in the gas phase. Indeed except for small polar molecules which will probably be packed in the crystal so that the dipoles are antiparallel, there is no reason why the internal field should always decrease the resonance frequency. To explain the almost universal

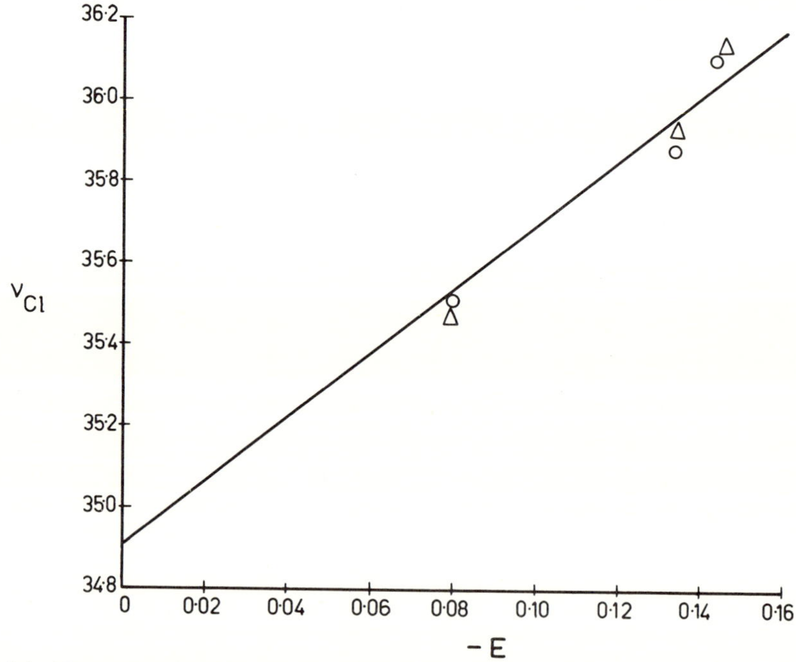

FIG 8.1. Observed chlorine resonance frequencies in 1,3,5-trichlorobenzenes (circles) versus calculated internal electric field. Crosses are observed Br^{81} frequencies in 1,3,5-tribromobenzene divided by 6·662. The Units of electric field are $\mu/\varepsilon \times 10^{24}$.

lowering of the frequency we must have recourse to the fact that the molecules in the crystal are not static but are vibrating. This point naturally leads us to a discussion of the effects of temperature on quadrupole coupling constants measured in the solid state.

A natural preface to a study of the effects of vibrations on the coupling constants of solids is a mention of the effect of vibrational state on the coupling constants of nuclei in molecules in the vapour phase. We will here be concerned with molecules forming essentially covalent bonds; such effects in ion-pair molecules such as the alkali halides can often be extremely large and will be discussed in Chapter 12. Table 3 shows the result of such a study on the methyl halides. The results shown are typical and serve to illustrate the point that simple linear vibrations which do not change the direction of the axes of the field-gradient tensor do not much change the magnitude of the quadrupole coupling constant.

TABLE 8.3. Variation of coupling constant with vibrational state in the methyl halides. Data taken from Ref. 11.

Molecule	Vibrational quantum number	Halogen coupling constant (Mc/s)
$C^{12}H_3Cl^{35}$	$v = 0$	$-74{\cdot}77 \pm 0{\cdot}10$
	$v_3 = 1$	$-74{\cdot}87 \pm 0{\cdot}10$
	$v_6 = 1$	$-74{\cdot}89 \pm 0{\cdot}20$
$C^{12}H_3Br^{79}$	$v = 0$	$577{\cdot}15 \pm 0{\cdot}10$
	$v_3 = 1$	$577{\cdot}32 \pm 0{\cdot}20$

In solids, however, the situation is different. The vibrations of the crystal as a whole will in general result in rapidly changing directions for the field-gradient axes. Consider the effect of small rotations θ_x, θ_y or θ_z about the X, Y and Z axes of the field-gradient tensor. Then, including terms up to the second order in θ_x, θ_y and θ_z, the new values of the field-gradient components, q'_{xx} etc., are related to the corresponding components, q_{xx} etc., in the unperturbed molecule by the expressions

$$\begin{aligned} q'_{xx} &= (1 - \theta_y^2 - \theta_z^2)q_{xx} + \theta_z^2 q_{yy} + \theta_y^2 \cdot q_{zz}, \\ q'_{yy} &= \theta_z^2 \cdot q_{xx} + (1 - \theta_x^2 - \theta_z^2)q_{yy} + \theta_x^2 \cdot q_{zz}, \\ q'_{zz} &= \theta_y^2 \cdot q_{xx} + \theta_z^2 \cdot q_{yy} + (1 - \theta_x^2 - \theta_y^2)q_{zz}, \\ q'_{xy} &= \theta_z \cdot q_{xx} + (\theta_x\theta_y - \theta_z)q_{yy} - \theta_x\theta_y q_{zz}, \\ q'_{yz} &= -\theta_y\theta_z q_{xx} + \theta_x q_{yy} + (\theta_y\theta_z - \theta_x)q_{zz}, \\ q'_{xz} &= -\theta_y \cdot q_{xx} - \theta_x\theta_z q_{yy} + (\theta_x\theta_z + \theta_y)q_{zz}. \end{aligned} \tag{1}$$

The observed coupling constant is then given by the average of this expression over the angular displacements caused by the crystalline vibrations. Now since the vibration is symmetrical about the equilibrium position the averages $\langle\theta_x\rangle$, $\langle\theta_y\rangle$ and $\langle\theta_z\rangle$ are zero. This has a consequence that the average value of the off-diagonal elements q'_{xy} etc. are also zero. The average value of the squares of the angular displacements $\langle\theta_x^2\rangle$, $\langle\theta_y^2\rangle$ and $\langle\theta_z^2\rangle$ are not however zero and manipulation of the above formulae show that the vibrationally-averaged coupling constant $e^2Qq'_{zz}$ and asymmetry parameter η' are related to those of the stationary molecule by the equations

$$\begin{aligned} e^2Qq'_{zz} &= e^2Qq_{zz}[1 - \tfrac{3}{2}(\langle\theta_x^2\rangle + \langle\theta_y^2\rangle) - \tfrac{1}{2}\eta(\langle\theta_x^2\rangle - \langle\theta_y^2\rangle)] \\ &\approx e^2Qq_{zz}[1 - \tfrac{3}{2}(\langle\theta_x^2\rangle + \langle\theta_y^2\rangle)] \end{aligned} \tag{2}$$

$$\begin{aligned} \eta' &= \frac{e^2Qq_{zz}}{e^2Qq'_{zz}}[\eta - \tfrac{3}{2}(\langle\theta_x^2\rangle - \langle\theta_y^2\rangle) - \tfrac{1}{2}\eta(\langle\theta_x^2\rangle + \langle\theta_y^2\rangle)] \\ &\approx \eta. \end{aligned} \tag{3}$$

Thus the main effect of vibrations is to lower the coupling constant since the averages $\langle\theta_x^2\rangle$ and $\langle\theta_y^2\rangle$ will always be positive. The directions of the field-gradients axes in the vibrating and in the stationary molecules are however identical. Since the vibration amplitudes are temperature-dependent, decreasing with decreasing temperature, the pure quadrupole resonance frequencies will increase with decreasing temperature.

An attempt to put this temperature-dependence on a quantitative basis was first made by Bayer[5] and extended by Kushida[6] and by Wang.[7] They supposed that the torsional vibrations about the three field-gradient axes could be replaced by harmonic oscillators. Then

$$\langle\theta_i^2\rangle = \frac{\hbar}{A_i\omega_i}\left\{\frac{1}{2} + \frac{1}{e^{\hbar\omega_i/kT} - 1}\right\} \tag{4}$$

where ω_i is the oscillator frequency, A_i is the moment of inertia of the system and T the absolute temperature. This expression may now be substituted in equation (2) to give an expression for the temperature dependence of the coupling constant and hence of the resonance frequency.

$$\nu = \nu_0\left\{1 - \frac{3}{2}\sum_i \frac{\hbar A_i}{\omega_i}\left(\frac{1}{2} + \frac{1}{\exp(\hbar\omega_i/kT) - 1}\right)\right\} \tag{5}$$

where ν_0 is the frequency for the rigid lattice. The summation is of course over all the different torsional modes about the X or Y axes. For frequencies such that $\omega < kT/\hbar$ the exponential term may be expanded in the conventional way so that the expression becomes

$$\nu = \nu_0 \left\{ 1 - \frac{3kT}{2} \sum_{i=1}^{M} \frac{A_1}{\omega_i^2} - \frac{\hbar^2}{8kT} \sum_{i=1}^{M} A_i + \Omega(T) \right\}$$

where

$$\Omega(T) = -\frac{3}{2} \sum_{M+1}^{N} \frac{\hbar A_i}{\omega_i} \left(\frac{1}{2} + \frac{1}{\exp(\hbar\omega_i/kT) + 1} \right). \tag{6}$$

The first two summations are over all frequencies for which $\omega < kT/\hbar$ and the summation in $\Omega(T)$ over frequencies for which $\omega > kT/\hbar$. Because of the term $1/\omega_i$, $\Omega(T)$ may in practice be neglected so that the Bayer equation has the form

$$\nu = \nu_0 \left\{ 1 + bT + \frac{c}{T} \right\}$$

where

$$b = -\frac{3}{2} k \sum_{i=1}^{M} \frac{A_i}{\omega_i^2} = -\frac{3k}{2} M \left\langle \frac{A}{\omega^2} \right\rangle$$

$$c = -\frac{\hbar^2}{8k} \sum_{i=1}^{M} A_i = -\frac{\hbar^2}{8k} M \langle A \rangle. \tag{7}$$

Experimental tests of this equation show that it does not adequately describe the behaviour of the resonance frequencies over a large frequency range, although qualitatively it does have the correct form. Kushida *et al.*[8] pointed out, however, that Bayer's theory assumes an independence of A_i and ω_i on temperature whereas these terms are certainly temperature dependent owing to the change of volume with temperature. Bayer's theory is essentially therefore a constant volume theory whereas the experimental measurements are usually made at constant pressure. Kushida *et al.*[8] therefore experimentally determined both the temperature and pressure dependence of the resonance frequencies of several compounds which, together with a knowledge of the compressibility and coefficients of thermal expansion, enabled them to construct a curve giving the resonance frequencies at constant volume. These curves agreed extremely well with the predictions of the Bayer theory.

The pressure-dependence of quadrupole resonance frequencies mentioned above had previously been observed by a number of workers[9,10] and has subsequently been studied on numerous occasions in conjunction with investigations on the effect of temperature, usually directed towards a further test of the Bayer theory as modified by Kushida *et al.*[8] It seems likely that the principal mechanism giving rise to pressure shifts is the modification of the internal electric field caused by the macroscopic displacement of the molecules in the crystalline lattice. The coefficients of pressure dependence should thus be susceptible to theoretical investigation by Bersohn's method[4] and indeed this author has predicted the sign of this term for 1,3,5-trichlorobenzene on the basis of his calculations of the internal electrostatic field. As far as the present author is aware, however, the pressure dependence of 1,3,5-trichlorobenzene has not yet been measured nor have any attempts been made on other compounds to calculate the sign and magnitude of the effect.

To conclude this section therefore it seems clear that the joint effect of an internal electrostatic field and torsional vibration are sufficient to account for the observed differences between solid and gas phase coupling constants. This implies that extrapolation of the solid state coupling constant to 0°K via the Bayer equation combined with an estimate of the internal field by Bayer's method should yield the coupling constant of the isolated molecule. This is obviously unpractical on a routine basis and the usual solution adopted is to compare solid state coupling constants at as low a temperature as is convenient, usually 77°K.

REFERENCES

1. H. C. Allen, *J. Phys. Chem.* **57**, 501 (1953).
2. S. Kondo and I. Nitta, *X-Rays* **6**, 53 (1950).
3. D. Biedenkapp and A. Weiss, *Z. Naturf.* **22a**, 1124 (1967).
4. R. Bersohn, *J. Appl. Phys.* Suppl. to vol. **33**, 286 (1962).
5. H. Bayer, *Z. Phys.* **130**, 227 (1951).
6. T. Kushida, *J. Sci. Hiroshima University* **A19**, 327 (1955).
7. T. C. Wang, *Phys. Rev.* **99**, 566 (1955).
8. T. Kushida, G. B. Benedek and N. Bloembergen, *Phys. Rev.* **104**, 1364 (1956).
9. D. Dautreppe and B. Dreyfus, *C.r. R. hebd. Acad. Sci. Paris* **241**, 795 (1955).
10. C. Dean and E. Lindstrand, *J. Chem. Phys.* **24**, 1114 (1956).
11. J. Kraitchman and B. P. Dailey, *J. Chem. Phys.* **22**, 1477 (1954).
12. R. C. Johnson, Q. Williams and T. L. Weatherly, *J. Chem. Phys.* **36**, 1588 (1962).
13. D. R. Lide and C. K. Jen, *J. Chem. Phys.* **38**, 1504 (1963).
    W. Zeil, W. Hüttner and W. Plein, *Z. Naturf*, **17a**, 823 (1962).
14. R. L. Poynter, *J. Chem. Phys.* **39**, 1962 (1963).
15. J. Sheridan and W. Gordy, *J. Chem. Phys.* **20**, 591 (1952).
16. W. Zeil, M. Winnewisser and W. Hüttner, *Z. Naturf.* **16a**, 1248 (1961).
17. E. Rosenthal and B. P. Dailey, *J. Chem. Phys.* **43**, 2093 (1965).
18. P. C. Fletcher and E. Amble, *Phys. Rev.* **110**, 536 (1958).
19. W. Gordy and J. Sheridan, *J. Chem. Phys.* **20**, 591 (1957).
20. J. Sheridan and W. Gordy, *Phys. Rev.* **79**, 513 (1950).
21. D. R. Lide and D. E. Mann, *J. Chem. Phys.* **28**, 572 (1958).
22. M. T. Weiss and M. W. P. Strandberg, *Phys. Rev.* **83**, 567 (1951).
23. G. O. Sørensen, *J. Mol. Spec.* **22**, 325 (1967).
24. R. Livingston, *J. Chem. Phys.* **19**, 1434 (1951).
25. R. Livingston, *J. Phys. Chem.* **57**, 496 (1953).
26. H. G. Dehmelt, *Naturwiss.* **37**, 398 (1950).
27. H. Zeldes and R. Livingston, *J. Chem. Phys.* **21**, 1418 (1953).

28. W. Zeil, H. Heel, H. Pförtner and A. Schmitt, *Z. Elektrochem.* **63**, 1009 (1959).
29. J. Hatton and B. V. Rollin, *Trans. Faraday Soc.* **50**, 358 (1954).
30. H. Robinson, H. G. Dehmelt and W. Gordy, *J. Chem. Phys.* **22**, 511 (1954).
31. G. A. Matzkanin, T. A. Scott and P. J. Haigh, *J. Chem. Phys.* **42**, 1646 (1965).
32. C. T. O'Konski and T. J. Flautt, *J. Chem. Phys.* **27**, 815 (1957).
33. S. S. Lehrer and C. T. O'Konski, *J. Chem. Phys.* **43**, 1941 (1965).
34. L. Guibé, *C. r. hebd. Acad. Sci. Paris* **250**, 3014 (1960).
35. H. Allen, *J. Am. Chem. Soc.* **74**, 6074 (1952).
36. H. C. Meal, *J. Am. Chem. Soc.* **74**, 6121 (1952).

CHAPTER 9

THE STARK EFFECT IN PURE QUADRUPOLE RESONANCE

A. THEORY

1. Introduction

There remains one further effect of which the reader should be aware before going on to a survey of experimental results but which could not properly be discussed before the theoretical evaluation of field-gradients had been dealt with. This is the Stark effect in pure quadrupole resonance.

There is no direct interaction between an electrostatic field and a nuclear quadrupole. In sufficiently high fields, however, the electrons around the nucleus under study become distorted so that the field-gradient to which they give rise alters from that in the unperturbed state. This effect was first predicted theoretically by Bloembergen[1] and was soon afterwards observed experimentally.[2,3,4] Phenomenologically Bloembergen *et al.*[3,5] have shown that the effect may be described by a traceless third order tensor whose elements R^*_{ijk} are defined as

$$R^*_{ijk} = e\left(\frac{\partial q_{jk}}{\partial E_i}\right) = \left(\frac{\partial V_{jk}}{\partial E_i}\right). \tag{1}$$

As the above authors have pointed out piezoelectric crystals deform under the influence of the applied field. Experimentally this cannot be avoided, i.e. the measurements are made under constant stress. More interesting from the theoretical point of view would be the corresponding tensor elements under constant strain, R_{ijk}. These two quantities are related by the expression[3]

$$R^*_{ijk} = R_{ijk} + \sum_{m,n} S_{jkmn} d_{imn}, \tag{2}$$

where S_{jkmn} is a fourth order tensor relating the change in field-gradient produced by an external mechanical strain while d_{imn} are the piezoelectric coefficients. Both these terms are accessible experimentally but it has been shown[6,7] that their effect compared with the term R_{ijk} is small. It is the usual practice therefore to equate the experimentally determined R^*_{ijk} with the tensor elements at constant stress, R_{ijk}.

The details of the Stark effect on pure quadrupole resonance frequencies are slightly different for nuclei with integral or half-integral spins. They are therefore discussed separately.

2. Half-Integral Spins. Axially Symmetric Field-Gradients

This case is particularly simple. The levels occur in degenerate pairs characterized by the value of $|m|$ and the electrostatic field does not remove the degeneracy. The unperturbed frequencies are given by the expression (see equation 108, Chapter 3)

$$\nu_{|m| \to |m|+1} = \frac{3e^2Qq_{zz}}{4I(2I-1)h}\{2|m| + 1\}. \tag{3}$$

Straightforward differentiation of this expression yields

$$\frac{\partial \nu}{\partial q} = \frac{3e^2Q}{4I(2I-1)h}\{2|m| + 1\} \tag{4}$$

$$= \left(\frac{\nu}{q_{zz}}\right)_0. \tag{5}$$

Hence the linear Stark shift is given by

$$\delta\nu = \left(\frac{\nu}{q_{zz}}\right)_0 \delta q_{zz}. \tag{6}$$

According to the definition of the tensor elements given above

$$\delta q_{zz} = \frac{1}{e}\sum_i E_i R_{izz}. \tag{7}$$

Hence

$$\frac{\delta\nu}{\nu} = \frac{1}{eq_{zz}}\sum_i E_i R_{izz}. \tag{8}$$

For axial symmetry, however, $R_{xzz} = R_{yzz} = 0$. Thus

$$\frac{\delta\nu}{\nu} = \frac{E_z R_{zzz}}{eq_{zz}}. \tag{9}$$

The Stark effect in this case is described by just one parameter R_{zzz}.

Equation (6) predicts a linear shift of the resonance frequency for any particular orientation of the field-gradient tensor and the applied field. Since E_z is of odd parity, however, the shift in any particular direction changes in sign when the field is changed in sign or when the inversion operation is performed on the crystal. Consequently for a polycrystalline powder the application of an electric field results in a symmetrical broadening of an initially symmetrical absorption line. The actual line-shape depends on the mutual orientation of the radio-frequency and electrostatic fields. In the usual experimental arrangement they are perpendicular and Dixon and Bloembergen[5] have shown that the idealized line-shape is that shown in Fig. 1. The value of R_{zzz} may thus be directly obtained from the separation between the extrema of the derivative of the absorption. The same authors have

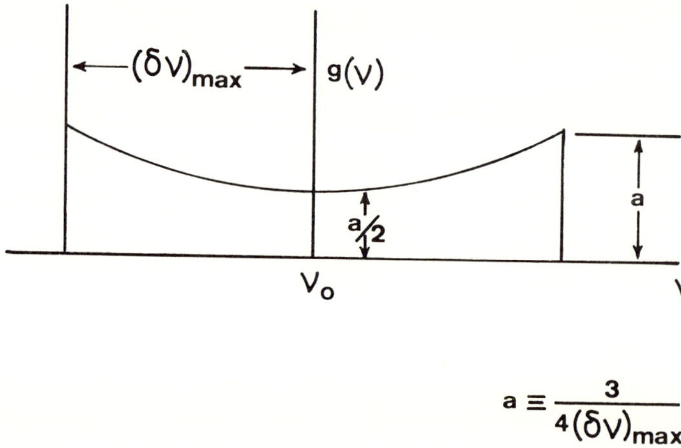

FIG. 9.1. Idealized line-shape for the Stark-broadened resonance of a polycrystalline powder for a nucleus with half-integral spin.

also shown that the *increase* in second moment, $\Delta\langle\Delta\nu\rangle^2$, for the broadened absorption curve is given by the expression

$$\Delta\langle\Delta\nu\rangle^2 = \tfrac{2}{5}(\delta\nu)^2_{\max}$$

where

$$(\delta\nu)_{\max} = \frac{3eQ\,.\,(2|m| - 1)}{4I(2I - 1)\,.\,h}\,.\,R_{zzz}\,.\,E \qquad (10)$$

and E is the magnitude of the electrostatic field. This expression provides a check on the direct measure of $\Delta\nu^0$ and a means of measuring it when the broadening is small.

3. Half-Integral Spins. Asymmetric Field-Gradients

Dixon and Bloembergen[5] have extended the above theory to the case of non-axial field-gradients. For nuclei with spin $I = \frac{3}{2}$, an analytical expression can be obtained for the energy levels of the unperturbed system (equation 115, Chapter 3). In order to achieve a proper separation of variables this must be re-expressed as

$$\nu = \frac{e^2Q}{2h}\left(q_{zz}^2 + \frac{\alpha^2}{3}\right)^{\frac{1}{2}} \tag{11}$$

where

$$\alpha \equiv \eta q_{zz} = q_{xx} - q_{yy}.$$

Then

$$\delta\nu = \left(\frac{\partial \nu}{\partial q_{zz}}\right)\delta q_{zz} + \left(\frac{\partial \nu}{\partial \alpha}\right)\delta\alpha. \tag{12}$$

From the definition of the tensor elements R_{ijk}

$$\delta q = \frac{1}{e}\sum_i E_i R_{izz} \tag{13}$$

$$\delta\alpha = \frac{1}{e}\sum_i E_i(R_{ixx} - R_{iyy}). \tag{14}$$

Partial differentiation of equation (11) and substitution of equations (13) and (14) yields

$$\frac{\delta\nu}{\nu_0} = \left[eq_{zz}\left\{1 + \frac{\eta^2}{3}\right\}\right]^{-1}\cdot\left[\sum_i E_i\left\{R_{izz} + \frac{\eta}{3}(R_{ixx} - R_{iyy})\right\}\right]. \tag{15}$$

The differences $(R_{ixx} - R_{iyy})$ are likely to be small, so that if the asymmetry is not too large the splitting is still dominated by the term R_{zzz}.

For half-integral spin nuclei with $I > \frac{3}{2}$ in asymmetric field-gradients analytical expressions are not possible for the unperturbed system and hence *a fortiori* not available for the Stark shift. Dixon and Bloembergen[5] have given the necessary basic equations and applied them to the results for the I^{127} nucleus in iodic acid.

The conclusions for polycrystalline powders given above and the correctness in detail of the theory has been checked by measurements on single crystals of sodium chlorate[7] and on *p*-dichlorobenzene.[8]

4. Integral Spins

The only nuclear spin of much interest here is $I = 1$. The system is characterized by a non-degenerate level $m = 0$ and a pair of levels arising from the $m = \pm 1$ states which are degenerate in axially symmetric field-gradients. The essential difference between integral and non-integral spins is that Kramers theorem states that the $\pm m$ degeneracy of the former can be lifted by an electric field but not of the latter. We must therefore treat the general case directly without presupposing axial symmetry.

Pure quadrupole resonance transitions are observed between the $m = 0$ and the $m = \pm 1$ states. The frequencies are given by

$$\nu_{\pm} = \frac{3e^2Q}{4h}\left\{q_{zz} \pm \frac{\alpha}{3}\right\}. \tag{16}$$

Proceeding as above yields for the Stark shift

$$\frac{\delta\nu}{\nu} = \left[e\left(1 \pm \frac{\eta}{3}\right)\right]^{-1}\left[\sum_i E_i\left\{R_{izz} \pm \frac{(R_{ixx} - R_{iyy})}{3}\right\}\right]. \tag{17}$$

Thus even for axially-symmetric field-gradients the shift depends on parameters other than R_{zzz}.

No experimental study of the Stark effect in a nucleus of spin $I = 1$ seems to have been made to date.

B. THEORETICAL INTERPRETATION OF THE STARK TENSOR ELEMENTS R_{ijk}

In this section we will concentrate our attention on the interpretation of the magnitudes of the tensor elements R_{ijk} for nuclei in covalent molecules in terms of the parameters describing the bonding.

All the experimental work to date has been confined to the halogens. It is therefore convenient to discuss as a specific example the change in field-gradient at the halogen atom in a diatomic molecule $A—X$ subject to an electrostatic field lying along the $A—X$ bond. The electrostatic field then effectively alters the coulomb integral of the atoms A and X. If the $A—X$ bond length is R and the potential gradient has its positive end at the X atom then the coulomb integral of X is increased by $ER/2$, while that of A is decreased by the same amount. This change in the relative electronegativities of A and X by ER results in the transfer of charge $\delta\rho$ from A to X. If the halogen bonding orbital is an *s-p* hybrid given by:

$$\phi_x = \alpha\phi_s + (1 - \alpha^2)\phi_p \tag{18}$$

then the change in field-gradient δq is given by

$$\delta q = -(1 - \alpha^2)\,\delta\rho. \tag{19}$$

The quantity $\delta\rho$ is given by the equation

$$\delta\rho = \frac{ER}{2}\{\pi_{A,X} - \pi_{A,A}\} \tag{20}$$

where the terms $\pi_{A,X}$ and $\pi_{A,A}$ are the atom-atom mutual polarizabilities defined by Coulson and Longuet-Higgins.[9] They may be calculated from assumed wave-functions for the system. For a single bond between two centres described by a bonding orbital ψ_b and an antibonding orbital ψ_a given in the Hückel approximation by

$$\begin{aligned} \psi_b &= a\phi_X + (1 - a^2)^{\frac{1}{2}}\phi_A \\ \psi_a &= (1 - a^2)^{\frac{1}{2}}\phi_X - a\phi_A \end{aligned} \tag{21}$$

the polarizabilities are given by

$$\pi_{A,A} = -\pi_{A,X} = \frac{4a^2(1 - a^2)}{E_a - E_b} \tag{22}$$

where E_a and E_b are the orbital energies of the antibonding and bonding orbitals respectively. Thus compounds forming weaker bonds should be more readily deformed by an electrostatic field. This tendency is reinforced by the presence of the term R since weak bonds are usually longer. Thus for a single bond the final expression for the change in field-gradient produced by an electric field is

$$\delta q = \frac{(1 - \alpha^2) \,.\, 4a^2(1 - a^2)RE}{E_a - E_b}. \tag{23}$$

Although this equation has been derived for a diatomic molecule it may be used for polyatomic molecules as in for example CCl_4 inasmuch as the individual carbon–chlorine bonds may be considered to be isolated from each other. An exactly analogous treatment may be made of a pi-bond and a similar formula obtained. In this case, however, since the field-gradient along the sigma axis produced by a pi-electron is opposite in sign to that produced by a sigma electron the Stark shift is in the opposite sense to that produced by the polarization of the sigma bond. In general therefore we have the important result that when the $A—X$ bond has multiple bond character then the Stark shifts are less than when the $A—X$ bond is entirely single.

There exists yet another mechanism by which the field-gradient at the halogen atom alters without, however, resulting in a drift of electrons towards A. This is simply a change in the extent of s-p hybridization brought about by the electrostatic field and is mainly a property of the isolated atom. Consider a pair of s-p hybrid orbitals, a bonding orbital containing one electron and a non-bonding orbital occupied by two electrons

$$\begin{aligned} \phi_1 &= \alpha\phi_s + (1 - \alpha^2)^{\frac{1}{2}}\phi_p \\ \phi_0 &= (1 - \alpha^2)^{\frac{1}{2}}\phi_s - \alpha\phi_p. \end{aligned} \tag{24}$$

In the presence of the electrostatic field these orbitals become

$$\begin{aligned} \phi'_1 &= \phi_1 + \varepsilon\phi_0 \\ \phi'_0 &= \phi_0 - \varepsilon\phi_1 . \end{aligned} \tag{25}$$

The field-gradients produced by the electrons in these two orbitals (occupied by one and two electrons respectively) are

$$\int \phi'_1\left(\frac{3\cos^2\theta - 1}{r^3}\right) . \phi'_1 \, d\tau + 2\int \phi'_0\left(\frac{3\cos^2\theta - 1}{r^3}\right)\phi'_0 \, d\tau$$

$$\equiv \int \phi'_1\mathscr{H}_q\phi'_1 \, d\tau + 2\int \phi'_0\mathscr{H}_q\phi'_0 \, d\tau. \tag{26}$$

This may be expanded in terms of the integrals over atomic orbitals, dropping terms in ε^2, to yield

$$\begin{aligned} \frac{q}{q_0} &= (1 - \alpha^2) - 2\varepsilon\alpha(1 - \alpha^2)^{\frac{1}{2}} + 2\alpha^2 - 4\varepsilon\alpha(1 - \alpha^2)^{\frac{1}{2}} \\ &= (1 + \alpha^2) + 2\varepsilon\alpha(1 - \alpha^2)^{\frac{1}{2}}. \end{aligned} \tag{27}$$

Hence

$$\delta q = 2\varepsilon\alpha(1 - \alpha^2)^{\frac{1}{2}} . q_0 . \tag{28}$$

The importance of this result is that it shows that there is no first order Stark shift due to *s-p* rehybridization unless the sigma bond is already an *s-p* hybrid. We have already seen that α^2 is in general small for halogens so that in general we would be justified in neglecting it.

The treatment given here differs somewhat from that given by Dixon and Bloembergen,[5] although the conclusions are essentially the same. These authors have used assumed values of bond-parameters to calculate the shifts in a variety of systems and compare the predicted shifts with their experimental values. Table 1 shows a selection of such data. The agreement with experiment is reasonably satisfactory with the notable exception of the Si—Cl bond and to a lesser extent the Cl—Cl bond. It is interesting to note that these two compounds are precisely those for which it has often been suggested that the lone-pair electrons on the chlorine atom form partial p_π—d_π bonds with the vacant 3d orbitals of the neighbouring atom. As seen above such multiple bonding would lead to a decrease in Stark shift and since Dixon and Bloembergen assumed that the A—X bonds were single such p_π—d_π bonding would, as indeed these authors point out, account for the experimentally observed low Stark shifts. Duchesne and his collaborators[10,11] have also measured Stark shifts in a number of covalent halides. They have preferred to relate their experimental values to experimental values of molecular polarizability.

The correlation between these two quantities is about as good as between the theoretical and experimental quantities.

TABLE 9.1

Bond	Theory $(1/q_{at})(dq/dE)$ $(kV/cm)^{-1}$	$d\Delta\nu/dE$ $[cps(kV/cm)^{-1}]$ Theory	Experiment
$C—^{35}Cl$	$1{\cdot}8 \times 10^{-6}$	95	40 to 70
$Si—^{35}Cl$	$2{\cdot}4 \times 10^{-6}$	130	23
$Sn—^{35}Cl$	$3{\cdot}7 \times 10^{-6}$	200	128
$Hg—^{35}Cl$	$5{\cdot}0 \times 10^{-6}$	270	117
$C—^{81}Br$	$2{\cdot}0 \times 10^{-6}$	640	610
$C—^{127}I$	$4{\cdot}35 \times 10^{-6}$	1430	930
$^{35}Cl_2$	$3{\cdot}7 \times 10^{-6}$	200	81
$^{81}Br_2$	$5{\cdot}7 \times 10^{-6}$	1800	1400
$^{127}I_2$	$8{\cdot}9 \times 10^{-6}$	2950	3000

REFERENCES

1. N. Bloembergen, *Science, N.Y.* **133**, 1363 (1961).
2. T. Kushida and K. Saika, *Phys. Rev. Lett.* **7**, 9 (1961).
3. J. Armstrong, N. Bloembergen and D. Gill, *Phys. Rev. Lett.* **7**, 11 (1961).
4. J. Armstrong, N. Bloembergen and D. Gill, *J. Chem. Phys.* **35**, 1132 (1961).
5. R. W. Dixon and N. Bloembergen, *J. Chem. Phys.* **41**, 1720, 1739 (1964).
6. D. Gill and N. Bloembergen, *Phys. Rev.* **129**, 2398 (1963).
7. F. A. Collins and N. Bloembergen, *J. Chem. Phys.* **40**, 3479 (1964).
8. P. Cornil and M. Read, *C.r. hebd. Acad. Sci. Paris*, **259**, 1833 (1964).
9. C. A. Coulson and H. C. Longuet-Higgins, *Proc. R. Soc.*, A **191**, 39 (1947).
10. J. Duchesne, M. Read and P. Cornil, *J. Phys. Chem. Solids* **24**, 1333 (1963).
11. M. Read, P. Cornil and J. Duchesne, *C.r. hebd. Acad. Sci. Paris* **256**, 5331 (1963).

PART 2

A SURVEY OF THE EXPERIMENTAL DATA

CHAPTER 10

THE HALOGEN COUPLING CONSTANTS OF ORGANO-HALOGEN COMPOUNDS

A. INTRODUCTION

With the exception of fluorine whose nuclear spin quantum number is only $\frac{1}{2}$(F^{19}), all the halogens have commonly occurring isotopes with a quadrupole moment. Cl^{35} (75·4 per cent), Cl^{37} (24·6 per cent), Br^{79} (50·6 per cent) and Br^{81} (49·4 per cent) all have spins of $\frac{3}{2}$ while I^{127} (100 per cent) has a spin of $\frac{5}{2}$. The values of e^2Qq_0 in the Townes and Dailey analysis of the coupling constants are given in Table 7/9. The compounds to be discussed in the present section represent the largest single group of compounds which have been studied by nuclear quadrupole resonance spectroscopy and they constitute a most suitable starting point for a survey of the experimental results and for a systematic application of simple methods, such as that due to Townes and Dailey, of using the data to derive information about molecular electronic structure. We will be concerned almost entirely with the variation in coupling constant of the halogen atom X in a compound R—M—X as the nature of R is varied. M will usually be a carbon atom with its remaining three valencies satisfied by other groups or atoms but we will also be interested in comparing the relative effects of two groups R_1 and R_2 for different atoms M_1 and M_2. This type of approach is one which comes very naturally to the organic chemist, who has available from a variety of other sources information about the electronic effects of different groups R. On the one hand, this initial exercise will therefore enable us to test some of the predictions of the Townes and Dailey analysis and on the other hand we hope to find out new information about the ways different atoms and groups interact.

Because chlorine is a less expensive element than either bromine or iodine and because its compounds are usually the more stable, there are more readily-available compounds of chlorine than of the other two halogens. Added to this, the separation between pure quadrupole levels in covalent chlorides (15–60 Mc/s) represents a technically more convenient frequency range than does that for the bromides (150–350Mc/s) or the iodides (150–700 Mc/s), so that the quadrupole coupling data for chlorine are at present by far the most abundant, not only of the halogens but of all nuclei. For this reason we discuss all the data for chlorine first and follow this discussion by a shorter section relating the results for bromine and iodine to those for chlorine.

The compounds discussed here all have singly bonded halogen atoms. It is therefore assumed that the principal field-gradient axis lies along the halogen single bond even when, as is usually the case, this is not required by the molecular symmetry. This assumption has been checked by either microwave measurements or single crystal quadrupole resonancc measurements in a variety of compounds. For the Cl^{35}, Cl^{37}, Br^{79} and Br^{81} nuclei with spin $\frac{3}{2}$ the pure quadrupole resonance spectrum of the polycrystalline powder yields only the quantity $\frac{1}{2}e^2Qq(1 + \eta^3/3)^{\frac{1}{2}}$ (equation 115, Chapter 5). For even quite large values of η, however, this is almost indistinguishable from $\frac{1}{2}e^2Qq$. For these two elements values of η must be obtained from single crystal measurements; for I^{127} where two resonance frequencies are available, both e^2Qq and η are given by the spectrum of the polycrystalline powder. Note that according to equations (22) and (23), in Chapter 7 and the value of e^2Qq in Table 7.9 the coupling constants of covalently bonded chlorides are negative.

B. CARBON–CHLORINE COMPOUNDS

One of the first examples to be presented to the student of organic chemistry of the way in which the reactivity of the carbon chlorine bond varies with a simple modification in structure is the series CH_3Cl, $CH_3 . CH_2Cl$, $(CH_3)_2CHCl$ and $(CH_3)_3C—Cl$ where the electron releasing abilities of the methyl group increase the ionicity of the carbon-chlorine bond to such an extent that, for example, the hydrolysis of *t*-butyl chloride is many times more rapid than that of methyl chloride and even proceeds by a different mechanism involving pre-ionization of the molecule to $(CH_3)_3C^+$ and Cl^- fragments. According to equation (22), Chapter 7, such an increase in ionicity should be accompanied by a decrease in the ^{35}Cl nuclear quadrupole coupling constant and Table 1 shows the pure quadrupole resonance frequencies for this series. Also shown are the coupling constants for the series CH_3Cl, CH_2Cl_2, $CHCl_3$ and CCl_4 in which by the same reasoning the electron withdrawing effect of the chlorine atom is expected to decrease the ionic character of the carbon–chlorine bond and to thus result in an increased ^{35}Cl nuclear quadrupole coupling constant. The coupling constants can be seen to behave in the expected manner and indeed Fig. 1 shows that the implicit expectation that the coupling constant would be a linear function of the number of chlorine or

TABLE 10.1. Cl^{35} Nuclear quadrupole resonance frequencies of simple chloro compounds (Ref. 16).

Compound	Cl^{35} Nuclear quadrupole resonance frequency at 77°K
CH_3Cl	34·029
CH_3CH_2Cl*	32·891, 33·080
$(CH_3)_2CHCl$	32·069
$(CH_3)_3CCl$	31·065
CH_2Cl_2	35·991
$CHCl_3$	38·354, 38·308
CCl_4	40·6†

* Measured at 20°K ; † Average of 16 separate resonances.

methyl substituents is also fulfilled. For the methyl-substituted series, however, the change in coupling constant is so small as to lend support to the suggestion that much of the reactivity of *t*-butyl chloride derives from the relief of steric strain afforded by formation of the planar carbonium ion.

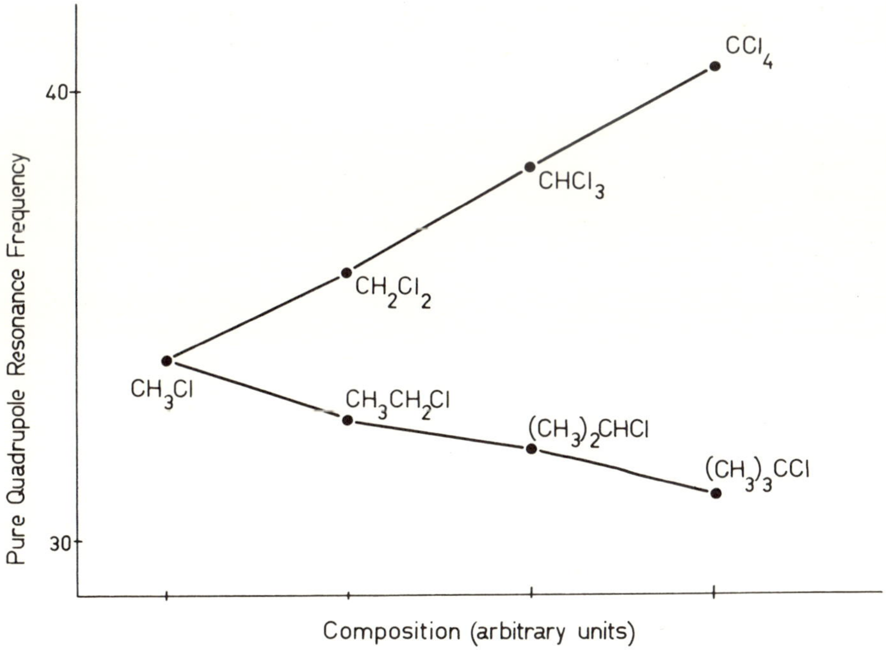

FIG. 10.1. ^{35}Cl Quadrupole resonance frequency (Mc/s at 77°K) of chloroalkanes.

1. "Inductive" Effects*

The organic chemist is used to considering two sorts of electronic displacement which affect electron distribution, the inductive and the conjugative effects. As far as chlorine is concerned the first corresponds to a movement of electrons within the carbon–chlorine sigma-bond while the second corresponds to participation of the lone-pair electrons of the chlorine atom in the bonding. It is obvious that in general a substituent R may act in both ways and many studies of the chemical reactivities of model systems have been made in order to separate these effects for various substituents and have led to the establishment of sets of parameters describing these two properties of the substituents. As far as the inductive effect is concerned the coupling constants of the series RCH_2Cl should be a measure of the inductive effect of the substituent R, since the presence of the methylene group prevents conjugation with the chlorine atom. The pure quadrupole resonance frequencies of such compounds are listed in Table 2 together with the values of the parameter σ^* derived from measurements of chemical reactivity and which is likewise

TABLE 10.2. Pure quadrupole resonance frequencies at 77°K (ν_Q) of compounds $R.CH_2Cl$ and the Taft polar substituent constant, σ^*. Δ is the difference between ν_Q for $R.CH_2Cl$ and ν_Q for $CH_3.CH_2Cl$

R	ν_Q	Δ	σ^*
$CH_3CH_2CH_2CH_2—$	33·103	0·4·	−0·130
$CH_3CH_2CH_2—$	33·255	0·50	−0·115
$CH_3CH_2—$	32·968	0·27	−0·100
$CH_3—$	32·70†	0	0
$CH_3CH{=}CH—$	33·455	0·75	0·36
$ClCH_2.CH_2—$	33·04†	0·34	0·385
$CH_2{=}CH—$	33·612	0·91	0·41
H—	34·029	1·33	0·49
$CH_3O.CH_2—$	33·453	0·75	0·52
$C_6H_5—$	33·627	0·93	0·60
$C_6H_5.O.CH_2—$	34·316	1·61	0·85
$HO.CO.CH_2—$	33·95	1·25	1·05
$ClCH_2—$	34·361	1·66	1·05
$CH_3.CO—$	35·28†	2·58	1·65
$H\overset{+}{N}(CH_3)_2.CH_2—$	35·065	2·36	1·90
$CH_3O.CO—$	35·962	3·26	2·00
$CF_3—$	37·98†	5·28	3·1

† Average of two resonances.

* No attempt has been made to give references to individual measurements of coupling constants in this section. Extensive tables are given in references 1 and 2 and more recently in I. P. Biryukov and M. G. Voronkov, *Coll. Czech. Chem. Comm.*, **32**, 830 (1967).

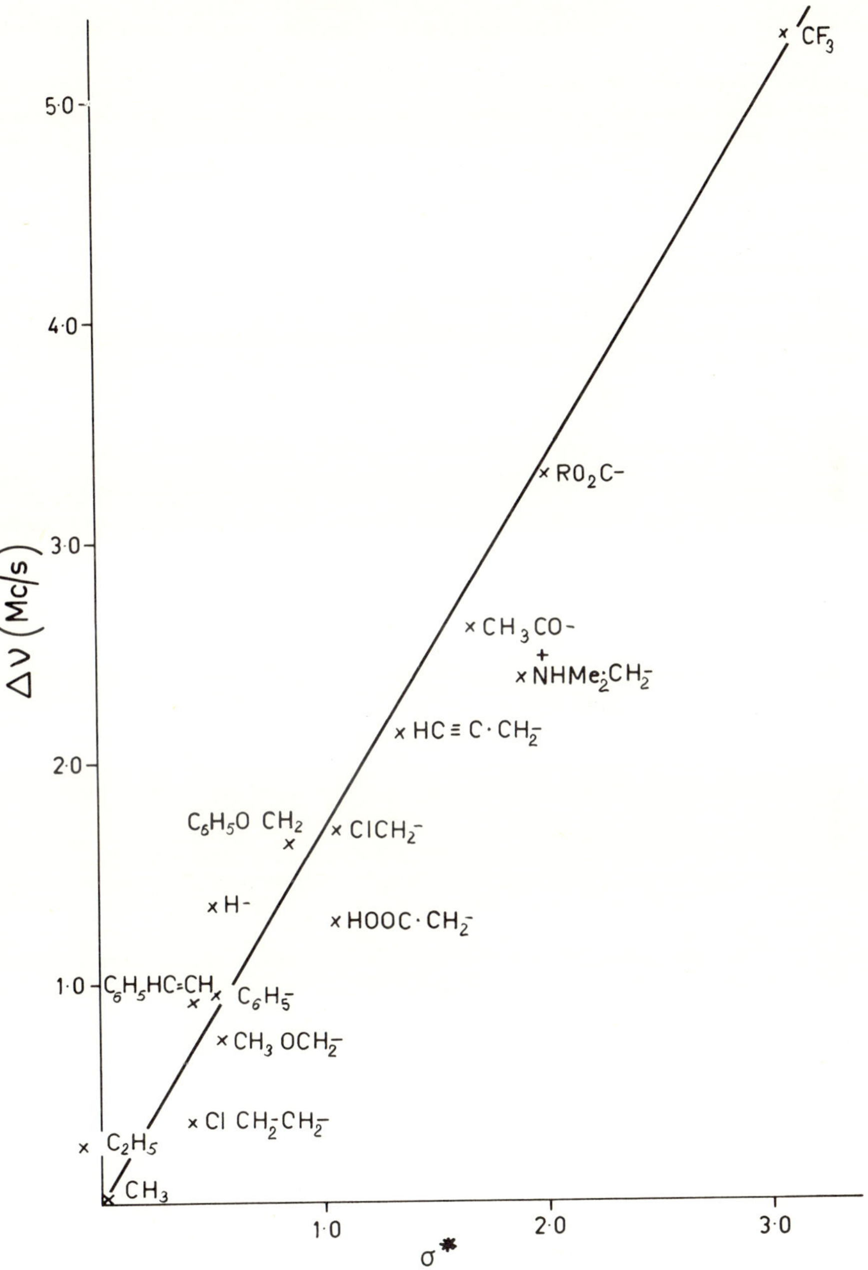

FIG. 10.2. ^{35}Cl Quadrupole resonance frequency (Mc/s at 77°K) of substituted chloroalkanes, RCH_2Cl and the Taft polarity parameter, σ^*, of the group R.

supposed to be a measure of the substituent inductive effect.* These data are plotted in graphical form in Fig. 2 and the fairly good straight line which can be drawn through these points demonstrates clearly the mutual compatibility of these two measures of the electronic effect of the substituent.

In a more general study of chloro alkanes Biryukov and Voronkov[1] have shown that the ^{35}Cl quadrupole resonance frequency of a substituted methyl chloride $R_1R_2R_3C\,.\,Cl$ is related to the Taft σ^* parameters of the substituents by the equation

$$\nu = 32{\cdot}05 + 1{\cdot}019 \sum_i \sigma_i^* \pm 0{\cdot}35 \text{ Mc/s}$$

$$= 32{\cdot}05(1 + 0{\cdot}0317 \sum \sigma_i^*) \pm 0{\cdot}35 \text{ Mc/s.} \qquad (1)$$

The coupling constants of the *n*-chloro alkanes are shown graphically in Fig. 3. The reason for the steady rise in resonance frequency on going from the ethyl to the butyl derivatives is interesting and the reason for it is not clear. It probably has little relation to the electronic structure of the chloro alkanes. The oscillations of the coupling constants for the higher alkyl halides is typical of the behaviour of many properties of solid derivatives of straight-chain alkanes and is probably mainly due to solid state effects. Hooper and Bray[2] have measured the coupling constants of the α–ω dichloro *n*-alkanes and the results are similarly plotted in Fig. 4. The rapidity with which the inductive effect of the chlorine atom is attenuated as the length of the carbon chain is increased is remarkable, and the coupling constants for the nitrogen mustards, shown in Table 3, are interesting in this respect.

* A number of reactivity parameters have been devised by chemists in order to express quantitatively the overall electronic behaviour of an atom or group, R. The two which will be most referred to here are the Hammett sigma constant, σ, which is defined for substituted benzene compounds by the relationship:

$$\sigma_R = \log\left(\frac{K_R}{K_0}\right);$$

K_R = acid dissociation constant of the substituted benzoic acid = $R.C_6H_4.COOH$;

K_0 = acid dissociation constant of benzoic acid.

This equation was originally restricted to only *para* or *meta* substituents since *ortho* substituents introduce additional steric effects. However the concept has been extended to include *ortho* substituents when the σ constants have been derived by indirect methods. The Hammett sigma constant has the disadvantage that both inductive and conjugative effects are mixed to an unknown extent. To overcome this difficulty R. W. Taft has devised the σ^* parameter which should be a measure of inductive effects only. This parameter is defined as the logarithm of the ratio of the acid dissociation constants of propionic acid, $CH_3CH_2.COOH$, the substituted acid $R_1CH_2.COOH$ and of propionic acid, $CH_3.CH_2.COOH$. It will be appreciated that since one is mainly interested in the difference in reactivity between two compounds the choice of reference compound is arbitrary. These parameters, based on *equilibrium* data, have proved extremely useful in systematizing the reactivities of organic compounds.

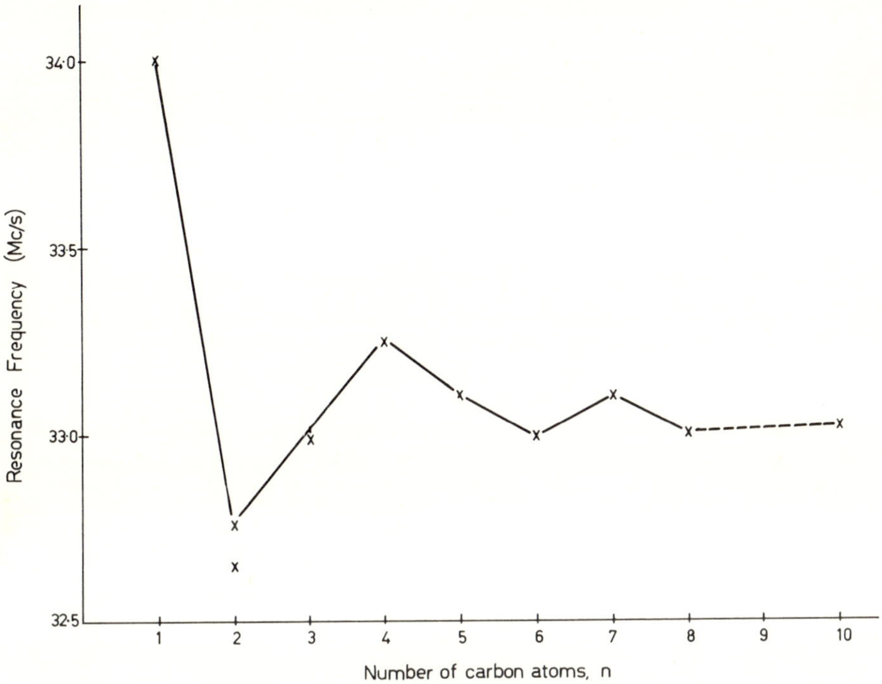

FIG. 10.3. ^{35}Cl Quadrupole resonance frequency (Mc/s at 77°K) of 1-chloro-*n*-alkanes, $CH_3(CH_3)_{n-1}Cl$.

TABLE 10.3. Cl^{35} quadrupole resonance frequencies at 77°K of nitrogen mustards (chloride salts). (Ref. 2)

$(CH_3)_2\overset{+}{N}H-CH_2-CH_2-Cl$	35·065
$(CH_3)_2\overset{+}{N}H-CH_2-CH_2-CH_2-Cl$	33·239

2. "Conjugative" Effects

If the chlorine atom is attached to an unsaturated carbon atom then the lone-pair electrons participate in the bonding. In resonance terms this amounts to contributions of structures of the type:

$$C = C - Cl \leftrightarrow C^- - C = Cl^+$$

to the ground state of the molecule. If the carbon atom is in the sp^2 hybrid state this partial double bond character of the carbon chlorine bond should

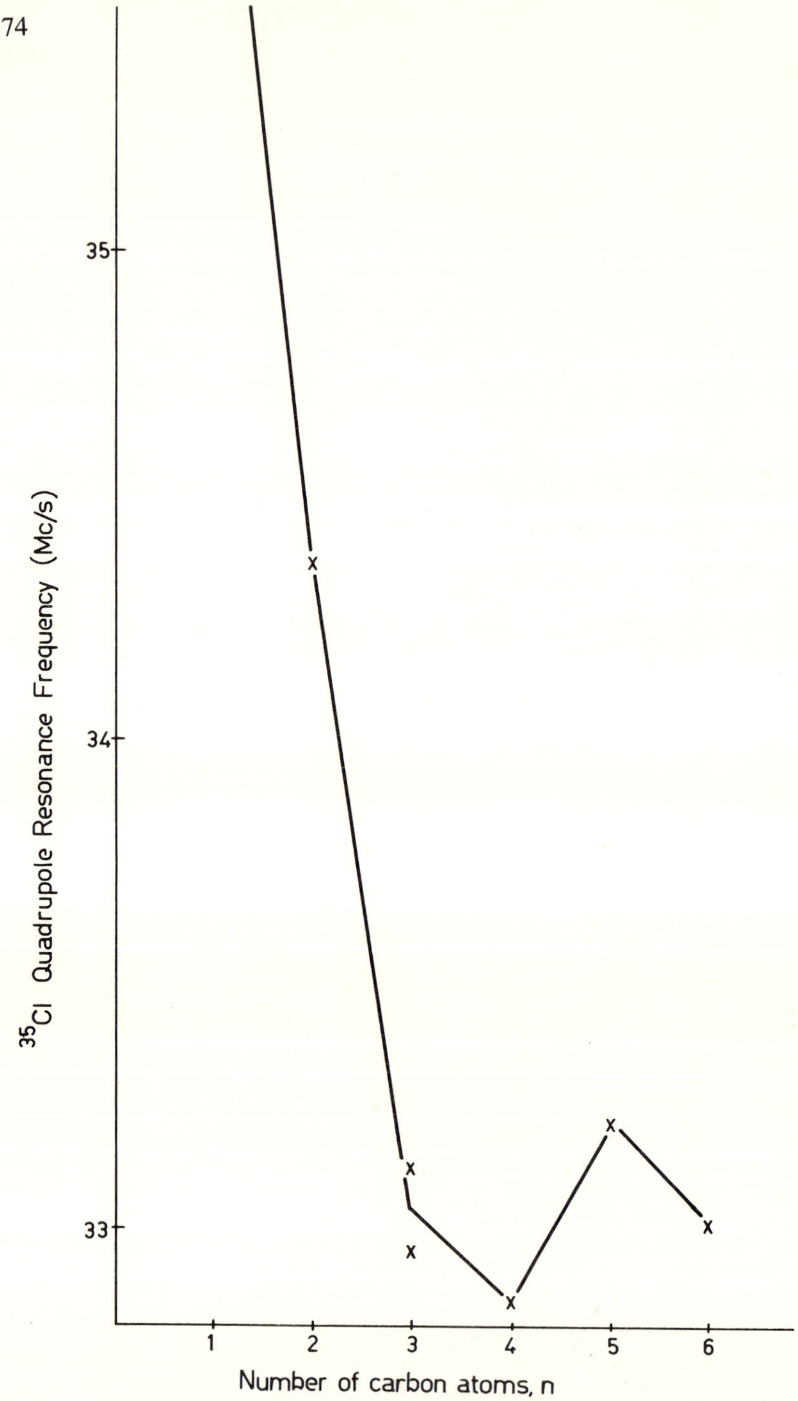

FIG. 10.4. ^{35}Cl Quadrupole resonance frequency (Mc/s at 77°K) of α–ω dichloro-n-alkanes, $Cl(CH_2)_nCl$.

reveal itself in a loss of cylindrical symmetry about the C—Cl bond and the consequent appearance of an asymmetry parameter in the quadrupole coupling tensor. For chlorine the detection of this asymmetry requires measurements of coupling constants either by microwave spectroscopy or by pure quadrupole resonance of single crystals accompanied by studies of the splitting produced by weak magnetic fields. Such measurements are therefore much less abundant than for the simple quadrupole coupling constants themselves.

Table 4 shows values of the asymmetry parameter for the ^{35}Cl nucleus directly attached to an sp^2-hybridized carbon atom. In some cases, particularly for *p*-dichlorobenzene which is undoubtedly the compound whose nuclear quadrupole resonance spectrum has been studied under the greatest variety of conditions, there are several values available and in these cases the one which appears to be the most accurate has been selected by the author. For *p*-dichlorobenzene two different values with mutually incompatible limits of accuracy are given. For comparison, Table 5 shows the asymmetry parameter for compounds in which the bond would be expected to be single and thus the asymmetry parameter zero even though this is not ensured by the molecular symmetry.

Bersohn[3] has shown (equation 28, Chapter 7) that the Townes and Dailey analysis of chlorine coupling constants leads to the following relationship between the fractional pi-bond character, ρ_π, and the coupling constants

$$\rho_\pi = \frac{2}{3}\left|\frac{e^2Qq}{e^2Qq_0}\right| . \eta. \tag{2}$$

The partial double-bond characters calculated according to this equation are also listed in Table 4.

Some of the details of this table require comment. In the first place the error in the asymmetry parameter for many of the compounds is rather large and prevents the drawing of firm conclusions. This is particularly true for the substituted chlorobenzenes. For some of the *p*-substituted chlorobenzenes accurate values are available and the relative values of η are those which would have been predicted on the simplest of theoretical bases. Thus η for *p*-chloroaniline is less than for *p*-dichlorobenzene owing to a contribution from resonance structures of the type

(I)

TABLE 10.4. Asymmetry parameters of the ^{35}Cl field-gradient tensor and partial double-bond character of the carbon halogen bond in conjugated organochlorine compounds. In the column headed "Method" the symbol M implies microwave spectroscopy, NQR/F is the field/frequency method due to Rehn (Ref. 14, Chapter 3) while NQR/G is the conventional zero-splitting locus method (Chapter 3, Section C2e).

Molecule	ν	η	$\frac{2}{3}\frac{e^2Qq}{e^2Qq_0}\cdot\eta(\%)$	$2-\left(\frac{e^2Qq}{e^2Qq_0}\right)\left(1+\frac{\eta}{3}\right)$	Method	Ref.
$H_2C{=}CHCl$	36·0 **	0·14 ± 0·03	6·18	1·314	M	22
$H_2C{=}CFCl$	36·35**	0·085 ± 0·015	3·8	1·313	M	23
$O{=}CCl_2$	36·23**	0·25 ± 0·03	11·06	1·285	M	24
$p.Cl_2C_6H_4$	34·25	0·075 ± 0·002	3·14	1·360	NQR/F	25
$p.Cl_2C_6H_4$	34·25	0·067 ± 0·001	2·81	1·362	NQR/G	26
$o.Cl_2C_6H_4$	34·84	0·100 ± 0·01	4·23	1·344	NQR/G	27
	35·12	0·100 ± 0·01	4·26	1·339		
1,3,5-$Cl_3C_6H_3$	35·020	0·093 ± 0·01	3·95	1·348	NQR/G	28
	35·269	0·111 ± 0·01	4·75	1·333		
	35·545	0·129 ± 0·01	5·56	1·327		
1,2,4,5-$Cl_4C_6H_2$	36.283	0·116 ± 0·002	5·14	1·313	NQR/F	25
$p.ClC_6H_4NH_2$	33·767	0·0489 ± 0·002	2·02	1·375	NQR/F	25
$p.ClC_6H_4OH$	34·367	0·058 ± 0·013	2·42	1·361	NQR/G	29
	34·145	0·223 ± 0·035	9·25	1·332		
$p.ClC_6H_4COOH$	~34·17	0·08 ± 0·03	3·32	1·360	NQR/G	30
$p.ClC_6H_4.CH_2Cl$	34·57	0·07 ± 0·02	2·94	1.356	NQR/G	31
$p.(OH)_2C_6Cl_4$	36·74	0·09 ± 0·03	4·03	1·310	NQR/G	32
	36·96	0·13 ± 0·03	5·86	1·298		

$(OH)C_6Cl_5$	36·98	0·163 ± 0·017	7·36	1·289	NQR/G	33
	37·71	0·164 ± 0·013	7·54	1·275		
	38·02	0·167 ± 0·010	7·75	1·269		
	37·75	0·185 ± 0·018	8·51	1·269		
	36·67	0·100 ± 0·007	4·48	1·310		
2,5-$Cl_2C_6H_3NH_2$	33·895†	0·079 ± 0·02	3·27	1·367	NQR/F	34
	34·011‡	0·059 ± 0·002	2·45	1·368		
2,5-$Cl_2C_6H_3NO_2$	37·23†	0·11	4·97	1·296	NQR/G	35
	35·48‡	0·09	3·87	1·335		
2,4-6-trichloro-s-triazine	36·300	0·23 ± 0·03	10·10	1·287	NQR/G	37
	36·317	0·26 ± 0·03	11·51	1·280		
2,6-dichloro-pyridine	33·850	0·118 ± 0·002	4·85	1·359	NQR/G	38
3,5-dichloro-pyridine	34·773	0·086 ± 0·002	3·63	1·348	NQR/G	38
Chloranil	36·79	0·21 ± 0·03	9·43	1·283	NQR/G	39
	36·86	0·21 ± 0·03	9·43	1·282		

* It is likely that the resonance with the higher asymmetry parameter is due to a chlorine atom hydrogen-bonded to the hydroxyl group of a neighbouring molecule; †The chlorine atom in the 2-position; ‡ The chlorine atom in the 5-position; ** Calculated from the observed quadrupole coupling constant.

TABLE 10.5. ^{35}Cl Asymmetry parameters in saturated organochlorine compounds

Compound	Formula	e^2Qq	η	Ref.
δ-Benzene hexachloride	$C_6H_6Cl_6$	37·268	0·00 ± 0·01	40
		36·818	0·034 ± 0·007	
		36·364	0·020 ± 0·005	
		36·256	0·037 ± 0·010	
		36·228	0·023 ± 0·008	
		36·948	0·015 ± 0·010	
γ-Benzene hexachloride	$C_6H_6Cl_6$	36·817	0·0166 ± 0·0014	41
		36·487	0·0231 ± 0·0031	
		36·230	0·0286 ± 0·0031	
		36·230	0·036 ± 0·010	
		35·693	0·0220 ± 0·0025	
		35·608	0·0311 ± 0·0031	
Monochloracetic acid	$ClCH_2:COOH$	34·8964	0·0	42
		35·4475	0·0	

In *p*-chlorophenol there are two distinct resonance frequencies and one of these has a very high asymmetry parameter. The crystal structure indicates that probably the chlorine atom having this high value of η is hydrogen-bonded to the hydroxyl group of a neighbouring molecule. For the other chlorine atom the asymmetry parameter is lower than in *p*-dichlorobenzene for the same reason as in *p*-chloroaniline. This conjugative effect of substituents is particularly well illustrated by the results for 2,5-dichloroaniline, where η for the chlorine atom *ortho* to the amino group is low for the reasons given above whereas in the *meta* position it is comparable with that of *p*-dichlorobenzene since no corresponding resonance structures can be written.

The very high value for the asymmetry parameter in $COCl_2$ (II) shows clearly the very considerable contribution of carbon-chlorine double bonded structures to the ground state of this molecule and indeed in general to the ground state of acid chlorides. Such structures seem to have a similar importance in chloranil (III) and of course in both cases this is due to the readiness with which the oxygen atom takes on a negative charge.

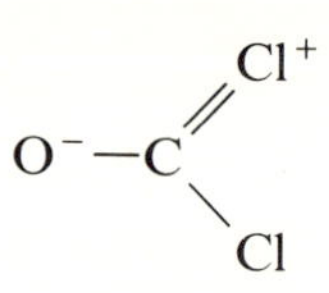

(II)

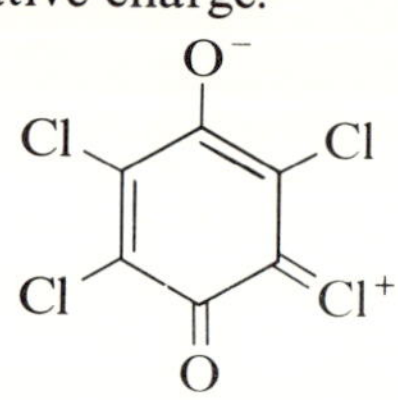

(III)

In the field of heterocyclic chemistry the difference between 2,6- and 3,5-dichloropyridine can clearly be ascribed to the fact that in the 2,6-compound double-bonded resonance structures can be written in which the negative charge is placed on the nitrogen atom. This is not possible in the 3,5 derivative

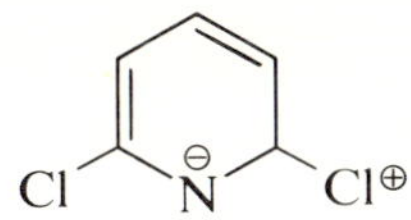

(IV)

This had been suspected for some time simply on the basis of the coupling constants themselves.[4,5] The importance of resonance structures such as (IV) is similarly and perhaps more strikingly illustrated by the difference between 1,3,5-trichlorobenzene and 2,4,6-trichloro sym-triazine (cyanuric chloride).

A straightforward manipulation of the field-gradients produced by the chlorine atom *p*-electron populations yields the following equation for the sigma-electron population, ρ_σ, for a chlorine atom bonded to an sp^2 hybridized carbon atom

$$\rho_\sigma = 2 - \left|\frac{e^2Qq}{e^2Qq_0}\right|\left(1 + \frac{\eta}{3}\right), \tag{3}$$

and the sigma-electron populations calculated by it are also shown in Table 4. The figure for vinyl chloride should be compared with that for ethyl chloride ($e^2Qq = -69{\cdot}1$ Mc/s) where the sigma bond population is 1·370 (the value obtained from microwave spectroscopy is taken from comparison with the microwave result for vinyl chloride). This demonstrates that as has been suggested on a variety of both theoretical and experimental grounds, the electronegativity of sp^2 hybridized carbon is greater than that of sp^3 hybridized carbon.

For chlorine attached to *sp* hybridized carbon, as in the chloroacetylenes, the cylindrical symmetry of the C—Cl bond is restored and measurements of η no longer reveal the extent of pi-bonding. The Townes-Dailey method yields

$$\left|\frac{e^2Qq}{e^2Qq_0}\right| = \rho_\pi - \rho_\sigma. \tag{4}$$

If ρ_π is assumed to have the same value as in vinyl chloride—and in view of the shorter C—Cl bond length

$$\left[\gt C{-}Cl, 1{\cdot}767\ \text{Å}, {=}C{-}Cl, 1{\cdot}719\ \text{Å};\ {\equiv}C{-}Cl,\ \ 1{\cdot}635\ \text{Å}\right]$$

it is not expected to be less—then the quadrupole coupling constant of chloroacetylene (−79·7 Mc/s. microwave) implies that $\rho_\sigma = 1{\cdot}213$, again in accordance with the expected sequence of electronegativity $sp^3 < sp^2 < sp$. It should be stressed, however, that these figures should not be taken too literally since the changing carbon–halogen bond length in the series paraffin–olefine–acetylene has the consequence that the terms neglected by the Townes–Dailey treatment are not approximately constant from one molecule to another. However the effect is certainly a genuine one as is illustrated by the data in Table 6 where on the one hand the presence of the methylene group prevents conjugation and on the other hand it maintains a uniform bond-length. The sigma electron densities in the remaining compounds of Table 4 all follow the evolution expected from electronegativities of the various atoms making up the molecule. Thus ρ_σ in phosgene, $COCl_2$, is greater than in vinyl chloride or 1,chloro-1,fluoroethylene and in cyanuric chloride or 2,6-dichloropyridine it is greater than in the corresponding chlorobenzenes.

TABLE 10.6. Pure quadrupole resonance frequencies at 77°K of chloroalkanes illustrating the effect of hybridization on the electronegativity of carbon atoms

Compound	Freq. (Mc/s)	Ref.
$CH_3.CH_2.Cl$	32·3*	43
$CH_2{=}CH.CH_2.Cl$	33·455	43
$CH{\equiv}CH.CH_2Cl$	34·812	43

* Extrapolated from measurement at 20°K.

In this connection equation (3) shows that when considerable pi-bonding is present the pure quadrupole resonance frequency is much less than would have been expected on the basis of the electron withdrawing effects of the substituents. This is the basis of the remark made above about the chloropyridines and we illustrate it by some of the data for these compounds (Table 7). The low frequency of 2-chloropyridine and 2-chloroquinoline are due to the resonance structures such as (IV) while for 3-chloropyridine, for which no such resonance structures can be written, the higher frequency is due to the inductive effect of the nitrogen atom leading to a lower C—Cl bond ionic character. For 7-chloroquinoline again no such structures can be written and the nitrogen atom is far enough away for its inductive effect to have been considerably attenuated so that the frequency is essentially that of chlorobenzene. For 4-chloropyridine and 6-chloroquinoline, however, the resonance structures (V) and (VI) can be written but the quadrupole resonance frequen-

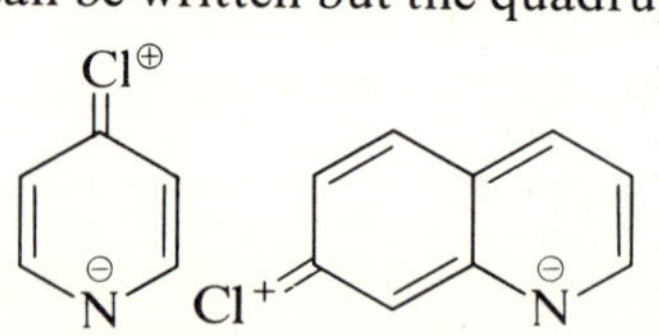

(V) (VI)

cies give no indication of additional double bonding in these compounds and direct measurements of asymmetry parameters would be interesting. This is a particularly important point in organic chemistry since the possibility of writing such structures involving considerable charge separation is often invoked to explain reaction kinetics or isomer ratios. Conclusions similar to

TABLE 10.7. Cl^{35} Quadrupole resonance frequencies (Mc/s at 77°K) of nitrogen-containing aromatic heterocycles

Compound	Quadrupole resonance frequency	Ref.
2-chloropyridine	34·194	4
3-chloropyridine	35·234	6
4-chloropyridine	34·739	6
2,6-dichloropyridine*	33·850	38, 44
3,5-dichloropyridine*	34·773	38, 44
2,4-dichloropyrimidine	35·310, 35·321	6
4,6-dichloropyrimidine	35·102, 35·132, 35·308, 35·425	6
2,4,6-trichloropyrimidine	35·702, 36·166, 36·250	5
2,4,5,6-tetrachloropyrimidine	36·281, 36·560, 37·068, 38·020, 38·420	6
2,4,6-trichloro *s*-triazine	36·738, 36·770	5, 45
2-chloroquinoline	33·271	4, 5
6-chloroquinoline	34·628	4, 5
7-chloroquinoline†	34·681	5
4,7-dichloroquinoline	34·764, 35·179, 35·591	4

* At room temperature; †At 86°K.

TABLE 10.8. Pure quadrupole resonance frequencies of acid chlorides, R COCl

R	Resonance frequency (Mc/s) at 77°K	Ref.
$ClCH_2-$	30·437, (37·517)*	46
Cl_2CH-	32·147, 32·962, (38·353, 38·521, 39·189, 39·386)*	46
Cl_3C-	33·721, (40·132, 40·474, 40·613)*	46
$ClCO-$	33·621	46
$ClCO(CH_2)_2-$	30·217	46
$ClCO(CH_2)_4-$	28·978	46
$ClCO(CH_2)_8-$	29·118	46
$ClCOCH{=}CH-$	30·380, 30·970	46
C_6H_5-	29·93	46
p-$CH_3O-C_6H_4-$	30·09	46
**C_4H_3O-		46
$(CH_3)_2N-$	31·8	46
$(CH_3CH_2)_2N-$	31·887	46
$(C_6H_5)_2N-$	35·2	47
$CH_3CH_2.O-$	33·858	46

* Resonance frequencies of chlorine nuclei other than those belonging to the —COCl groups.
** α-furyl.

the above have been obtained as a result of a fairly extensive investigation of the quadrupole resonance frequencies of substituted pyrimidines.[6]

The acid chlorides are another example of the phenomenon of strong double bonding lowering the coupling constant and some representative data are shown in Table 8. The figures for the related carbamoyl chlorides and chloroformates are also given but the measurement for diphenyl carbamoyl chloride should be repeated since the difference between it and the dialkyl carbamoyl chlorides implies a very considerable difference in electronic structure.

A measure of the total electronic effect of a substituent in a benzene ring is given by the so-called Hammett sigma parameter, defined for the particular substituent in a particular position (*meta* or *para*) by the difference between the pK_a of the monosubstituted benzoic acid and the pK_a of benzoic acid itself. An extensive investigation[7] of substituted chlorobenzenes has shown that a good correlation exists between the pure quadrupole resonance frequencies and the Hammett sigma constant of the substituent, given by

$$f = 1{\cdot}024\sigma + 36{\cdot}826 \text{ Mc/s} \pm 0{\cdot}36 \text{ Mc/s}. \tag{5}$$

The standard deviation of the above correlation represents about 7 per cent of the total spread in frequency for chlorobenzenes. As we have seen above the chlorobenzenes all have an asymmetry parameter but since this is usually not much greater than 0·1 the coupling constant is almost indistinguishable from twice the resonance frequency. However, the Hammett sigma constant is proportional to the total electronic effect of the substituent so that a correlation between sigma and the total *p*-electron population on the chlorine atom might have been expected. This total population is given by the expression

$$\sum \rho_p = 6 - \left|\frac{e^2Qq}{e^2Qq_0}\right|(1 + \eta) \tag{6}$$

and for the relatively few chlorobenzenes for which η is available the total *p*-electron defect is plotted against the sigma constant in Fig. 5. The correlation seems to be no better than that for the resonance frequency with the sigma constant. Further evidence that the scatter in the correlation with the Hammett sigma function does not arise from a neglect of double bonding is afforded by the data in Table 9 for the chlorobenzenes and the corresponding benzyl chlorides which show that in general although the frequencies of the benzyl chlorides are, as expected, less sensitive to the substituent than the chlorobenzenes, the frequency shifts produced by corresponding substituents are roughly proportional.

For the multichlorobenzenes (Table 10) Scrocco[8] has pointed out that the resonance frequency of a particular chlorine atom is well represented by an

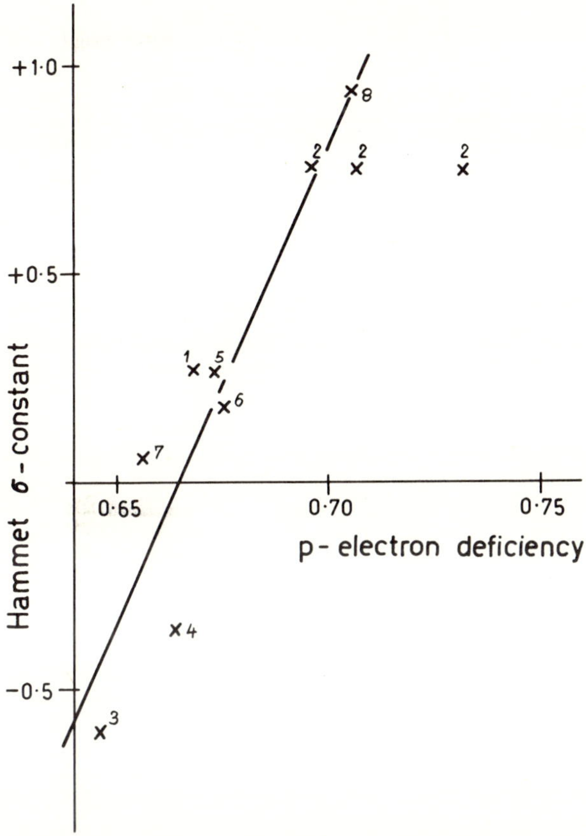

FIG. 10.5. *p*-Electron deficiency (equation 6) and Hammett sigma constant (σ) for chlorobenzene. (1) *p*-dichlorobenzene; (2) 1,3,5-trichlorobenzene; (3) *p*-chloroaniline; (4) *p*-chlorophenol; (5) *p*-chlorobenzoic acid; (6) *p*-chlorobenzyl chloride; (7) 2,5-dichloroaniline; (8) 2,5-dichloronitrobenzene.

equation of the form

$$f(^{35}\text{Cl}) = f(\text{chlorobenzene}) + a\Delta f_{ortho} + b\Delta f_{meta} + c\Delta f_{para}$$

where

$$\Delta f_{ortho} = 1{\cdot}30 \text{ Mc/s}; \quad \Delta f_{meta} = 0{\cdot}45 \text{ Mc/s}; \quad \Delta f_{para} = 0{\cdot}25 \text{ Mc/s}; \quad (7)$$

and *a*, *b* and *c* are the numbers of chloro atoms *ortho*, *meta* and *para* respectively to the chlorine atom whose frequency is being calculated. Figure 6 illustrates this correlation which perhaps not surprisingly is much better than for the more generally substituted chlorobenzenes. On the basis of the "separated electron pair" description of molecular electronic structure, Scrocco[8] has also shown in a very simple and thought-provoking manner how the relative values of the frequency increments may be estimated by calculating the perturbation produced in one carbon–chlorine bond by an equivalent

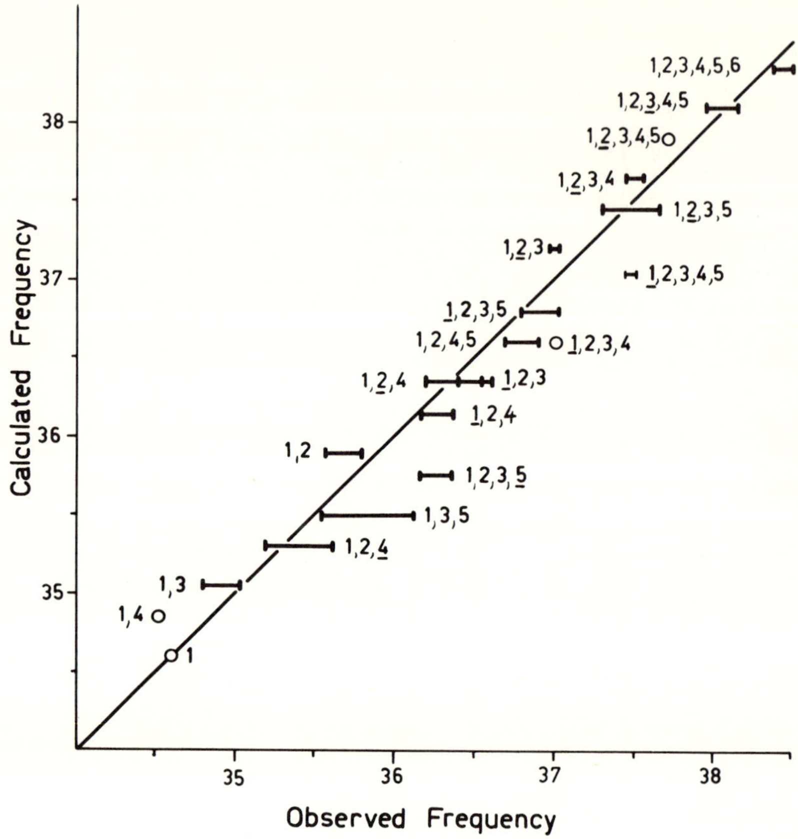

FIG. 10.6. Calculated (equation 7) and observed ^{35}Cl pure quadrupole resonance frequencies (Mc/s at 77°K) of the chlorobenzenes.

dipole located in the other (Fig. 7). Simple electrostatics combined with the known geometries of the chlorobenzenes leads to the ratios

$$\Delta f_{ortho} : \Delta f_{meta} : \Delta f_{para} = 1 : 0{\cdot}27 : 0{\cdot}20$$

compared with the experimental 1 : 0·35 : 0·19.

A consequence of this theory, which neglects all changes in double bond character of the C—Cl bond, is that for any monoatomic substituent the ratios of the frequency increments for substitution in the *o*, *m* and *p* positions should also be in approximately the same ratio. The extension of this result to polyatomic substituents is not really justified although the data collected in Table 11 show that in most cases the frequency shifts follow this pattern.

The coupling constants of the chloro derivatives of the condensed aromatic hydrocarbons would have been an interesting subject for study if the chlorine

TABLE 10.9. ^{35}Cl pure quadrupole resonance frequencies of chlorobenzenes, XC_6H_4Cl and benzychlorides, $XC_6H_4.CH_2Cl$ at 77°K

X	ν(Mc/s) Chlorobenzene	$\nu_X - \nu_H$	ν(Mc/s) Benzylchloride	$\nu_X - \nu_H$	Ref.
H	34·622	0·0	33·627	0·0	2
p-Cl	34·779	0·157	33·754	0·023	48
m-Cl	34·809		34·426		48
	34·875	0·283*	34·622	1·02*	
	35·030		34·683		
			34·875		
o-Cl	35·824		34·225		48
	35·755	1·11*	34·560	0·90*	
	35·755		34·795		
	35·580				
p-$ClCH_2$	34·567	−0·055	33·863	0·236	2
m-NO_2	35·457	0·835	34·216	0·589	2

* Average difference.

atom had shown a greater tendency towards double bond formation since it is an easy matter to calculate what the relative bond-forming tendencies in the various positions should be. As Table 12 shows, however, the coupling constants of 1- and 2-chloronaphthalene are not significantly different from each other or from that of chlorobenzene and an essentially similar situation prevails for 1,4-dichloronaphthalene, 9,10-dichloroanthracene and *p*-dichlorobenzene. Nevertheless, much more data on this point are available for the iodo compounds and these results are discussed below.

TABLE 10.10. ^{35}Cl pure quadrupole resonance frequencies (Mc/s at 77°K) of the chlorobenzenes (Ref. 49)

Compound	Pure quadrupole resonance frequency
Chlorobenzene	34·622
o-dichlorobenzene	35·580, 35·755, 35·824
m-dichlorobenzene	34·809, 34·875, 35·030
p-dichlorobenzene	34·779
1,2,3-trichlorobenzene	36·214, 36·238, 36·268, 36·563, 36·973, 37·031
1,2,4-trichlorobenzene	35·189, 35·617, 36·173, 36·380, 36·400, 36·623
1,3,5-trichlorobenzene	35·545, 35·894, 36·115
1,2,3,4-tetrachlorobenzene	37·013, 37·455, 37·557
1,2,3,5-tetrachlorobenzene	36·185, 36·386, 36·816, 36·872, 37·020, 37·289, 37·483, 37·657
1,2,4,5-tetrachlorobenzene	36·702, 36·738, 36,843, 36·898
Pentachlorobenzene	37·466, 37·504, 37·720, 37·959, 38·153
Hexachlorobenzene	38·381, 38·452, 38·493

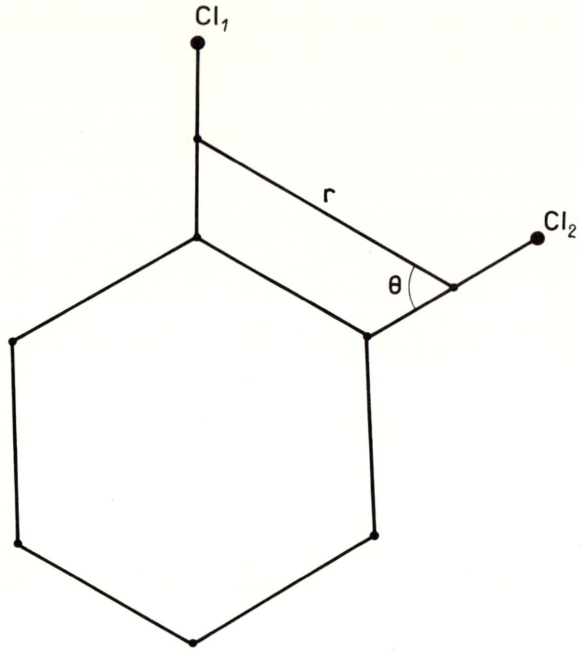

FIG. 10.7. Calculation of the perturbation of the electrons on atom Cl_1 and in the CCl_1 bond by a dipole situated in the CCl_2 bond.

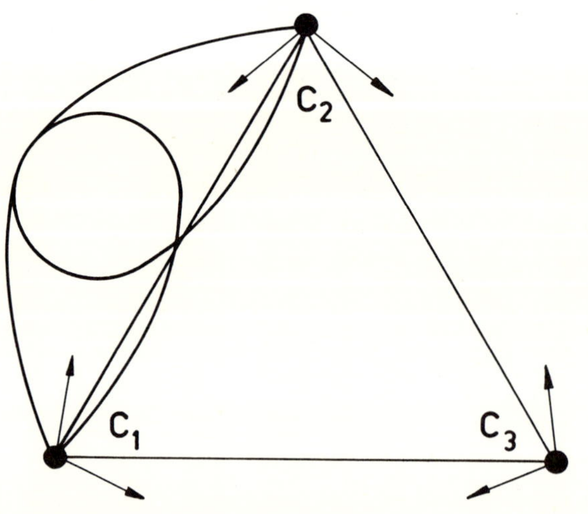

FIG. 10.8. Orbital overlap in cyclopropanes.

TABLE 10.11. ^{35}Cl Pure quadrupole resonance frequencies (Mc/s at 77°K) of *o*, *m* and *p* substituted chlorobenzenes (Ref. 7)

Substituent	*o*		*m*		*c*	
	Frequency	Δf	Frequency	Δf	Frequency	Δf
—COOH	36·305	1·663	35·227	0·585	34·673	0·031
—OH			34·766		34·700	
	35·37	0·728	34·925	0·204*	34·925	0·170*
—NHCOCH$_3$	35·150	0·508	34·735	0·093	34·792	0·150
—NCO	34·415	−0·227	34·653	+0·011	35·259	0·617
—NO$_2$	37·260	1·618	35·457	0·815		
—CF$_3$	35·633	0·991	35·073	0·431		

* Average difference.

TABLE 10.12. 35 Cl quadrupole resonance frequencies (Mc/s at 77°K) of chloronaphthalenes

Compound	Frequency	Ref.
1-chloronaphthalene	34·606	2
2-chloronaphthalene	34·69	50
1,4-dichloronaphthalene	35·125	51
1,5-dichloronaphthalene	34·818	51
1,4,5,8-tetrachloronaphthalene	35·238	51
	36·768	
Octachloronaphthalene	38·189	52
	38·161	
	38·158	
	37·847	
	37·826	
	37·769	
	37·721	
9,10-dichloroanthracene	34·899	2

The foregoing discussion serves to build up our confidence that the ^{35}Cl quadrupole coupling constants of organo chlorine compounds behave in a reasonable manner and that their values reflect significant changes in the electronic configuration of the chlorine atom. We now discuss therefore the coupling constants of compounds where the electronic behaviour of the substituents is less certain and try to use them to find out exactly what is going on.

3. Derivatives of Cyclopropane

Because it is impossible to construct *s-p* hybrid orbitals with an acute interorbital angle, any localized bond structure for cyclopropane must involve bond-bending. Coulson and coworkers[9] have shown that maximum overlap between the various orbitals is achieved with an interorbital angle of 104° for the orbitals forming the carbon-carbon bonds (Fig. 8); the $\widehat{HCH}$ bond angles predicted by this model agree reasonably well with those observed. On the other hand Walsh[10] has suggested that structures such as

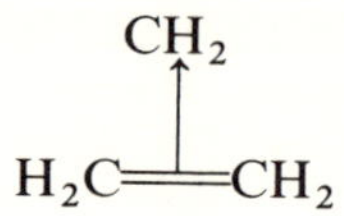

(VII)

contribute to the ground state of cyclopropane and indeed from the purely chemical point of view cyclopropane behaves as an unsaturated system.

Flygare *et al.*[11] have measured the microwave spectrum of 1,1-dichlorocyclopropane and by means of high resolution studies determined the complete quadrupole coupling tensor in the inertial axis system. They were thus able to compare the directions of the principal axes of the coupling tensor for the two chlorine atoms with the directions of the C—Cl internuclear vectors and found that they differed by not more than 15′ and that the asymmetry parameter had the value $\eta = 0{\cdot}032 \pm 0{\cdot}004$. In the analogous methylene chloride the bond vectors and principal axis directions did not differ by more than 1° and $\eta = 0{\cdot}036 \pm 0{\cdot}003$. In a similar comparison of cyclopropyl chloride and 2-chloropropane in which, however, only the diagonal elements of the coupling tensors in the inertial axis systems could be obtained, Schwendeman and his co-workers[12] arrived at the conclusion that in both molecules the two directions were essentially coincident and that the asymmetry parameter was not much greater than 0·03. Both these investigations therefore lead to the conclusion that C—Cl bonds in the chlorocyclopropanes are essentially single so that the cyclopropane ring does not act as a conjugative electron acceptor. Note however that this conclusion does not imply that it cannot act as a conjugative donor and it is indeed to this situation, i.e. the ready formation of cyclopropylmethyl carbonium ions, $C_3H_5{-}C^+{-}(R)_2$, that the majority of the chemical evidence for the "unsaturated" nature of the cyclopropyl ring refers.

Todd *et al.*[13] have measured the pure quadrupole resonance frequencies of several chlorocyclopropanes and have sought to explain the universally higher frequencies of the chlorocyclopropanes compared with their unstrained analogues (Table 13). Using the orbitals proposed by Coulson *et al.*[9] they calculated the orbital electronegativity of the sigma orbital used by the carbon atom to bond with chlorine and obtained good agreement with the observed coupling constants. On these grounds they likewise rejected any suggestion of

TABLE 10.13. ^{35}Cl Pure quadrupole resonance frequencies (Mc/s at 77°K) of chlorocyclopropanes and their open-chain or macrocyclic analogues (Ref. 13)

Compound	Resonance frequency at 77°K
Cyclopropyl chloride	34·063
1-Chloropropane	32·968
1,1-Dichlorocyclopropane	36·610
1,1-Dichloro-2,2-dimethylcyclopropane	36·797
2,2-Dichloropropane	34·883
1,1-Dichlorocyclohexane	34·904; 34·989
1,1-Dichlorocyclopentane	34·984; 35·060

partial double bond character of the carbon–chlorine bond which, as seen above, would decrease the resonance frequency.

An interesting system related to the above molecules is the trichlorocyclopropenium ion. For this molecule it proved also possible to measure the asymmetry parameter[14] and, as most people would no doubt have expected, it is very large and is indeed the largest measured for any chlorine atom bonded to carbon ($e^2Qq = 78{\cdot}8$ Mc/s, $\eta = 0{\cdot}35$). The usual analysis of the quadrupole resonance data shows that the partial double bond character of the C—Cl bonds is 0·16 so that half the positive charge on the three-membered ring has been distributed among the three chlorine atoms.

4. First-Row Substituents

In the early part of this section, we saw that the quadrupole coupling constants of compounds of the type RCH_2Cl behaved much in the manner which the known electronic properties of the substituent would lead one to expect. On the grounds, therefore, that fluorine is much more electronegative than chlorine it would likewise be expected that successive replacement of chlorine by fluorine would lead to a steadily increasing quadrupole coupling constant of the remaining chlorine atoms. In fact the coupling constants decrease. A similar phenomenon is observed with the alkoxy substituent and relevant data are shown in Table 14. The only obvious explanation of this phenomenon would seem to be a contribution to the ground state of structures such as

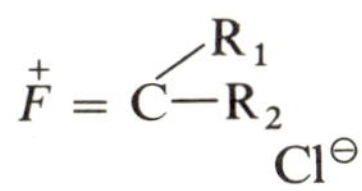

(VIII)

In molecular orbital terms this amounts to hyperconjugation due to interaction of the C—Cl bond with the lone-pair electrons on the fluorine or oxygen atoms (IX)

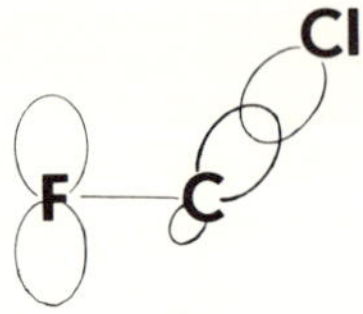

(IX)

and the magnitudes of the various overlap integrals, calculated from the tables of Mulliken *et al.*[15] for Slater orbitals,[15] show that this suggestion is quite a reasonable one. From the point of view of chemical reactivity this observation is quite an interesting one since it has always been supposed that the high rate of hydrolysis of chloromethylmethylether was due to the low energy of the resonance stabilized cation $CH_3.O^+{=}CH_2$. The nuclear quadrupole coupling constants suggest that the carbon-chlorine bond is already highly ionic in the molecular ground state.

TABLE 10.14. Pure quadrupole resonance frequencies (Mc/s at 77°K) of chloroalkanes with first-row substituents

Compound	Pure quadrupole resonance frequency at 77°K	Ref.
CCl_4	40·6	16
$CFCl_3$	39·161, 39·151, 39·703	16
CF_2Cl_2	38·450	16
CF_3Cl	38·089	16
CCl_3COOH	39·96, 40·165, 40·240	53
$CFCl_2COOH$	39·594	43
$CF_2ClCOOH$	37·485	43
CH_2Cl_2	35·991	16
$CH_3O.CH_2Cl$	30·181	43
$ClCH_2O.CH_2Cl$	32·381, 32·587	16
$CH_3S.CH_2Cl$	33·104	43
$ClCH_2.S.CH_2Cl$	34·526, 34·749	43

C. CHLORINE BONDED TO ATOMS OTHER THAN CHLORINE

In this section we will be concerned with presenting the facts rather than attempting any detailed discussion of their significance. This will take place in the next chapter where the general problem of the relationship between the chlorine quadrupole coupling constant and the nature of the bond to the chlorine atom is considered.

1. Silicon-Chlorine Compounds

The coupling constants of the chlorosilanes (Table 15) are, as the relative electronegatives of silicon and carbon would lead one to expect, much lower than those of the corresponding carbon compounds.

Apart from one or two initial measurements by Livingston,[16] most systematic work in this field has been carried out by Hooper and Bray[2] and by Biryukov *et al.*[17] The latter workers have shown that the resonance frequencies in Mc/s at 77°K of compounds $RSiCl_3$—a series for which the greatest amount of data is available—are well-represented by the equation

$$\nu = 18{\cdot}919 + 0{\cdot}457\sigma^* \pm 0{\cdot}099 \text{ Mc/s} \tag{8}$$

TABLE 10.15. Quadrupole resonance frequencies at 77°K of ^{35}Cl nuclei directly bonded to a silicon atom (Refs. 2 and 17)

Compound	Quadrupole resonance frequency (Mc/s at 77°K)
$SiCl_4$	20·463, 20·415, 20·408, 20·273
CH_3SiCl_3	19·220, 19·095, 19·004
$(CH_3)_2SiCl_2$	17·800, 17·762
$(CH_3)_3SiCl$	16·465
$CH_3CH_2.SiCl_3$ (α-phase)	18·930, 18·890, 18·750, 18·642, 18·625
(β-phase)	18·870, 18·850, 18·760
$(CH_3)_2CH.SiCl_3$	18·950, 18·890, 18·805, 18·770, 18·746
$H.SiCl_3$ (α-phase)	19·292, 19·132, 19·082, 19·025, 18·930, 18·882
$H.SiCl_3$ (β-phase)	19·146, 19·024, 18·915
$ClCH_2.SiCl_3$	19·55, 19·47
$Cl_2CH.SiCl_3$	19·55
$C_6H_5SiCl_3$	19·160, 19·130, 19·050
m-$ClC_6H_4SiCl_3$	19·295, 19·187, 19·45
2,5-$Cl_2.C_6H_3SiCl_3$	19·62
$CH_2{=}CH.SiCl_3$	19·310, 19·115, 18·960
$CH_2{=}CCl.SiCl_3$	20·361, 19·401
$Cl_3Si.SiCl_3$	19·306, 19·274
$Cl_3SiO.SiCl_3$	19·56, 19·860
$Cl_3Si.CH_2.SiCl_3$	19·15
$Cl_3Si(CH_2)_2SiCl_3$	19·030, 18·914
$F_3C.(CH_2)_2SiCl_3$	19·380, 19·170, 19·030
$NC.(CH_2)_2SiCl_3$	19·21, 19·076
$ClCH_2(CH_2)_2SiCl_3$	18·86
$(ClCH_2)(CH_3)SiCl_2$	18·19, 17·98
$(Cl_2CH)(CH_3)SiCl_2$	18·74
$(CH_3CH_2)_2SiCl_2$	17·735, 17·521
$(CH_3CH_2)H.SiCl_2$	18·179, 17·954
$(CH_2{=}CH)(CH_3)SiCl_2$	17·190
$(ClCH_2)(CH_3)_2SiCl$	17·090, 17·022

where σ^* is the Taft polarity parameter. The frequency of the most generally substituted chlorosilane, $R_1R_2R_3SiCl$, is however best given by a line of somewhat different slope

$$\nu = 16{\cdot}649 + 0{\cdot}399 \sum_{R} \sigma^*_R \pm 0{\cdot}128 \text{ Mc/s} \tag{9}$$

which can be re-expressed in the form

$$\nu = 16{\cdot}649\left(1 + 0{\cdot}0239 \sum_{R} \sigma^*_R\right), \tag{10}$$

and comparison with equation (1) shows that the frequencies of the chlorosilanes are much less sensitive to the nature of the substituents than the chloroalkanes, both in a relative as well as an absolute sense. The chlorosilanes are thus not very suitable for studies of substituent effects since changes in frequency produced by substituents tend to disappear in the "noise" induced by solid state effects. A possible reason for this low sensitivity to substituent effects is discussed in Section B of Chapter 12.

2. Germanium-Chlorine Compounds

Apart from germanium tetrachloride there only appears to be available the resonance frequencies of three organochloro germanes (Table 16). Despite the paucity of the data it seems clear that the frequencies of the chlorogermanes are much more affected by the nature of the substituent than are the

TABLE 10.16. ^{35}Cl quadrupole resonance frequencies in Mc/s at 77°K of chlorogermanes (Ref. 18)

$GeCl_4$	25·745, 25·736, 25·714, 25·449
$CH_3.CH_2GeCl_3$	22·795, 22·300, 22·160
$(CH_3)_2GeCl_2$	20·322, 20·180
$F_3C.(CH_2)_2GeCl_3$	23·395, 23·285, 23·260

chlorosilanes and on the basis of the frequencies of the four compounds in question Voronkov and Biryukov[18] propose the relationship

$$\nu = 17{\cdot}420 + 0{\cdot}937 \sum_{R} \sigma^*_R \tag{11}$$

between the pure quadrupole resonance frequency, ν, in Mc/s at 77°K of the compound $R_1R_2R_3GeCl$ and the polarity parameter σ^*. Thus although in an absolute sense the frequencies of the chlorogermanes are slightly less affected by the substituent than the chloroalkanes, relatively speaking they appear to be much more sensitive as is shown by the written form of equation (11)

$$\nu = 17{\cdot}420\left(1 + 0{\cdot}0538 \sum_{R} \sigma^*_R\right) \tag{12}$$

The coupling constant of GeH_3Cl has been determined by microwave spectroscopy[19] as $-46{\cdot}95 \pm 0{\cdot}015$ Mc/s which is in essential agreement with the results given above. However it should be emphasized (see below) that no structural information is available for the solid compounds and the tetrahedral unit may not persist.

3. Tin-Chlorine Compounds

The ^{35}Cl data for this group of compounds are reasonably abundant (Table 17) and in addition the Sn^{119m} coupling constants for these and other organo-tin compounds are available from Mössbauer spectroscopy (Section B2 of Chapter 12). However as discussed in the section just mentioned it is by no means certain that all or indeed any of these compounds form molecular crystals in which a tetrahedral $R_1R_2R_3SnCl$ molecule retains its identity. This doubt is reinforced by the unusually large sensitivity of the ^{35}Cl resonance frequency to the nature of the other groups attached to the central atom and by the anomalous positive temperature coefficient of the resonance frequencies of some of these compounds. Without the necessary structural information it seems rather pointless to discuss the data much further.

TABLE 10.17. ^{35}Cl pure quadrupole resonance frequencies of the chlorostannanes (Megacycles/sec. at 77°K)

Compound	Pure quadrupole resonance frequencies (Mc/s) at 77°K	Ref.
$SnCl_4$	23·719, 24·140, 24·226, 24·294	54
n-$C_4H_9SnCl_3$	20·012, 20·448, 20·521	55
$C_6H_5SnCl_3$	20·112, 20·632, 21·279	55
$(CH_3)_2SnCl_2$†	15·466	55
$(C_2H_5)_2SnCl_2$	15·520, 15·740	55
$(n$-$C_3H_7)_2SnCl_2$†	14·930	55
$(n$-$C_4H_9)_2SnCl_2$†	15·982,* 15·992*	55
$(n$-$C_8H_{17})_2SpCl_2$	15·136*	55
$(C_6H_5)_2SnCl_2$	17·440, 17·902, 18·020, 18·772	55
$(C_6H_5)_3SnCl$	16·750,* 16·985*	55

* Measured at 303°K; † The resonance frequency has a *positive* temperature coefficient.

4. *N*-Chloro Compounds

Table 18 shows the data for the six compounds for which results are available, and the resonance frequencies are remarkable for the wide variation with structural type. Particularly interesting is the difference between chloramine-*T* and dichloramine-*T* (5·9 Mc/s); for the carbon-chlorine bond

TABLE 10.18. ^{35}Cl pure quadrupole resonance frequencies (Mc/s at 77°K) of *N*-chloro compounds

Compound	Formula	Frequency (Mc/s at 77°K)	Ref.
Chloramine-*T*	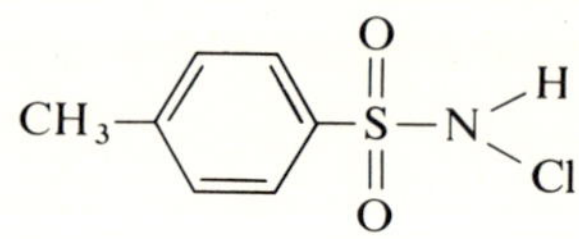	45·724	2
Di-chloramine-*T*	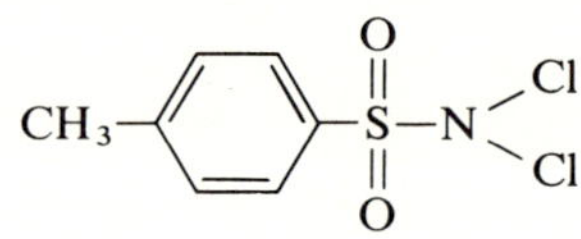	51·600	2
N-chlorosuccinimide	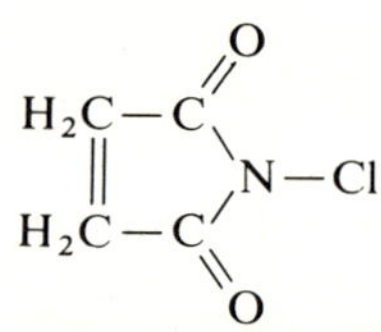	54·100	56, 4
1,3-Dichloro 5,5-di-methyl hydantoin	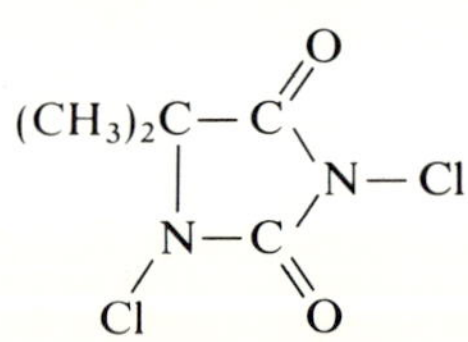	55·950 53·627	2
N-chloro-*p*-quinoneimide	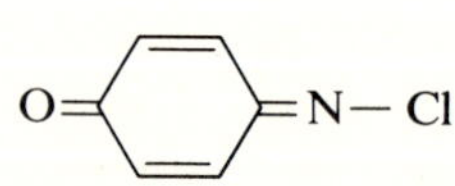	44·992	4
N-chloropiperidine	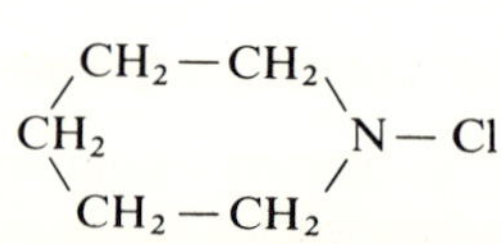	43·889	56

the average frequency difference between the compounds R_1R_2CHCl and $R_1R_2CCl_2$ is 2·3 Mc/s. This is the inverse phenomenon to that observed for chlorine bonded to phosphorus and silicon.

In *N*-chloro *p*-quinoneimide there is undoubtedly considerable partial double bond character to the N—Cl bond and this of course reduces the resonance frequency. The asymmetry parameter of this chlorine atom does not appear to have been measured.

The very high value of the coupling constant for all these compounds reflects of course the high electronegativity of nitrogen. For *N*-chlorosuccinimide the coupling constant indicates an essentially pure covalent bond and it is interesting in this respect to note that to the organic chemist *N*-chlorosuccinimide is well-known as a source of chlorine atoms Cl˙ for free-radical chlorinations or as a source of "positive chlorine".

5. The Phosphoryl Chlorides

The pure quadrupole resonance frequencies for these compounds, $R_1R_2P(O)Cl$ are collected in Table 19, and Fig. 9 shows the relationship of the

TABLE 10.19. ^{35}Cl pure quadrupole resonance frequencies of phosphoryl chlorides, $R_1R_2P(O)Cl$ (Ref. 57)

R_1	R_2	Quadrupole resonance frequency at 77°K
Cl	Cl	28·989
		28·935
CH_3	Cl	26·729
		26·426
CH_3CH_2	Cl	26·332
		26·194
$ClCH_2$	Cl	26·872
		26·172
		35·050*
C_6C_5	Cl	25·357
CH_3O	Cl	27·468
		27·099
C_6H_5O	Cl	27·700
$(C_2H_5)_2N$	Cl	27·034
		26·097
$(C_2H_5)_2N$	CH_3	27·695
C_2H_5O	CH_3	26·744
		26·445

* Resonance frequency of the chlorine atom on the chloromethyl group.

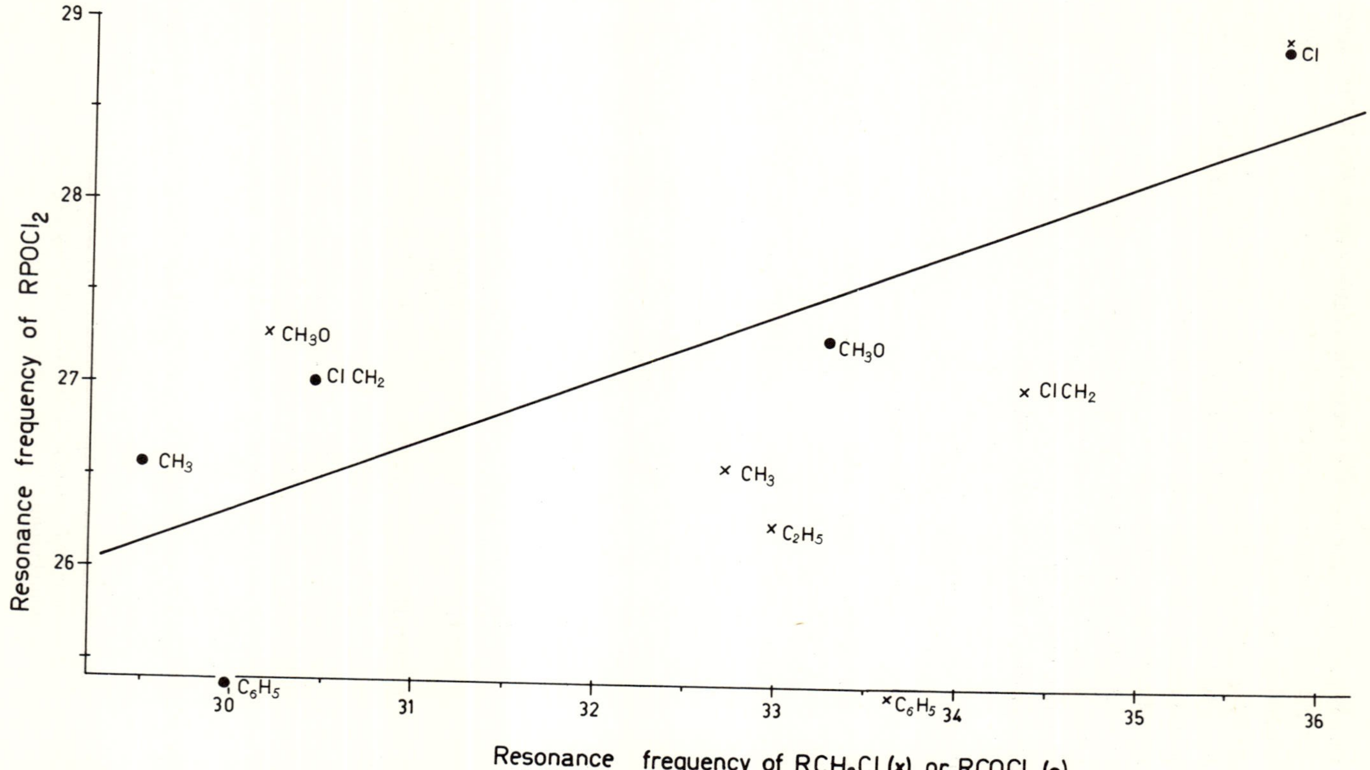

FIG. 10.9. Correlation of the ^{35}Cl pure quadrupole resonance frequencies (Mc/s at 77°K) of the phosphoryl chlorides, $RPOCl_2$, with those of the correspondingly alkyl chlorides, RCH_2Cl (X) or carbonyl chlorides, RCOCl (●).

resonance frequencies of the dichloro compounds, $RP(CO)Cl_2$, to those of the corresponding acid chlorides, RCOCl. This correlation with the frequencies of the acid chlorides was taken to imply that contrary to the conclusions from other physical and physicochemical measurements, there is some degree of conjugation between the halogen and the substituent *via* the phosphorus atom. The results, however, clearly show that, in conformity with other evidence, the effect of a substituent is transmitted much less efficiently through a phosphorus atom than through a carbon atom.

$$\nu_{CH_3CH_2Cl} - \nu_{ClCH_2\cdot CH_2Cl} = 1{\cdot}66 \text{ Mc/s};$$

$$\nu_{CH_3P(O)Cl_2} - \nu_{ClCH_2\cdot POCl_2} = 0{\cdot}46 \text{ Mc/s}.$$

6. The Sulphonyl Chlorides

Data for the sulphonyl chlorides, RSO_2Cl, are collected in Table 20. Figure 10 shows that there is no correlation between these resonant frequencies and those of the corresponding alkyl chlorides, RCH_2Cl. Particularly striking is

TABLE 10.20. ^{35}Cl pure quadrupole resonance frequencies at 77°K of sulphonyl chlorides, $R.SO_2Cl$

R	Quadrupole resonance frequency (Mc/s at 77°K)	Ref.
Ethyl $CH_3CH_2.$	32·519	58
n-Butyl $CH_3CH_2.CH_2.CH_2$	32·759	58
Phenyl C_6H_5	32·470, 32·538, 32·892	58
p-Nitrophenyl *p*-$NO_2C_6H_4$	33·438	59
m-Nitrophenyl *m*-$NO_2C_6H_4$	34·337	59
p-Chlorophenyl *p*-$Cl.C_6H_4$	32·832	59
p-Bromophenyl *p*-BrC_6H_4	32·895	59
p-Tolyl *p*-$CH_3.C_6H_4\cdot$	32·460	60
m-Carboxyphenyl *m*-$HOOC.C_6H_4$	33·672	2
2-Thienyl 2-C_4H_3S	33.151	58

the difference between the *p*-nitrophenyl and *m*-nitrophenyl derivatives and this leads one to suspect the possibility of there being strong conjugation with the sulphonyl group in the *para*-compound. The value of the ^{35}Cl asymmetry parameter would be interesting in this respect but it does not seem that any measurements of this sort have been made for the sulphonyl chlorides.

A better series for comparison would undoubtedly be the acid chlorides, RCOCl, but there is unfortunately almost no overlap between the groups, R, for these last compounds and those of the sulphonyl chlorides for which data are available.

D. BROMINE RESONANCES

The coupling constants for the corresponding organic halides are approximately linearly related. For the two bromine isotopes this is illustrated by the data in Table 21 and by Fig. 11, where the data for the chlorides and bromides have been converted to a common scale by dividing by their respective atomic coupling constants. In agreement with the sign of e^2Qq_0 (Table 9) and equations (22) and (23) all in Chapter 7, both Br^{79} and Br^{81} coupling constants of covalent bromides are positive.

In order to compare the extents to which the bromides and chlorides are affected by the nature of the group, R, it is convenient to suppose that the coupling constant of an organic, halide, R—X, will be given by an equation of the form:

$$(e^2Qq)_{RX} = (e^2Qq)_0(1 + f_X f_R) \tag{13}$$

f_R is thus a measure of the polarizing power of the group R and f_X of the polarizability of the halogen X. If f_R for some simple reference compound is set equal to zero, and the methyl group is the obvious choice, we then have for the corresponding bromides and chlorides

$$\frac{(e^2Qq)_{CH_3Br} - (e^2Qq)_{R-Br}}{(e^2Qq)_{CH_3Br}} = \left(\frac{f_{Br}}{f_{Cl}}\right)\frac{(e^2Qq)_{CH_3Cl} - (e^2Qq)_{RCl}}{(e^2Qq)_{CH_3Cl}}. \tag{14}$$

Figure 12 shows this relationship between these two quantities for the compounds in Table 21 and the reasonably good straight line which can be drawn through these points shows that equation (13) and the derived equation (14) are fairly good representations of the observed frequencies. The slope of this line, 1.18, indicates that, as might be expected, the polarizability of the C—Br bond is greater than that of the C—Cl bond. A further example of this phenomenon is shown in the data presented in Table 22 for the halogenonitroalkanes.

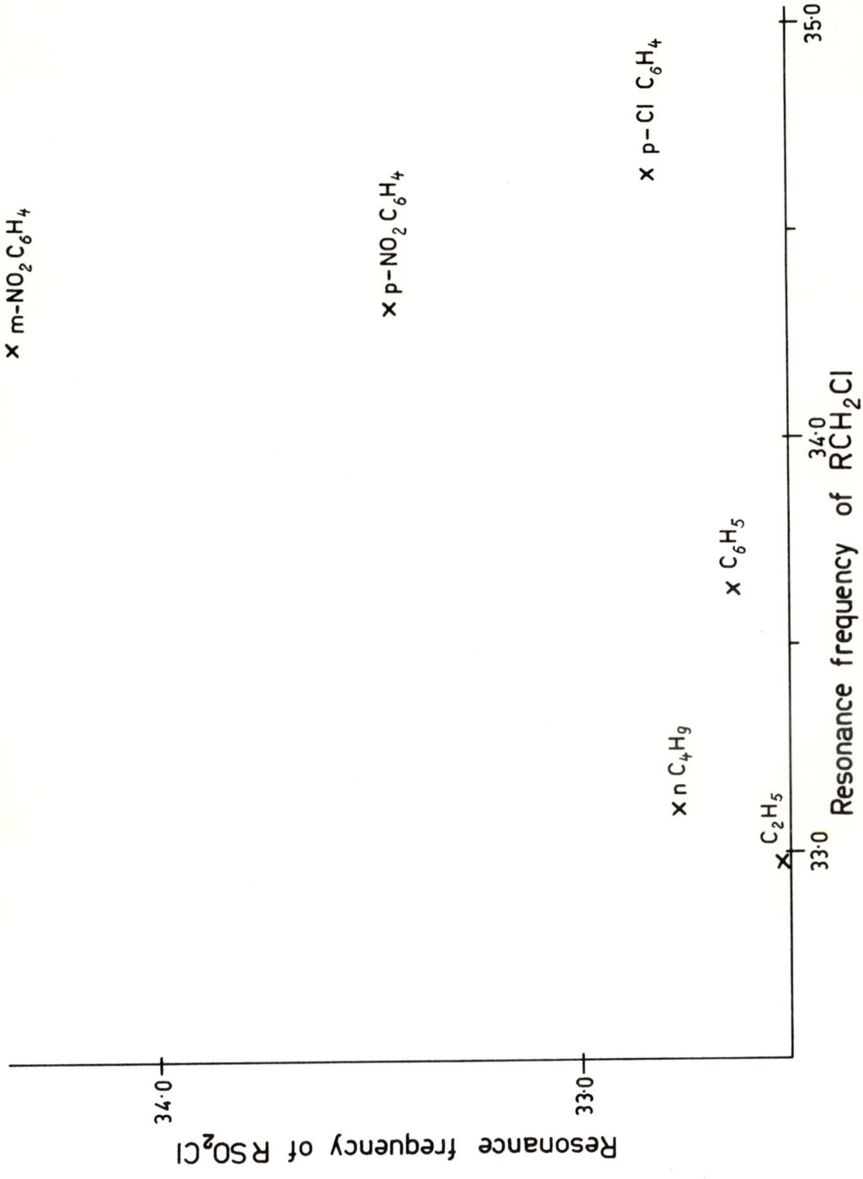

FIG. 10.10. Correlation of the ^{35}Cl pure quadrupole resonance frequencies (Mc/s at 77°K) of the sulphonyl chlorides, RSO$_2$Cl, with those of the corresponding alkyl chlorides, RCH$_2$Cl.

TABLE 10.21. Pure quadrupole resonance frequencies (Mc/s at 77°K) of organo bromine and chlorine compounds. The quantity ν_0 is one half the corresponding value of e^2Qq_0 Table 7.9)

R	νBr^{79}	νBr^{81}	Ref.	$(\nu/\nu_0)Br$	νCl^{35}	Ref.	$(\nu/\nu_0)Cl$	Point on figure
$-CH_2COOH$	286·74 284·55		61	0·744 0·739	36·429 36·131	53	0·664 0·658	1
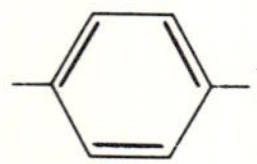$-NHCOCH_3$		226·805	62	0·705	34·792	7	0·634	2
$-CH_3$ $-C_2H_5$		220·981 207·790	63 63	0·687 0·646	34·029 32·649 32·759	16 16	0·620 0·595 0·597	3 4
$-C_3H_7$ $-CF_3$		210·277 252·263	63 63	0·654 0·784	32·968 38·089	16 16	0·601 0·694	5 6
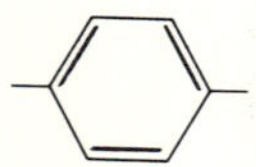$-NH_2$		221·862	61, 62	0·689	34·146	31	0·622	7
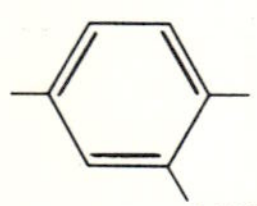NH_2		222·614	61, 62	0·692	34·388, 34·468	7	0·626, 0·628	8
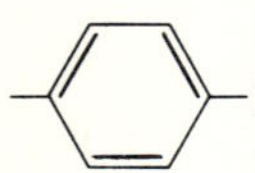$-NH_3Cl$		232·043	61	0·722	35·448	66		
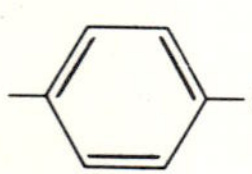$-OMe$		226·690	61, 62	0·705	34·753	31	0·633	9
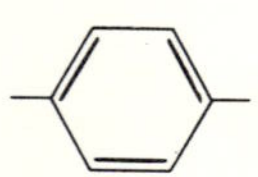$-OCH_2-CH_2-$	262·548		2	0·682	34·316	47	0·625	10

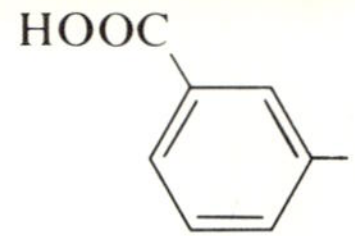		226·434	20	0·704	35·227	31	0·642	11
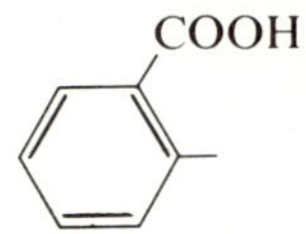		234·185	20	0·729	36·305	31	0·662	12
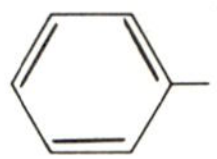	268·856		62	0·698	34·622	16	0·631	13
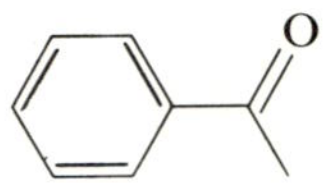		191·868	64	0·597	29·93	60	0·545	14
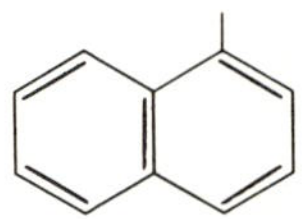	267·894		64	0·695	34·606	2	0·631	15
H—		224·0 ± 0·7	65	0·696	35·15 ± 1·4	65	0·641	16
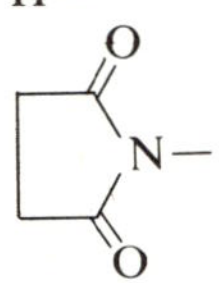		358·73	56	1·115	54·100	4	0·985	17
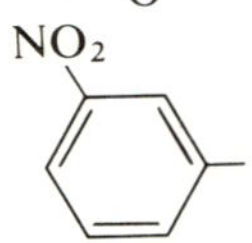		233·630	20	0·727	35·457	31	0·646	18
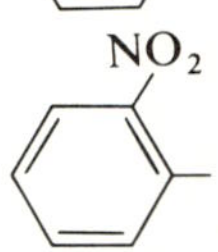		247·486	20	0·769	37·260	31	0·678	19

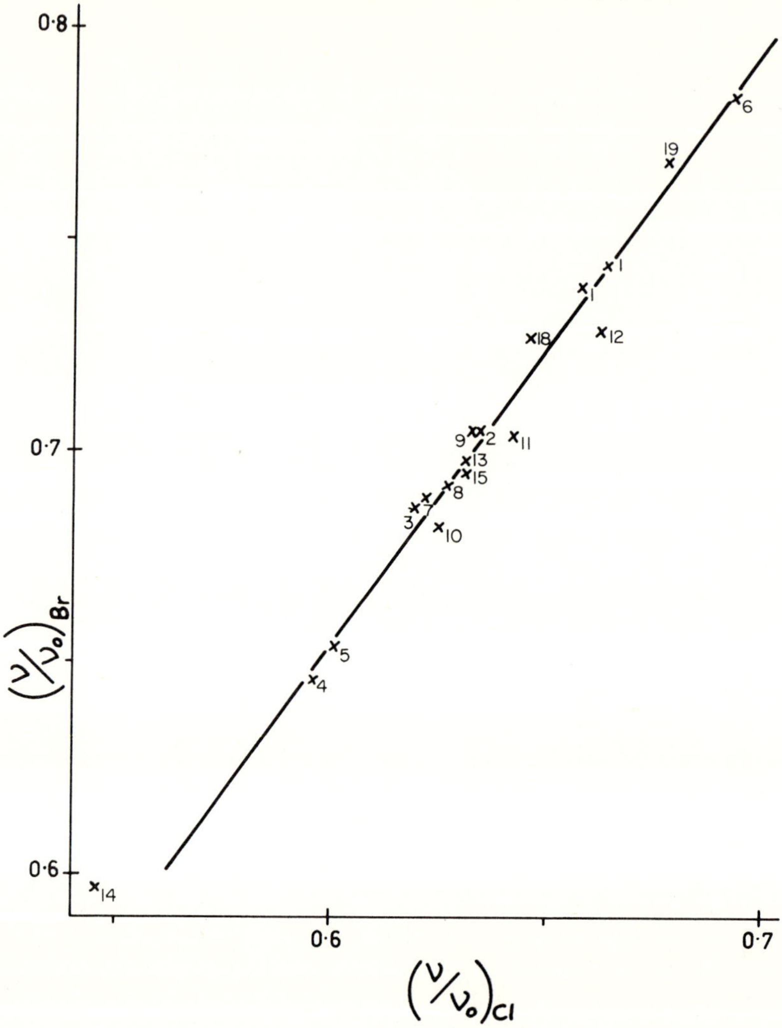

FIG. 10.11. Reduced quadrupole resonance frequencies, ν_{RX}/ν_{atomic}, of corresponding organo-bromine and organochlorine compounds.

The quadrupole resonance frequencies of the n-alkyl bromides[2] are plotted in Fig. 13 as a function of the chain length and their behaviour resembles qualitatively that of the corresponding chlorine compounds. The rise in coupling constant from the propyl to the butyl derivative observed in the chlorides does not occur here.

The pure quadrupole resonance frequencies of the bromobenzenes[2] show a relationship to the Hammett sigma constants of the substituents similar to

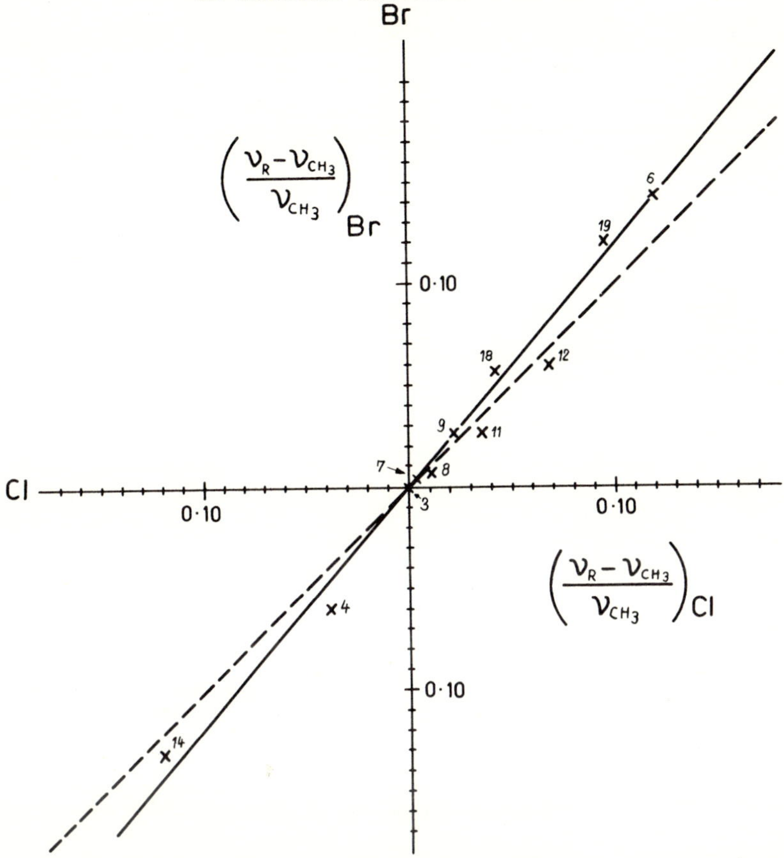

FIG. 10.12. Relative pure quadrupole resonance frequency shifts $(\nu_{RX} - \nu_{CH_3X})/\nu_{CH_3X}$ for organochlorine and organobromine compounds.

TABLE 10.22. Quadrupole resonance frequencies at 77°K for the halogenonitroalkanes, (^{35}Cl and ^{79}Br isotopes) (Ref. 67)

Compound	X = Cl ν	$\nu - \nu_{CX_4} \equiv \Delta\nu$	$\Delta\nu / \nu_{CX_4}$	X = Br ν	$\nu - \nu_{CX_4} \equiv \Delta\nu$	$\Delta\nu / \nu_{CX_4}$
CX_4	40·64*	0·00	—	320·66*	0·00	—
CX_3NO_2	41·003					
	41·235	0·62	0·0152	—	—	—
	41·546					
$CX_2(NO_2)_2$	42·133	1·57	0·0386	344·913	25·25	0·0756
	42·287					
$CX(NO_2)_3$	42·949	2·31	0·0568	359·394	38·75	0·1207

* Average frequency

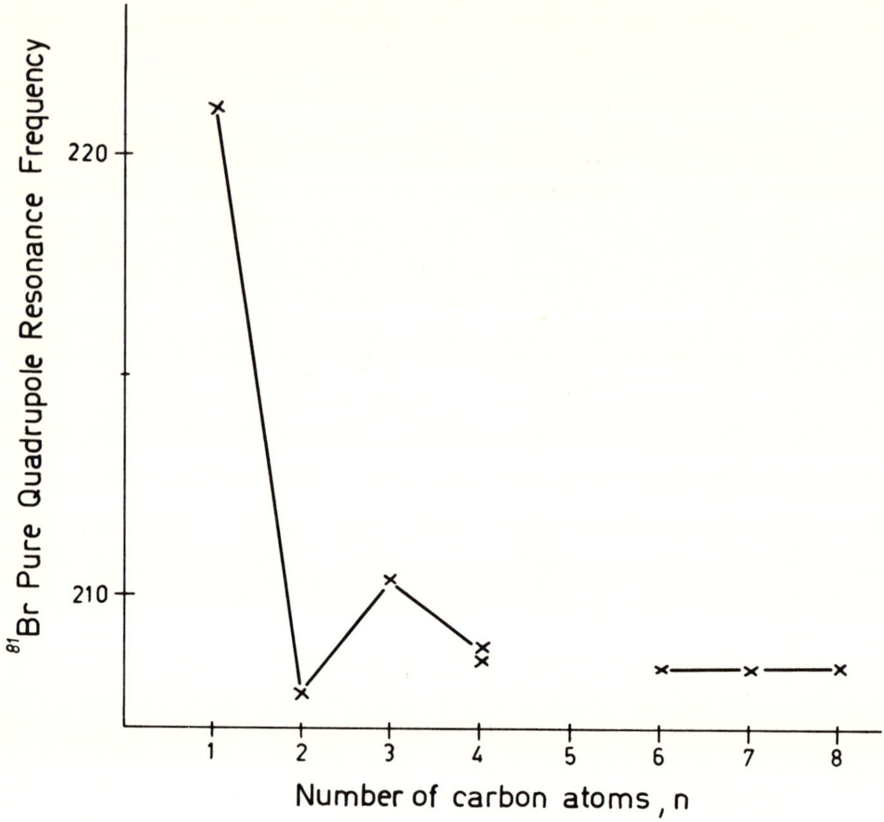

FIG. 10.13. ^{81}Br pure quadrupole resonance frequencies of the *n*-bromoalkanes, $CH_3(CH_2)_{n-1}Br$.

that observed for the chlorobenzenes. Fewer compounds are available for the correlation but the pure quadrupole resonance frequency for ^{81}Br in the substituted bromobenzenes is given by

$$\nu(^{81}\text{Br}) = 226{\cdot}932 \text{ Mc/s} + 7{\cdot}639\sigma \pm 2{\cdot}98 \text{ Mc/s}. \tag{15}$$

This equation and the corresponding one for the chlorobenzenes may be expressed as

$$\nu(^{35}\text{Cl}) = 34{\cdot}826(1 + 0{\cdot}0287\sigma) \text{ Mc/s} \tag{16}$$

$$\nu(^{81}\text{Br}) = 226{\cdot}932(1 + 0{\cdot}0336\sigma) \text{ Mc/s}. \tag{17}$$

TABLE 10.23. Asymmetry parameters of the field-gradient tensor at the bromine nuclei in conjugated organobromine compounds.

Compound	Nucleus	Frequency (Mc/s)	η	Method	Ref.
p-dibromobenzene			$0{\cdot}045 \pm 0{\cdot}002$	Zeeman	68
1,3,5-tribromobenzene	Br^{81}	230·350	$0{\cdot}062 \pm 0{\cdot}002$	Zeeman	69
		231·476	$0{\cdot}062 \pm 0{\cdot}003$		
		231·695	$0{\cdot}084 \pm 0{\cdot}002$		
p-bromophenol	Br^{79}	264·69	$0{\cdot}07 \pm 0{\cdot}03$	Zeeman	70
p-bromoaniline	Br^{81}	221·86	0·029	Zeeman	71
2,6-dibromoaniline	Br^{81}	226·58	0·069	Zeeman	71
2,4,6-tribromoaniline	Br^{81}	230·42	0·074	Zeeman	71
		233·56	0·040	Zeeman	
Acetyl bromide	Br^{81}	464*	0·21	Microwave	72

* Quadrupole coupling constant.

It is again evident that the bromine atom is more susceptible to changes in sigma than the chlorine atom, that is to say that the carbon-bromine bond is more polarizable.

Few measurements of bromine asymmetry parameters are available and these have been collected in Table 23. In every case where data for corresponding systems are available the bromine asymmetry parameter is less than that of the chlorine compound. Despite the greater polarizability of the bromine atom this result is to be expected in view of the length of the carbon–bromine bond ($=C-Br = 1{\cdot}85$ Å) compared with the carbon–chlorine bond ($=C-Cl = 1{\cdot}70$ Å) and the more diffuse nature of 4*p* compared with 3*p* orbitals. The displacement of the point corresponding to the benzoyl derivatives in Fig. 12 can also be explained if the asymmetry parameter in benzoyl bromide is less than that in benzoyl chloride and the result for acetyl bromide supports this hypothesis.

TABLE 10.24. ^{81}Br pure quadrupole resonance frequencies (Mc/s at 77°K) of the bromotropolones (Ref. 73).

Compound	Quadrupole resonance frequency
3-Bromotropolone	232·05
4-Bromotropolone	226·96
5-Bromotropolone	226·15
3,5-Dibromotropolone	233·17, 240·86
3,7-Dibromotropolone	237·21, 237·86
3,5,7-Tribromotropolone	241·40

TABLE 10.25. ^{81}Br quadrupole coupling constants at 77°K of *N*-bromo compounds (Ref. 56)

Compound	Formula	Frequency (Mc/s)
N-bromoacetamide	$CH_3CONHBr$	344·1
N-bromo-2-pyrrolidineone	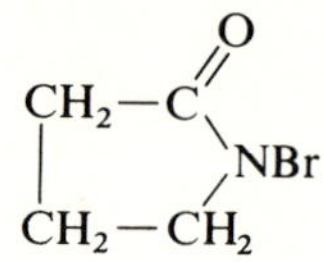	343·3
N-bromo caprolactam	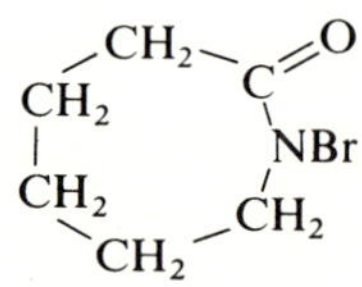	341·9
N-bromosuccinimide	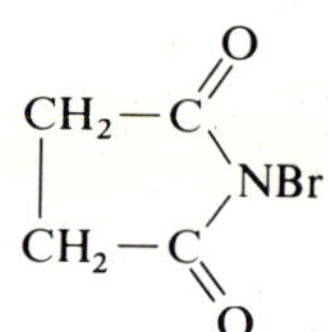	358·73
N-bromophthalimide	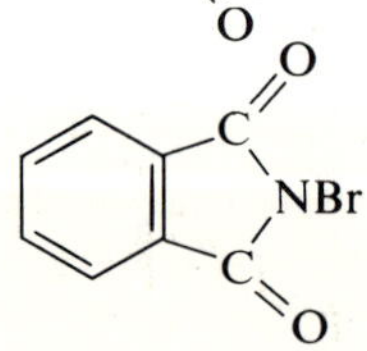	370·8

An interesting series of compounds studied by ^{81}Br pure quadrupole resonance is the bromotropolones (Table 24), the skeletal formula of tropolone being shown in (X).

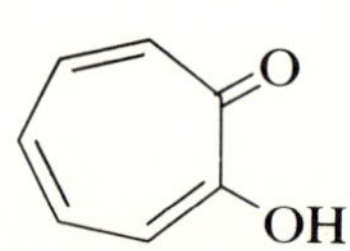

(X)

The closeness of the two frequencies for the two formally different bromine atoms in 3,7-dibromotropolone was taken to indicate the presence of an almost symmetrical intramolecular hydrogen bond (XI).

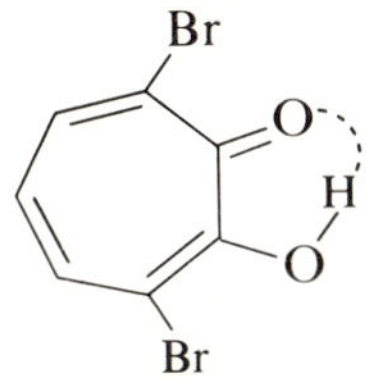

(XI)

This conclusion must however be considered to be based on extremely flimsy evidence, particularly in view of the absence of values of the asymmetry parameters for these two bromine atoms.

1. *N*-bromo Compounds

The coupling constants of the *N*-bromo compounds (Table 25), like those of the analogous chloro compounds, are so high as to indicate that the carbon-bromine bond is essentially covalent, even having a slight positive charge on the halogen atom. Again *N*-bromo succinimide is well-known as a source of atomic bromine for free-radical bromination.

E. IODINE RESONANCES

The data for this nucleus are even more sparse than for the two bromine isotopes, nevertheless the results for iodo derivatives are particularly interesting inasmuch as the spin of $I^{127}(5/2)$ makes the asymmetry parameters immediately available from pure quadrupole resonance measurements. Tables 26, 27, 28 and 29 list I^{127} pure quadrupole resonance frequencies of organo iodine compounds and, where both $m = \pm\frac{1}{2} \leftrightarrow m = \pm\frac{3}{2}$ and $m = \pm\frac{3}{2} \leftrightarrow m = \pm\frac{5}{2}$ transitions have been observed, the derived coupling constants and asymmetry parameters. The coupling constants of covalent iodides are negative (see equations (22) and (23) and Table 9 in Chapter 7).

Like the organic bromides the coupling constants of the organic iodides are linearly related to those of the corresponding chlorides. This is illustrated in Fig. 14 where once again the coupling constants have been divided by the appropriate atomic coupling constants. Figure 15 shows a plot of the equation analogous to (14), i.e.

$$\frac{(e^2Qq)_{CH_3I} - (e^2Qq)_{RI}}{(e^2Qq)_{CH_3I}} = \left(\frac{f_I}{f_{Cl}}\right)\frac{(e^2Qq)_{CH_3Cl} - (e^2Qq)_{RCl}}{(e^2Qq)_{CH_3Cl}} \tag{18}$$

TABLE 10.26. I^{127} nuclear quadrupole resonance frequencies and coupling constants (Mc/s) of aliphatic iodo compounds

Compound	Formula	ν_1	ν_2	T(°K)	e^2Qq	η	Ref.
Methyl iodide	CH_3I	265·102	529·670	77	1765·846	0·0278	74
Methylene iodide	CH_2I_2	286·883	568·362	77	1897·368	0·0858	74
Carbon tetraiodide	CI_4	319·549	—	77	—	—	74
Trifluoromethyl iodide	CF_3I	310·275	—	83	—	—	74
Ethyl iodide	CH_3CH_2I	247·115 247·374	494·058	83	1649	0·029	63
n-Propyl iodide	$CH_3CH_2CH_2I$	250·810 250·962	499·810	83	1671	0·052	63
i-Propyl iodide	$(CH_3)_2CH.I$	238·946	—	77	—	—	63
n-Butyl iodide	$CH_3CH_2CH_2CH_2I$	249·062	497·539	83	1660	0·028	75
trans-Acetylene diiodide	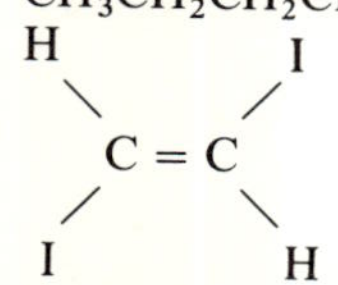	277·1	—	83	—	—	76

TABLE 10.27. I^{127} nuclear quadrupole resonance frequencies and coupling constants (Mc/s) of iodobenzenes

Compound	Formula	ν_1	ν_2	T(°K)	e^2Qq	η	Ref.
p-Di-iodobenzene	IC_6H_4I	275·89 ± 0·15	550·16 ± 0·30	301	1834·7	0·0477	20
		276·31 ± 0·15	551·68 ± 0·30		1839·4	0·0365	
o-Iodobenzoic acid	IC_6H_4COOH	294·58 ± 0·15	573·43 ± 0·30	300	1919·6	0·1462	20
m-Iodobenzoic acid	IC_6H_4COOH	275·67 ± 0·15	548·56 ± 0·30	301	1830·0	0·0627	20
p-Iodobenzoic acid	IC_6H_4COOH	272·89 ± 0·04	544·18 ± 0·30	300	1814·8	0·0476	20
Methyl *m*-iodobenzoate	$IC_6H_4COOCH_3$	276·24 ± 0·01	550·16 ± 0·02	297	1835·0	0·0571	20
Methyl *p*-iodobenzoate	$IC_6H_4COOCH_3$	284·03 ± 0·03	564·21 ± 0·04	297	1882·7	0·0726	20
Sodium *o*-iodobenzoate	IC_6H_4COONa	294·81 ± 0·04	—	—	—	—	20
Sodium *m*-iodobenzoate	IC_6H_4COONa	272·73 ± 0·03	545·17 ± 0·04	298	1817·4	0·0203	20
		273·76 ± 0·02	547·42 ± 0·03	—	1824·8	0·0127	20
Sodium *p*-iodobenzoate	IC_6H_4COONa	277·59 ± 0·10	—	—	—	—	20
		280·82 ± 0·15					
		282·25 ± 0·15					
o-Iodonitrobenzene	$IC_6H_4NO_2$	315·19 ± 0·20	613·88 ± 0·40	299	2055·0	0·1447	20
		315·97 ± 0·20	615·26 ± 0·40		2059·9	0·1454	
m-Iodonitrobenzene	$IC_6H_4NO_2$	285·63 ± 0·02	570·29 ± 0·30	301	1901·5	0·0505	20
		286·66 ± 0·02	571·43 ± 0·30		1905·8	0·0365	
p-Iodonitrobenzene	$IC_6H_4NO_2$	284·36 ± 0·15	565·82 ± 0·30	301	1887·7	0·0630	
o-Iodophenol	IC_6H_4OH	277·60 ± 0·02	551·80 ± 0·05	299	—	—	20
p-Iododiphenyl	$IC_6H_4C_6H_5$	275·40 ± 0·15	—	—	—	—	20
m-Iodoaniline	$IC_6H_4NH_2$		—	—	—	—	20
p-Iodoaniline	$IC_6H_4NH_2$	261·14 ± 0·03	521·25 ± 0·05	301	1841·1	0·0691	20
p-Iodoaniline hydrochloride	$IC_6H_4NH_3Cl$	288·28 ± 0·02	572·15 ± 0·04	296	1909·5	0·0771	20
p-Iodoacetanilide	$IC_6H_4NHCOCH_3$	273·74 ± 0·15	546·06 ± 0·30	301	1820·9	0·0449	20

TABLE 10.27 *continued*

Compound	Formula	ν_1	ν_2	T(°K)	e^2Qq	η	Ref.
3,5-Di-iodo-2-hydroxybenzoic acid	$I_2C_6H_2(OH)COOH$	284·19 ± 0·15 295·14 ± 0·15	—	—	—	—	20
Tetraiodophthalic anhydride	$I_4C_6(CO)CO$ (CO groups bridged by O)	313·63 ± 0·20 316·20 ± 0·20 324·03 ± 0·20 328·53 ± 0·20	3 lines at about 604, 609, 616	—	—	—	77
o-Iodoanisole	*o*-$I.C_6H_4OCH_3$	287·216	—	77	—	—	47
m-Iodoanisole	*m*-$I.C_6H_3OCH_3$	274·593	—	77	—	—	47
p-Iodoacetophenone	*p*-$I.C_6H_4COCH_3$	285·17	564·27	298	1884·1	0·0913	21
p-Iodotrimethylanilinium iodide	*p*-$I.C_6H_4.\overset{+}{N}(CH_3)I^-$	287·37	573·40	298	1912·0	0·0426	21
m-Iodoaniline	*m*-$IC_6H_4NH_2$	270·491	—	77	—	—	77
2-Iodo *p*-xylene	2-I,1,4$(CH_3)_2C_6H_3$	270·770	—	77	—	—	77
2-Iodo *m*-xylene	2-I,1,3$(CH_3)_2C_6H_3$	273·353	—	77	—	—	77
2,5-Di-iodo-α,α,α,-trifluorotoluene	2,5-I_2, $ICF_3.C_6H_3$	287·2	—	293	—	—	78

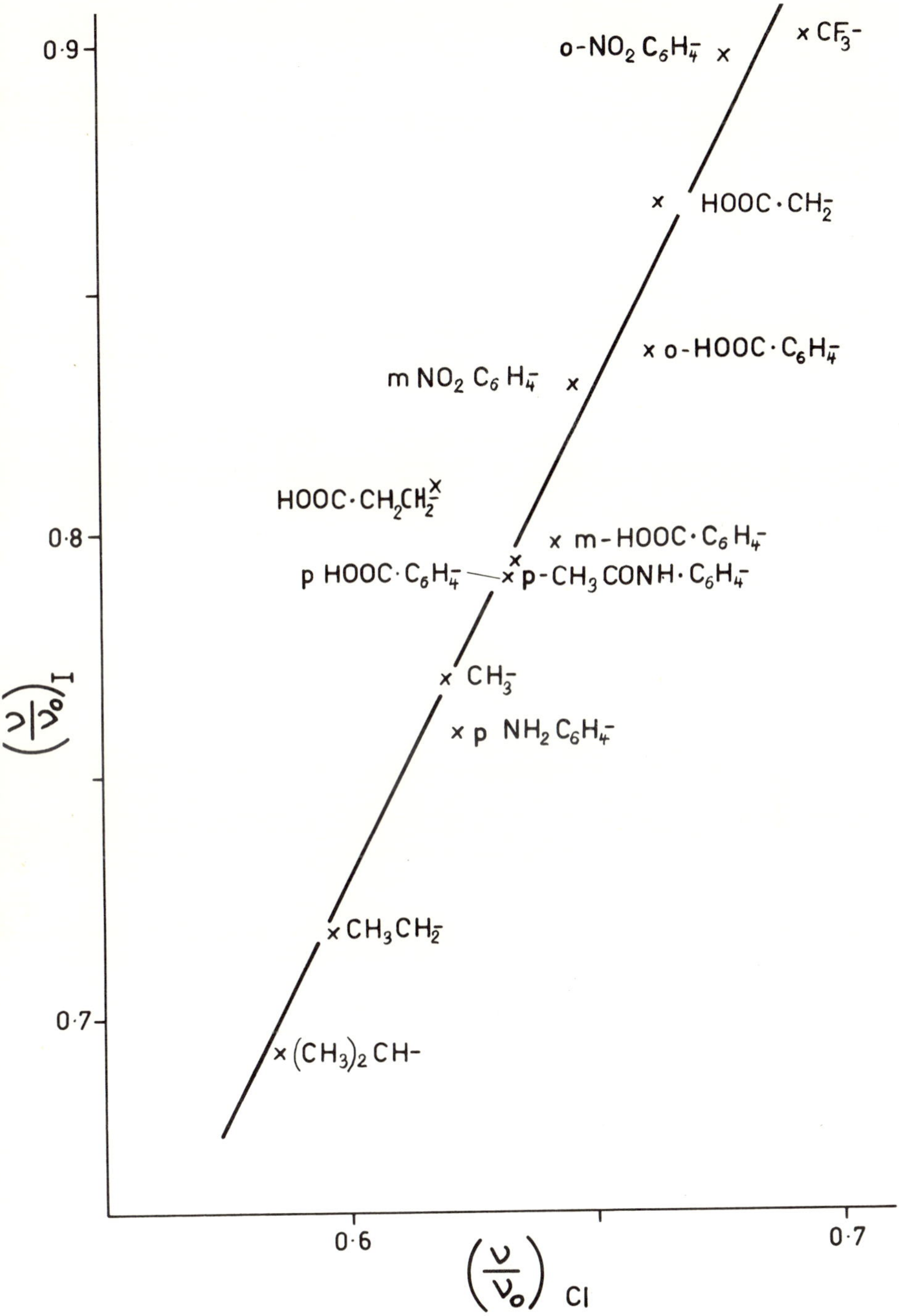

FIG. 10.14. Reduced quadrupole resonance frequencies, ν_{RX}/ν_{atomic}, of corresponding organo-iodine and organochlorine compounds.

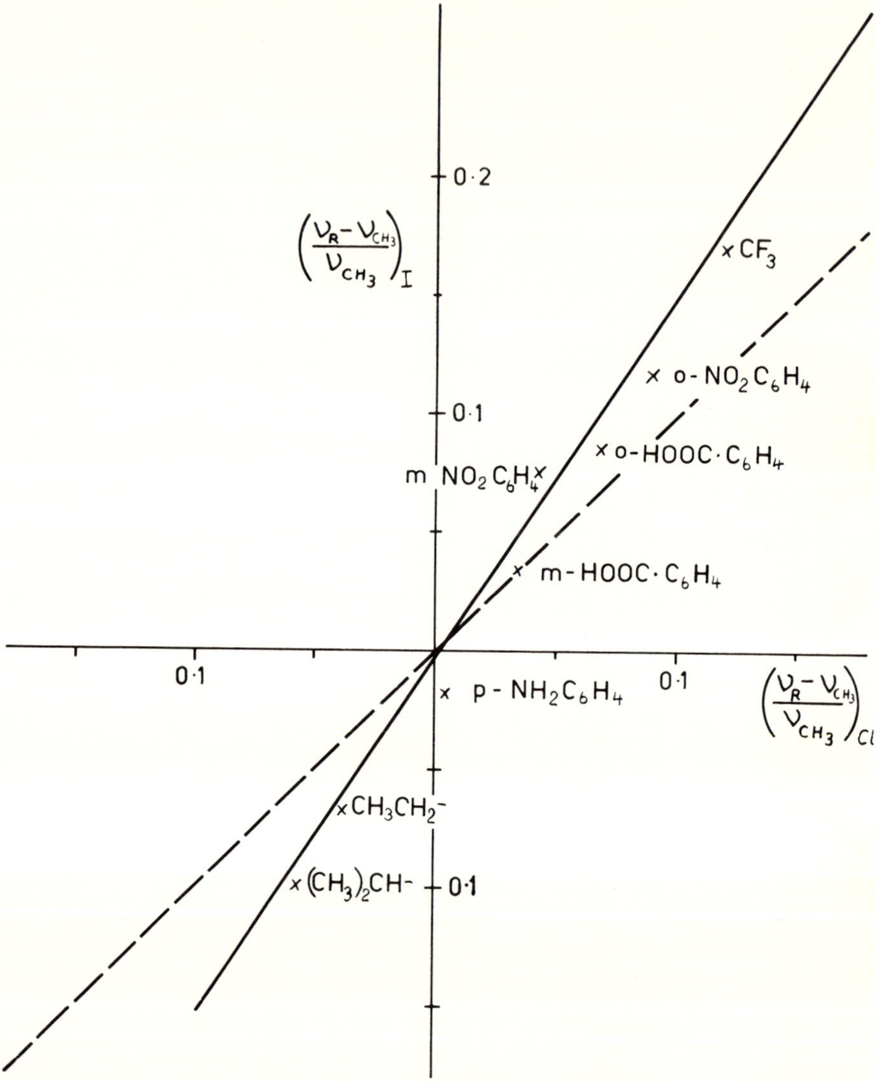

FIG. 10.15. Relative pure quadrupole resonance frequency shifts $(\nu_{RX} - \nu_{CH_3X})/\nu_{CH_3X}$ for organochlorine and organoiodine compounds.

and the expected greater polarizability of the iodine atom relative to both the chlorine and bromine atoms is clearly visible ($f_I/f_{Cl} = 1{\cdot}53$).

For the substituted iodobenzenes Ludwig[20] has related the frequencies of the $m = \pm\frac{1}{2} \leftrightarrow m = \pm\frac{3}{2}$ transition to the values of the Hammett sigma constant. The linear correlation is given by approximately

$$\nu = 267{\cdot}0 + 1{\cdot}424\sigma \text{ Mc/s} \tag{19}$$

TABLE 10.28. I^{127} pure quadrupole resonance data (Mc/s) for iodo-derivatives of condensed hydrocarbons (Ref. 21)

Compound	Formula	ν_1	ν_2	T°K	e^2Qq	η
1,5-Di-iodonaphthalene	$C_{10}H_6I_2$	278·51	551·58	298	1841·4	0·0873
		279·68	553·55		1848·2	0·0901
2,6-Di-iodonaphthalene	$C_{10}H_6I_2$	276·85	551·01	298	1838·1	0·0615
9,10-Di-Iodoanthracene	$C_{14}H_8I_2$	279·48	550·81	298	1840·3	0·1070
		285·30	562·40		1879·0	0·1063
2,7-Di-iodophenanthrene	$C_{14}H_8I_2$	272·7	541·0	298	1811	0·06
		273·6	544·6		1814	0·07
1,3-Di-iodoazulene	$C_{10}H_6I_2$	275·71	549·24	298	1832·0	0·0554
		278·38	554·13		1848·4	0·0605

and re-expressing this in the form

$$\nu = 267{\cdot}0\,(1 + 0{\cdot}0533\sigma) \tag{20}$$

illustrates the more polarizable nature of the iodine atom or more precisely the carbon-iodine bond compared with the carbon-chlorine or carbon-bromine bonds (cf. Equations 16 and 17).

Like the bromobenzenes, the asymmetry parameters for the iodobenzenes are in general less than those of corresponding chloro compounds, at least when the substituent is in the *meta* or *para* position. For *ortho* substituted derivatives direct interactions between the substituent and the iodine atom bromine bonds (cf. Equations 16 and 17).

TABLE 10.29. I^{127} pure quadrupole resonance data (Mc/s) for iodo derivatives of heterocyclic compounds (Ref. 21)

Compound	ν_1	ν_2	T(°K)	e^2Qq	η
2,iodo,2 acetylamino-thiophene	293·2	585·8	293	1953	0·03 ± 0·03
2,iodo,5 carboxythiophene	292·02	581·63	298	1940·0	0·0565
	295·65	590·33		1968·3	0·0355
2,5 diiodothiophene	296·57	590·32	298	1969·2	0·0360
	297·94	594·89		1983·6	0·0608
2,iodo,5 carboxymethyl thiophene	307·82	615·44	298	2051·6	0·0159
2,iodo,5 acetylthiophene	307·93	615·67	298	2052·3	0·0155
2,iodo,5 nitrothiophene	313·62	624·95	298	2084·4	0·0529
2,3,4,5 tetraiodothiophene	298·0	—	298	—	—
	312·0	—	298	—	—
2,5 diiodofuran	296·6	—	298	—	—
2,3,4,5 tetraiodopyrrole	288·9	—	298	—	—
	305·1	—		—	—
4,iodo pyrazole	293·73	586·90	298	1957	0·116
3,5 diiodopyridine	279·11	557·64	298	1859	0·03

make somewhat doubtful the relationship of the asymmetry parameters to the double-bond character of the carbon-iodine bond. Table 30 shows a selection of double bond characters of *m*- and *p*-substituted iodobenzenes calculated according to equation (2) and for those compounds for which a direct comparison is possible a lower value is found for this quantity than in the corresponding chloro compounds.

TABLE 10.30. Partial double bond character of carbon halogen bonds in substituted iodobenzenes

Compound	e^2Qq (Mc/sec)	η (%)	$\frac{2}{3}\frac{e^2Qq}{e^2Qq_0} \cdot \eta$ (%)
p-Di-iodobenzene	1834·7 ± 1·0	4·77 ± 0·40	2·1
IC_6H_4I	1839·4 ± 1·0	3·65 ± 0·50	1·6
m-Iodobenzoic acid	1830·0 ± 1·0	6·27 ± 0·30	2·7
IC_6H_4COOH			
p-Iodobenzoic acid	1814·8 ± 0·3	4·76 ± 0·20	2·1
IC_6H_4COOH			
Methyl *m*-iodobenzoate	1835·0 ± 0·1	5·71 ± 0·02	2.5
$IC_6H_4COOCH_3$			
Methyl *p*-iodobenzoate	1882·7 ± 0·2	7·26 ± 0·05	3.3
$IC_6H_4COOCH_3$			
Sodium *m*-iodobenzoate	1817·4 ± 0·2	2.03 ± 0·20	0·9
IC_6H_4COONa	1824·8 ± 0·2	1·27 ± 0·20	0·5
m-Iodonitrobenzene	1901·5 ± 0·2	5·05 ± 0·20	2·3
$IC_6H_4NO_2$	1905·8 ± 0·2	3·65 ± 0·25	1·6
p-Iodonitrobenzene	1887·7 ± 1·0	6·30 ± ·030	2·8
$IC_6H_4NO_2$			
p-Iodoaniline	1738·1 ± 0·2	3·91 ± 0·10	1·6
$IC_6H_4NH_2$			
p-Iodoaniline hydrochloride	1909·5 ± 0·2	7·71 ± 0·04	3·5
$IC_6H_4NH_3Cl$			
p-Iodoacetanilide	1820·9 ± 1·0	4·49 ± 0·45	1·9
$IC_6H_4NHCOCH_3$			

Bogaard[21] has attempted, with fair success, to correlate the asymmetry parameters of iodo-substituted condensed hydrocarbons with the predictions of simple molecular orbital theory for these compounds. It is, however, very unfortunate that the nucleus which yields the asymmetry parameter with the minimum of effort is the one which is the least susceptible to pi-bonding effects.

For saturated compounds where there is no threefold axis of symmetry through the carbon-iodine bond a non-zero value for the asymmetry parameter is theoretically possible. This possibility is realized for the molecules

CH_2I_2, CH_3CH_2I, $CH_3CH_2CH_2I$, and $CH_3CH_2CH_2CH_2I$.* The small but non-zero value for CH_3I presumably reflects a lack of cylindrical symmetry of the C—I bond within the crystal lattice.

REFERENCES

1. I. P. Biryukov and M. G. Voronkov, Latrijas P. S. R. *Zinat. Akad. Vest. Kim.* (1965) 115; ibid. *Coll. Czech. Chem. Comm.* **32**, 830 (1967).
2. H. O. Hooper and P. J. Bray, *J. Chem. Phys.* **33**, 334 (1960).
3. R. Bersohn, *J. Chem. Phys.* **22**, 2078 (1954).
4. S. L. Segel, R. G. Barnes and P. J. Bray, *J. Chem. Phys.* **25**, 1286 (1956).
5. M. J. S. Dewar and E. A. C. Lucken, *J. Chem. Soc.* 2653 (1958).
6. P. J. Bray, S. Moskowitz, H. O. Hooper, R. G. Barnes and S. L. Segel, *J. Chem. Phys.* **28**, 99 (1958).
7. P. J. Bray and R. G. Barnes, *J. Chem. Phys.* **27**, 551 (1957).
8. E. Scrocco, *Advances in Chemical Physics* **5**, 319 (1963).
9. C. A. Coulson and T. H. Goodwin, *J. Chem. Soc.* 2851 (1962).
10. A. D. Walsh, *Nature, Lond.* **159**, 165; 712 (1947).
11. W. H. Flygare, A. Narath and W. G. Gwinn, *J. Chem. Phys.* **36**, 200 (1962).
12. F. L. Tobiason and R. H. Schwendeman, *J. Chem. Phys.* **40**, 1014 (1964).
13. J. E. Todd, M. A. Whitehead and K. E. Weber, *J. Chem. Phys.* **39**, 404 (1963).
14. E. A. C. Lucken and C. Mazeline, *J. Chem. Soc.* A (1968). In press.
15. R. S. Mulliken, C. A. Rieke, D. Orloff and H. Orloff, *J. Chem. Phys.* **17**, 1248 (1949).
16. R. Livingston, *J. Phys. Chem.* **57**, 496 (1953).
17. I. P. Biryukov, M. G. Voronkov, I. A. Safin; *Teor. Eksperim. Khim. Akad. Nauk Ukr. S.S.R.* **1**, 373 (1965).
18. M. G. Voronkov and I. P. Biryukov, *Teor. Experim. Khim. Akad. Nauk Ukr. S.S.R.* **1**, 124 (1965).
19. J. M. Mays and C. H. Townes, *Phys. Rev.* **81**, 940 (1951); *J. Chem. Phys.* **20**, 1695 (1952).
20. G. W. Ludwig, *J. Chem. Phys.* **25**, 159 (1956).
21. M. P. Bogaard, Thesis, University of Sydney (1967).
22. J. H. Goldstein, *J. Chem. Phys.* **24**, 106 (1956).
23. J. K. Bragg, T. C. Madison, A. H. Sharbaugh, *Phys. Rev.* **77**, 148 (1950).
24. G. W. Robinson, *J. Chem. Phys.* **21**, 1741 (1953).
25. V. Rehn, *J. Chem. Phys.* **38**, 749 (1963).
26. P. Bucci, P. Cecchi and E. Scrocco, *Ric. Sci.* **34**, 129 (1964).
27. H. Hartmann and H. Haas, *Zeit. Naturforschung* **18a**, 977 (1963).
28. Y. Morino and M. Toyama, *J. Phys. Soc. Japan* **15**, 288 (1960).
29. M. Suhara, T. Yonemitsu and T. Tonomura, *Bull. Chem. Soc. Japan* **38**, 2205 (1965).
30. S. Ogawa and K. Ohi, *J. Phys. Soc. Japan*, **15**, 1064 (1960).
31. H. C. Meal, *J. Am. Chem. Soc.* **74**, 6121 (1952).
32. T. Sakurai, *Acta Cryst.* **15**, 443 (1962).
33. T. Sakurai, *Acta Cryst.* **15**, 1164 (1962).

* These asymmetries seem on the whole larger than those of the chlorine compounds and may reflect the high polarizability of the iodine atom: however the sizes of the two populations available for comparison is too small for it to be possible to attach much importance to this point at present.

34. D. Giardino, Thesis, University of Pittsburgh (1961).
35. N. Nagarajan and C. R. K. Murty, *Curr. Sci. India* **32**, 254 (1963).
36. N. Nagarajan and C. R. K. Murty, *Ind. J. Pure Appl. Phys.* **2**, 80 (1964).
37. Y. Morino, T. Chiba, T. Shimozawa and M. Toyama, *J. Phys. Soc. Japan* **13**, 869 (1958).
38. P. Bucci, P. Cecchi and A. Colligiani, *J. Am. Chem. Soc.* **87**, 3027 (1966).
39. S. Chu, G. A. Jeffrey and T. Sakurai, *Acta Cryst.* **15**, 661 (1962).
40. Y. Morino, M. Toyama and K. Itoh, *Acta Cryst.*, **16**, 129 (1963).
41. G. Soda, M. Toyama and Y. Morino, *Bull. Chem. Soc. Japan* **38**, 1965 (1965).
42. V. Grechishkin, *Zh. Strukt. Khim.* **3**, 597 (1962).
43. E. A. C. Lucken, *J. Chem. Soc.* (1959), 2954.
44. E. A. C. Lucken and C. Mazeline, *Proc. XIII Colloque Ampère*, North Holland Publishing Co., Amsterdam 1965.
45. H. Negita and S. Satou, *J. Chem. Phys.* **27**, 602 (1957).
46. P. J. Bray, *J. Chem. Phys.* **23**, 703 (1955).
47. P. J. Bray, R. G. Barnes and S. L. Segel, Unpublished data quoted in the Catalog of Nuclear Quadrupole Interactions and Resonance Frequencies in Solids, Pt. II.
48. S. Saito, *J. Chem. Phys.* **36**, 1397 (1962).
49. P. J. Bray, G. O'Keefe and R. G. Barnes, *J. Chem. Phys.* **25**, 792 (1956).
50. R. M. Myasnikov, V. I. Robas and G. K. Semin, *Zh. Strukt. Khim.* **6**, 474 (1965).
51. V. S. Grechishkin and G. B. Soifer, *Zh. Strukt. Khim.* **5**, 914 (1964).
52. V. S. Grechishkin, *Zh. Strukt. Khim.* **6**, 162 (1965).
53. H. C. Allen, *J. Phys. Chem.* **57**, 501 (1953).
54. P. J. Green and J. D. Graybeal, *J. Am. Chem. Soc.* **89**, 4305 (1967).
55. E. Swiger and J. D. Graybeal, *J. Am. Chem. Soc.* **87**, 1464 (1965).
56. H. Kashiwagi, D. Nakamura and M. Kubo, *Tetrahedron* **21**, 1095 (1965).
57. E. A. C. Lucken and M. A. Whitehead, *J. Chem. Soc.* 2459 (1961).
58. P. J. Bray and D. Esteva, *J. Chem. Phys.* **22**, 570 (1954).
59. P. J. Bray and P. J. Ring, *J. Chem. Phys.* **21**, 2226 (1953).
60. D. W. McCall and H. S. Gutowsky, *J. Chem. Phys.* **21**, 1300 (1953).
61. P. J. Bray, *J. Chem. Phys.* **22**, 950 (1954).
62. P. J. Bray and R. G. Barnes, *J. Chem. Phys.* **22**, 2023 (1954).
63. H. Zeldes and R. Livingston, *J. Chem. Phys.* **21**, 1418 (1953).
64. P. J. Bray, *J. Chem. Phys.* **22**, 1787 (1954).
65. G. Jones and W. Gordy, *Phys. Rev.* **136**A, 1229 (1964).
66. M. J. S. Dewar and E. A. C. Lucken, *J. Chem. Soc.* 426 (1959).
67. G. K. Semin and A. A. Fainzil'berg, *Zhur. Strukt. Khim* **6**, 213 (1965).
68. P. Bucci, P. Cecchi and A. Colligiani, *J. Am. Chem. Soc.* **86**, 2513 (1963).
69. P. Bucci, P. Cecchi and A. Colligiani, *Ric. Sci. Rend. Sez.* A**8**, 1144 (1965).
70. K. V. S. Rama Rao and C. R. K. Murty, *Curr. Sci.* **34**, 660 (1965).
71. P. Bucci, P. Cecchi, A. Colligiani and R. Meschia, *Bull. Sci. Fac. Chim. Ind. Univ. Bologna* **23**, 307 (1965).
72. L. C. Krisher, *J. Chem. Phys.* **33**, 1237 (1960).
73. R. Ikeda, D. Nakamura, S. Itô, H. Sugiyama and M. Kubo, *Tetrahedron* **20**, 909 (1964).
74. H. Robinson, H. G. Dehmelt and W. Gordy, *J. Chem. Phys.* **22**, 511 (1954).
75. S. Kojima, K. Tsukada, S. Ogawa and A. Shimauchi, *J. Chem. Phys.* **21**, 1415 (1953).
76. H. G. Dehmelt, *Z. Phys.*, **130**, 356 (1951).
77. S. Segel, Unpublished data in Catalog of Nuclear Quadrupole Resonance Frequencies.
78. J. Hatton and B. V. Rollin, *Trans. Faraday Soc.*, **50**, 358 (1954).

CHAPTER 11

N^{14} QUADRUPOLE COUPLING CONSTANT

A. INTRODUCTION

The N^{14} nucleus is present to the extent of 99·6 per cent in naturally-occurring nitrogen atoms. Its spin is unity and its quadrupole moment is rather small, about $+0{\cdot}016 \times 10^{-24}\,\text{cm}^2$. As a consequence the coupling constants are small, seldom more than 6 Mc/s. Most of the experimental results to date have been obtained by pure quadrupole resonance but data obtained from other sources will be mentioned when necessary.

The low coupling constant of N^{14} is the main reason for data for this nucleus being obtained so slowly. The intensities of the pure quadrupole resonance lines are very low and extremely sensitive spectrometers and much patience are required to get a result. For a nucleus of unit spin, however, the pure quadrupole resonance frequencies are given by the expression

$$\nu_{\pm} = \frac{3}{4}e^2Qq\left(1 \pm \frac{\eta}{3}\right) \tag{1}$$

$$\nu_0 = \tfrac{1}{2}\eta \cdot e^2Qq$$

so that, unlike the nuclei with spin $I = \frac{3}{2}$, all the components of the field-gradient tensor can be obtained in a pure quadrupole resonance experiment on a polycrystalline powder. Naturally, however, the directions of the field-gradient axes in the molecular framework must be obtained from a study of single crystals of known internal structure, although in some cases all or some of these are determined by the molecular symmetry. The frequency ν_0 is usually small and rarely observed but can be useful in the case of multiple resonances in correctly pairing the $\nu_{\pm}$ resonances.

Finally, as far as the theoretical interpretation of the coupling constants is concerned, it has not so far proved possible to obtain an experimental value for a coupling constant in some state or other of atomic nitrogen. In its ground state ($1s^22s^22p^3$, ^4S) the spherical symmetry of the nitrogen atom precludes the observation of a coupling constant, and the coupling constant in some excited state with non-spherical symmetry is too small for the hyperfine structure of atomic spectral lines to be observable with present techniques. The coupling constant yielded by an electron in a valence $2p$ orbital of ^{14}N, $(e^2Qq_0)_N$ is therefore not accessible to direct measurement, as is the case for the halogens, but must be obtained by studies of N^{14} couplings in molecules. As will be seen below a value in the region $-8{\cdot}0$ to $-10{\cdot}0$ Mc/s seems fairly appropriate.

Quadrupole coupling constants are completely defined if not only the values of the principal components are known but also their orientation in the molecular framework. For the singly coordinated nuclei we have been considering up to now this presents no problem since the principal direction of the field-gradient tensor lies along the bond direction. For polycoordinate nuclei, however, no such easy simplification is in general possible. For symmetrical trivalent derivatives such as ammonia or the linear cyanides these directions are defined by molecular symmetry. In this case the principal field-gradient axis coincides with one of the angular momentum axes so that the data obtained by microwave spectroscopy are immediately available in the principal axis system. For less symmetrical molecules, however, (e.g. pyridine, methylamine) the directions of the principal axes may only be partially defined by the molecular symmetry and in extreme cases not at all. In these circumstances there arises an important difference between data acquired by microwave methods and those by pure quadrupole resonance. In the latter the principal *values* are given by the resonance frequencies and the *directions* can in principle be obtained by single crystal studies, although only one measurement of this sort is at present known to the author. Even if they cannot be measured experimentally they can often be assigned by analogy or on the basis of elementary ideas about electronic structure. For microwave measurements, however, the coupling data are referred to the angular momentum axes, which do not in general coincide with the field-gradient axes, and since ^{14}N coupling constants are small the off-diagonal elements in the angular momentum axis-system are usually not measurable, thus making it impossible to obtain the principal values of the coupling tensor. Occasionally the field-gradient axes can be determined by isotopic substitution, for example with nitric acid. In general, therefore, measurements by microwave spectroscopy must be very precise and detailed to be of use to anyone interested in electronic structure.

B. TRICOORDINATE NITROGEN COMPOUNDS

Compounds with the formula NR_3 have at least C_{3v} symmetry with the nitrogen atom lying on the symmetry axis so that the asymmetry parameter is necessarily zero. The s-p^3 hybrid orbitals which elementary theory uses to describe the bonding are given by

$$\phi_1 = \left\{\frac{-\cos\alpha}{1-\cos\alpha}\right\}\psi_s - \sqrt{\frac{2}{3}}\psi_x + \left\{\frac{1+2\cos\alpha}{3(1-\cos\alpha)}\right\}^{\frac{1}{2}}\psi_z$$

$$\phi_2 = \left\{\frac{-\cos\alpha}{1-\cos\alpha}\right\}\psi_s - \sqrt{\frac{1}{6}}\psi_x + \frac{1}{\sqrt{2}}\psi_y + \left\{\frac{1+2\cos\alpha}{3(1-\cos\alpha)}\right\}^{\frac{1}{2}}\psi_z$$

$$\phi_3 = \left\{\frac{-\cos\alpha}{1-\cos\alpha}\right\}\psi_s + \sqrt{\frac{1}{6}}\psi_x - \frac{1}{\sqrt{2}}\psi_y + \left\{\frac{1+2\cos\alpha}{3(1-\cos\alpha)}\right\}^{\frac{1}{2}}\psi_z$$

$$\phi_4 = \left\{\frac{-\cos\alpha}{1-\cos\alpha}\right\}\psi_s - \left\{\frac{-3\cos\alpha}{1-\cos\alpha}\right\}^{\frac{1}{2}}\psi_z \qquad (2)$$

where ψ_s, ψ_x, ψ_y and ψ_z are the $2s$, $2p_x$, $2p_y$ and $2p_z$ orbitals of the nitrogen atom and α is the $R\hat{N}R$ angle. The orbital ψ_4 is directed along the C_{3v} axis and contains the non-bonding pair of electrons while the orbitals ϕ_1, ϕ_2 and ϕ_3 are the bonding sigma orbitals occupied by a electrons each. Then the simple Townes and Dailey analysis yields

$$\frac{e^2Qq}{e^2Qq_0} = -\frac{3\cos\alpha}{1-\cos\alpha}(2-a). \qquad (3)$$

Since α is always greater than 90°, equation (3) predicts that the coupling constants of amines are negative, as is observed. However the data cannot be analysed without a knowledge of α to which equation (3) is unfortunately rather sensitive. In the neighbourhood of the tetrahedral angle (where the

TABLE 11.1. N^{14} coupling constants (Mc/s) and σ-bond populations in compounds NR_3. e^2Qq_0 has been assumed to have the value $-9{\cdot}0$ Mc/s and the coupling constants have been derived from microwave measurements

	e^2Qq	e^2Qq/e^2Qq_0	α	$\frac{3\cos\alpha}{1-\cos\alpha}$	a	Ref.
NH_3	−4·084	0·454	107°	0·679	1·33	23
$N(CH_3)_3$	−5·47	0·578	108·7 ± 1·0°	0·728 ± 0·03	1·21 ± 0·04	24
NF_3	−7·07	0·785	102·5 ± 1·5°	0·534 ± 0·05	0·53 ± 0·13	25

angular term is equal to 0·75) a change of 4° in α produces a 15 per cent change in the value of $3\cos\alpha/1-\cos\alpha$. Table 1 shows values of a calculated from Equation 3 for three simple compounds for which measurements of α are available; if the angular term is maintained constant at 0·75 the corresponding values of a are 1·39, 1·23 and 0·95, respectively. It may well be that the explicit inclusion of the angular term in such an approximate method of analysing the data is unjustifiable and indeed it does seem that its inclusion somewhat overemphasises the low value of a in NF_3. This point arises again in section 11.D. However, in both cases the results are qualitatively much as would be expected: thus in NF_3 a is less than unity owing to the higher electronegativity of fluorine compared to nitrogen while ammonia and trimethylamine have similar values of a, both greater than unity, that of ammonia being as expected somewhat greater than that of trimethylamine.

To the organic chemist used to thinking of the methyl group as being an electron-releasing group (viz. the relative basicities of trimethylamine and ammonia are $pK = 4{\cdot}26$ and $4{\cdot}75$, respectively, in water at 25°C, this last result is perhaps somewhat strange. Of course the methyl group may be electron-releasing by a hyperconjugative mechanism and the effect of this on the sigma populations a depends very critically on the model used to describe the hyperconjugation. Furthermore such hyperconjugation would very much depend on the relative orientations of the hydrogens of the methyl group and the lone-pair of electrons on the nitrogen atom. There is some evidence that such an effect is occurring for the coupling constants of triethylene diamine and hexamethylene tetramine (Table 2) where the rigid cage structure of these molecules imposes the configuration illustrated in I.

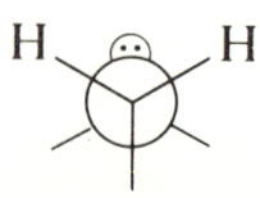

(I)

The bond angles in these two compounds are essentially tetrahedral and the higher coupling constant of hexamethylene tetramine is presumably due to the electron attracting effect of the three adjacent nitrogen atoms. Clearly, however, in amines there are a number of factors involved which are not yet

TABLE 11.2. N^{14} coupling constants (Mc/s) in rigid amines NR_3. The bond angle α has been assumed to have the tetrahedral value and e^2Qq_0 equal to $-9{\cdot}0$ Mc/s.

	$\lvert e^2Qq \rvert$	$\lvert e^2Qq/e^2Qq_0 \rvert$	a	Ref.
Triethylene diamine 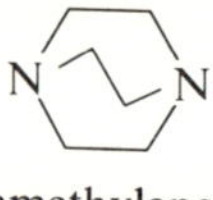	3·695	0·410	1·45	26
Hexamethylene tetramine 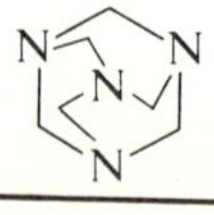	4·543	0·505	1·33	27

clear and which make the interpretation of the coupling constants rather difficult.

The value given for the coupling constant of ammonia above is taken from microwave spectroscopy. The value obtained by pure quadrupole resonance, i.e. in the solid phase, is much lower than this, 3·1607 Mc/s at 77°K. This exceptionally large difference is almost certainly due to hydrogen bonding. This fact and the loss of axial symmetry in such compounds, makes an interpretation of the coupling constants of primary and secondary amines susceptible to considerable uncertainty. Nevertheless if it is assumed that the field-gradient principal axis lies along the direction of the lone-pair orbital and that all the bond angles are identical, then the coupling constant data for compounds NR'_2R'' may be analysed by the equations

$$\frac{e^2Qq}{e^2Qq_0} = -\frac{3\cos\alpha}{1-\cos\alpha}\{2 - \tfrac{1}{3}(2b_1 + b_2)\} \tag{4}$$

$$\eta \cdot \frac{e^2Qq}{e^2Qq_0} = \tfrac{1}{2}|b_1 - b_2| \tag{5}$$

where b_1 is the population in the NR′ sigma bond and b_2 is the population of the NR″ bond.

Table 3 shows the results of the analyses of the coupling constants of methylamine and dimethylamine. Also included are the results for trimethylamine and the solid state coupling for ammonia. It seems likely that the gradual decrease in the calculated values of a in the series $NH_3 \rightarrow N(CH_3)$ may be taken as evidence of decreasing H-bonding in the series (in solid NH_3 each nitrogen atom is H-bonded to three equivalent protons). Most interesting, however, is the constancy of the difference ($b_{NH} - b_{NC}$) deduced from the asymmetry parameter and its similarity to the difference shown in Table 1. This difference is probably much less affected by hydrogen bonding than are the actual values of a themselves.

TABLE 11.3. N^{14} coupling constants (Mc/s) in amines. e^2Qq_0 is assumed equal to $-9{\cdot}0$ Mc/s and the solid state coupling constants for ammonia have been employed. α has been assumed to be the tetrahedral angle, 109° 28′.

	e^2Qq	η	b_{NH}	b_{NC}	$b_{NH} - b_{NC}$	Ref.
.NH_3	3·161	0·0	1·53	—	—	28
CH_3NH_2	3·986	0·186	1·44	1·36	0·08	29
$(CH_3)_2NH$	4·650	0·169	1·37	1·28	0·09	29
$(CH_3)_3N$	5·194	0·0	—	1·23	—	30

Hydrazine has the distinction of having the largest asymmetry parameter yet observed[1] ($e^2Qq = 4{\cdot}82$, $\eta = 0{\cdot}807$) and in fact the original investigation mistook a ν_0 resonance at around 1·8 Mc/s for a ν_- resonance. This high asymmetry is readily explicable from equation (5) since the population difference $|b_1 - b_2|$ will here be extremely large. Also however the principal field-gradient axis will now strongly bend away from the lone-pair axis so that equation (5) is quantitatively rather incorrect.

At first sight it would seem that these difficulties for unsymmetrical amines would disappear if microwave spectroscopic measurements yielded the orientation of the field-gradient tensor in the molecular framework. Instead, however, such measurements only reveal the inadequacy of the Townes and Dailey theory for dealing with such complicated situations. This is mainly due to the sensitivity to the angular term, and, for example such a study has recently been performed for dimethylamine.[2] The directions of the field-gradients for this molecule are shown in Fig. 1. (Note that the labelling of the axes is not the same as in the references[2] since we have maintained the convention $|e^2Qq_{zz}| > |e^2Qq_{yy}| > |e^2Qq_{xx}|$). Our first stumbling block in the

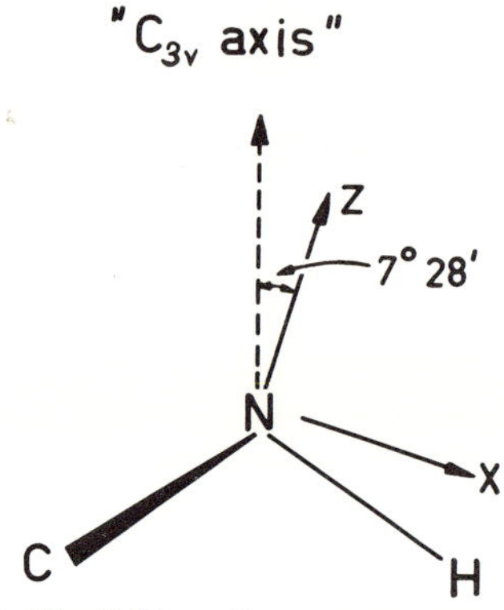

FIG. 11/1. The field-gradient axes in dimethylamine.

Townes and Dailey analysis is that the bond angles in this molecule are not known very accurately but are all in the neighbourhood of the tetrahedral angle, 109° 28′. The sp^3 hybrid orbitals for a regular tetrahedron are

$$\phi_1 = \tfrac{1}{2}\psi_s + \frac{\sqrt{3}}{2}\psi_z;$$

$$\phi_2 = \tfrac{1}{2}\psi_s + \sqrt{\frac{2}{3}}\,\psi_x - \frac{1}{2\sqrt{3}}\cdot\psi_z$$

$$\phi_3 = \tfrac{1}{2}\psi_s + \frac{1}{\sqrt{2}}\psi_y - \frac{1}{\sqrt{6}}\psi_x - \frac{1}{2\sqrt{3}}\psi_z;$$

$$\phi_4 = \tfrac{1}{2}\psi_s - \frac{1}{\sqrt{2}}\psi_y - \frac{1}{\sqrt{6}}\psi_x - \frac{1}{2\sqrt{3}}\psi_z. \qquad (6)$$

The orbital ϕ_1 is assigned to the lone-pair electrons while we let ϕ_2 correspond to the N—H bond with population $\equiv b_2$ and ϕ_3 and ϕ_4 to the two N—C bonds with population $\equiv b_1$. The Z axis lies along the lone-pair orbital and the X axis lies in the molecular symmetry plane. We have used capital letters for subscripts in order to distinguish our arbitrary axis system from the field-gradient axes for which we use lower-case symbols.

The first success of the Townes and Dailey theory as regards the orientation of the axes is that we have seen from the results for ammonia and trimethylamine that $b_2 > b_1$. Simple manipulation of the above orbitals and their populations shows that the Townes and Dailey theory indeed then predicts that the field-gradient X-axis should lie in the molecular symmetry plane if $b_2 > b_1$. (Note that the labelling of the p-orbitals is arbitrary and has been chosen with the foreknowledge that the field-gradient axes should, if the Townes and Dailey theory predicts correctly, lie close to the chosen axes).

In order to proceed further we must express our field-gradients and p-orbitals in the same axis system. Which system we choose is in principle immaterial but it is easier to transform the field-gradients to the p-orbital axes since we can then use directly equations (4) and (5). We have however one extra piece of information, namely the off-diagonal element of the field-gradient tensor in the XYZ system. These elements are expectation values of the off-diagonal operators:

$$\mathscr{H}_{xz} = -\int \frac{3\sin\theta\cos\theta\cos\phi}{r^3}\,\mathrm{d}\tau \tag{7}$$

$$\mathscr{H}_{yz} = -\int \frac{3\sin\theta\cos\theta\sin\phi}{r^3}\,\mathrm{d}\tau \tag{8}$$

$$\mathscr{H}_{xy} = -\int \frac{3\sin^2\theta\sin\phi\cos\phi}{r^3}\,\mathrm{d}\tau. \tag{9}$$

For p-orbitals it is easy to show that the following relationships hold:

$$\int p_i \,.\, H_{ij} \,.\, p_i\,\mathrm{d}\tau = 0 \tag{10}$$

$$\int p_i \,.\, H_{ij} \,.\, p_k\,\mathrm{d}\tau = 0 \tag{11}$$

$$\int p_i \,.\, H_{ij} \,.\, p_i\,\mathrm{d}\tau = \frac{3}{4}\int p_i \,.\, H_{ii} \,.\, p_i\,\mathrm{d}\tau \equiv \tfrac{3}{4}\,q_0 \tag{12}$$

where the generalized subscript, i, j, and k represent one or other of the subscripts x, y, z. For the above tetrahedral orbitals the only off-diagonal element is of course q_{xz} and the reader may easily verify that it is given by the equation

$$q_{xz} = \left(\frac{b_1 - b_2}{2\sqrt{2}}\right) q_0. \tag{13}$$

This can be equated to the formal off-diagonal element obtained from the angle between the XYZ and x, y, z axis systems

$$q_{XZ} = (q_{xx} - q_{zz}) \sin\theta \cos\theta \tag{14}$$

The reader will note that the orbital populations are already given by equations (4) and (5) where the field-gradient parameters in the XYZ system are used, i.e.

$$\eta' = \frac{q_{XX} - q_{YY}}{q_{ZZ}}. \tag{15}$$

Equations (13) and (14) above thus allow a check on the adequacy of the Townes and Dailey theory. In the first place we note that since θ is a negative angle the off-diagonal element defined in equation (14) is negative. Comparison of this with equation (13) shows that this implies $b_2 > b_1$, as of course we already know from the results for the symmetrical amines. Thus again the Townes–Dailey theory predicts the correct relative orientations of the axes. However when it comes to a more quantitative estimate of the angle we find that while equation (5) indicates that $(b_2 - b_1) = 0{\cdot}39$, equations (13) and (14) yield 0·24 for this value. Putting it another way, whereas the experimental angle between the Z and the z axes is 7° 28′, equations (13) and (14) yield a value of 12° 33′. Furthermore the population difference between the sigma orbitals in ammonia and in trimethylamine, as deduced from the quadrupole resonance data (Table 1) is only around 0·1. Obviously in such a complex situation as this one just cannot push the Townes and Dailey analysis too far and should be well content that the orientation of the axes is given correctly.

Results similar to the above have been obtained by Pierce *et al.*[3] in an attempt to analyse their data for methyldifluoramine, CH_3NF_2, and Lide's[4] data for difluoramine, HNF_2.

In amides, $RCONH_2$, a variety of studies have shown that the amido group is planar, presumably because of the gain in conjugation energy between the lone-pair electrons of the nitrogen and the carbonyl group which thus results. In this case the sigma orbitals are general *s-p* hybrids given by

$$\phi_1 = \psi_z;$$

$$\phi_2 = \cot^2\gamma \,.\, \psi_s + (1 - \cot^2\gamma)^{\frac{1}{2}}\psi_y$$

$$\phi_3 = \frac{1}{\sqrt{2}}\{(1 - \cot^2\gamma)^{\frac{1}{2}}\psi_s - \cot\gamma\psi_y + \psi_x\}$$

$$\phi_4 = \frac{1}{\sqrt{2}}\{(1 - \cot^2\gamma)^{\frac{1}{2}}\psi_s - \cot\gamma\psi_y - \psi_x\} \tag{16}$$

where 2γ is the $\widehat{HNH}$ bond angle. The orbital ϕ_1 contains the lone-pair of electrons but to which we can no longer give the population 2 because of the conjugation, while the orbital ϕ_2 corresponds to the NC bond and ϕ_3 and ϕ_4 to the two NH bonds. Then if the populations of orbitals ϕ_1, ϕ_2 and $\phi_3(=\phi_4)$ are a, b and c respectively it may readily be shown that the coupling constants are given by

$$\frac{e^2Qq}{e^2Qq_0}\left(1-\frac{\eta}{3}\right) = (a-c) \tag{17}$$

$$\frac{e^2Qq}{e^2Qq_0}\,.\,\eta = \tfrac{3}{2}(c-b)(1-\cot^2\gamma). \tag{18}$$

These equations assume that the z axis is perpendicular to the molecular plane and this assumption has been checked experimentally for thiourea.[5] It is also assumed that the Y-axis coincides with the CN bond; this is not exactly so for thiourea, but of course in amides the two N—H bonds are not strictly equivalent.

The $\widehat{HNH}$ bond angle in urea is 122·1 ± 0·8°, thus being very close to 120°. If 2γ is assumed to be 120° in both urea and thiourea and the NH bond population assumed to be the same as in ammonia (1·33) then the coupling constant data can be analysed to yield the values of the other two parameters a and b. The results are shown in Table 4. The similarity of the values of the

TABLE 11.4. N^{14} coupling constants (Mc/s) in amides. e^2Qq is assumed equal to −9·0 Mc/s and the HNH angle equal to 120°

	e^2Qq	η	a	b	c	Ref.
$CO(NH_2)_2$; urea	3·507	0·323	1·68	1·21	(1·33)	31
$CS(NH_2)_2$; thiourea	{3·1216	0·3954	1·63	1·20	(1·33)	5
	{3·0996	0·3930				

N—C bond population to those obtained in amines is most satisfactory. The main difference between the two compounds resides in the different values of the N^{14} pi-electron population. The relative values are in the sense which most people would have anticipated—thus the barrier to rotation about the C—N bond is higher in thioamides than in amides[6]—although the difference between them is perhaps rather small. These differences may be described in resonance terms to be a greater contribution to the ionic structure (II) in thioamides than in amides.

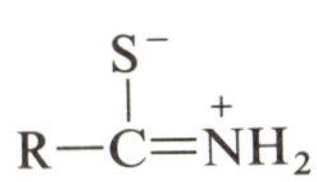

(II)

In aromatic amines, typified by aniline, there is the problem of the configuration of the amino group. Conjugation between the amino group and the benzene ring would tend to produce a planar configuration as in amides. However, Coulson[7] has advanced theoretical reasons for supposing that in aniline the molecule would be non-planar. A complete determination of the directions of the field-gradient tensor in the molecular framework would of course provide an answer to this question, since for the planar configuration the z-axis would be perpendicular to the benezene ring. This again has not been attempted. Table 5 shows results for various anilines. Remembering that for the planar configuration with $2\gamma = 120°$, the product $e^2Qq \, . \, \eta/e^2Qq_0$ is equal to $\frac{1}{2}(\Delta\sigma)$ (where $\Delta\sigma$ is the difference between the NC and NH populations) while for the tetrahedral configuration it is $\Delta\sigma$, the observed couplings are slightly more in favour of a non-planar configuration.

TABLE 11.5. N^{14} couplings (Mc/s) in anilines; e^2Qq_0 is assumed equal to $-9{\cdot}0$ Mc/s

Compound	e^2Qq	η	$e^2Qq \, . \, \eta/e^2Qq_0$	Ref.
aniline	3·977	0·263	0·117	32
p-bromoaniline	4·135	0·231	0·106	31
p-chloroaniline	4·117	0·243	0·111	31
p-phenylenediamine	3·910	0·264	0·115	31

To complete this section on tricoordinate derivatives of nitrogen there remain the organic nitro compounds, RNO_2, which of course have a planar configuration, and the related nitrate ion, NO_3^-. For the nitro compounds no exact experimental measurements of coupling constants are available, although Moniz and Gutowsky[8] have estimated the values 1·35 Mc/s and 2·6 Mc/s for nitromethane and nitrobenzene respectively from measurements of N^{14} relaxation times in nuclear magnetic resonance, and these low values are certainly in agreement with the expected electronic structure of the nitro group. For the nitrate ion the quadrupole coupling constant of the isolated ion is uncertain owing to the rather large contributions to the field-gradient which can arise from the various point charges in an ionic crystal, nevertheless it is unlikely to be far removed from 1·0 Mc/s[9].

C. TRIPLY BONDED NITROGEN COMPOUNDS

1. Molecular Nitrogen

The hybrid orbitals adapted to triple bond formation by a nitrogen atom have the general form

$$\phi_1 = \psi_x; \qquad \phi_2 = \psi_y$$

$$\phi_3 = \alpha\psi_s + \sqrt{1-\alpha^2}\psi_z; \qquad \phi_4 = \sqrt{1-\alpha^2} \, . \, \psi_s - \alpha\psi_z. \tag{19}$$

For molecular nitrogen if ϕ_4 is the lone-pair orbitals of occupation number 2, molecular symmetry requires that the occupation number of the first three orbitals be each unity and the usual analysis then yields

$$\frac{(e^2Qq)_{N_2}}{e^2Qq_0} = \alpha^2. \tag{20}$$

Pauling's hybridization postulate of maximum bond-forming strength implies that $\alpha^2 = \frac{1}{2}$. Unlike the hybridization of the amines however the value of α is here undetermined by the molecular geometry and indeed a variety of fairly exact calculations of the electronic structure of molecular nitrogen indicate that the lone-pair orbital has more s-character than is implied by the above. Nevertheless, for the purposes of the Townes and Dailey analysis the experimental value of the coupling constant (4·7 Mc/s)[11,12] and the assumption that $\alpha^2 = \frac{1}{2}$ implies that $e^2Qq_0 = -9{\cdot}4$ Mc/s which as stated previously seems to be fairly appropriate. Any further discussion of this point leads us at once back to the problem of the exact calculation of nuclear quadrupole coupling constants which has been discussed in Chapter 8.

2. The Cyanides

For the cyanides the general form of the orbitals given for molecular nitrogen (equation 19) still obtains of course but the populations of the three bonding orbitals ϕ_1, ϕ_2 and ϕ_3 now take on the more general values *a*, *a* and *b* respectively. This yields for the coupling constant

$$\frac{e^2Qq}{e^2Qq_0} = 2\alpha^2 + b(1 - \alpha^2) - a. \tag{21}$$

If $\alpha^2 = \frac{1}{2}$ this reduces to

$$\frac{e^2Qq}{e^2Qq_0} = 1 + \frac{b}{2} - a. \tag{22}$$

Since nitrogen is the more electronegative partner both *a* and *b* are expected to be greater than unity and provided that 2*a* is not less than (1 + *b*) equation (11) implies that, as observed, the coupling constant of, say, acetonitrile is less than that of molecular nitrogen. Equation (22) predicts, as is observed, that the coupling constants of nitriles are negative.

In the light of equations (21) and (22) we may enquire as to the probable effect of the nature of the substituent R on the coupling constant of the nitriles, RCN. Even though a particular substituent may be unable to conjugate with the nitrile group it will have an indirect effect on both *a* and *b*. Depending on whether *b* or *a* is more affected, the quadrupole coupling constant may be either increased or decreased by, say, an electron-withdrawing substituent. The possibility that the extent of *s*-hybridization of the sigma bond, α^2, may also change with the nature of the substituent adds a further complication. Clearly it will be difficult to say what will happen in any particular case and the results for HCN, CH_3CN, $(CH_3)_3C.CN$ and $Cl_3C.CN$ (see Table 6) illustrate the rather random manner in which the coupling constants vary. Thus, for example, although the Cl_3C and $(CH_3)_3C$ groups are respectively electron-withdrawing and electron-releasing with respect to the methyl group, the coupling constants of both trichloroacetonitrile and of pivalonitrile are higher than that of acetonitrile. Likewise the coupling constant of Germyl cyanide is about 20 per cent greater than that of acetonitrile even though the electronegativity of Germanium is much less than that of carbon. One could of course "explain" this last fact by bringing in acceptor *d*-orbitals on Germanium which would lead to a decrease in the pi-electron population—and hence to an increase in e^2Qq_0 by participation of resonance structures such as (III) to the ground state of the molecule.

$$\overset{-}{R}=C=\overset{+}{N}$$

(III)

When a substituent can conjugate with the CN group naturally yet another complication is added. If the substituent has lone-pair electrons conjugation will result in contribution of structures such as (IV) to the ground state of the molecule, i.e. the pi-electron population, *a*, will

$$\overset{+}{R}=C=\overset{-}{N}$$

(IV)

increase. This will lead to a decrease in the coupling constant (equation 22) and such a phenomenon is almost certainly responsible for the extremely low coupling in FCN and to a lesser extent of the other halogen cyanides (Table 7). When the substituent has two p_π orbitals of course such conjugation does not lead to a loss of cylindrical symmetry about the C=N bond; however, if only one pi-orbital is present the conjugation should result in the appearance of an asymmetry parameter. The very marked asymmetries of the thiocyanates (Table 7) is perhaps the strongest evidence that such conjugative interactions are occurring.

This point brings us to a further phenomenon exhibited by the nitrile group which contributes yet another complication to the difficulties of rationalizing the quadrupole coupling constants. We have seen that for halogen compounds the asymmetry parameter of singly-bonded halogen nuclei not attached to any conjugative group is essentially zero even when the molecular symmetry does not impose a cylindrically symmetrical field-gradient, as for example in the hexachlorocyclohexanes (see Table 10.5). A

TABLE 11.6. ^{14}N pure quadrupole resonance frequencies in Megacycles/sec of organic nitriles at 77°K

R	Resonance frequencies (Mc/s)	e^2Qq/h (Mc/s)	η	Ref.
H—	3·0223, 3·0052	4·0183	0·0085	33
CH_3—	2·8078, 2·7992	3·738	0·0046	33
CH_3CH_2—	2·8513, 2·8121	3·7756	0·0208	34
$(CH_3)_2CH$—	2.8804, 2·8296	3·8066	0·0266	35
$(CH_3)_3C$—	2·8956, 2·8845	3·8543	0·0058	35
$CH_3CH_2.CH_2$—	2·9255, 2·8322	3·8451	0·0537	35
$CH_3(CH_2)_2CH_2$—	2·9118, 2·8090	3·8138	0·0539	35
$CH_3(CH_2)_3CH_2$—	2·8953, 2·8038	3·7994	0·0482	35
$(CH_3)_2CH.CH_2$—	2·9075, 2·8080	3·8103	0·0522	35
$(CH_3)_2CH.CH_2CH_2$—	2·9118, 2·8332	3·8300	0·0273	35
CH_2Cl—	3·0582, 2·7832	3·8943	0·141	36
$CHCl_2$—	2·9986, 2·9895, 2·9343, 2·9173	3·947*	0·0344*	35
CCl_3—	3·0444, 3·0337	4·0521	0·0053	37
$C_6H_5.CH_2$—	2·9085, 2·8811	3·8597	0·0141	35
$C_6H_5.CH_2.CH_2$—	2·8848, 2·8267	3·8073	0·0305	35
N≡C—	3·225, 3·178	4·269	0·022	26
$NC.CH_2$—	3·0154, 2·8670	3·9216	0·0757	34
$NC.CH_2CH_2$—	2·9957, 2·8447	3·8936	0·0775	35
$NC.(CH_2)_2CH_2$—	2·9231, 2·7995	3·8150	0·0648	35
$NC.(CH_2)_3CH_2$—	2·8615, 2·8190	3·787	0·0224	35
$NC.(CH_2)_4CH_2$—	2·8814, 2·8246	3·8040	0·0298	35
$NC.(CH_2)_7CH_2$—	2·8962, 2·8063	3·8017	0·0473	35
$CH_3O.CH_2$—	3·1122, 2·7552	3·9116	0·1825	35
$CH_3O.CO.CH_2$—	2·9731, 2·8075	3·8537	0·0859	35
CH_2=CH—	2·9063, 2·8085, 2·8852, 2·8015	3·800*	0·0477*	35
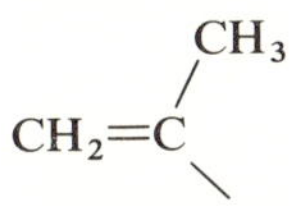	2·9157, 2·8303	3·8307	0·0466	35
$NC.C(CH_3)_2.N$=$N.C(CH_3)_2$—	3·0146, 2·9468	3·9743	0·0341	35
$NC.CH_2.NH.CH_2$—	2·9328, 2·745	3·7852	0·0922	35
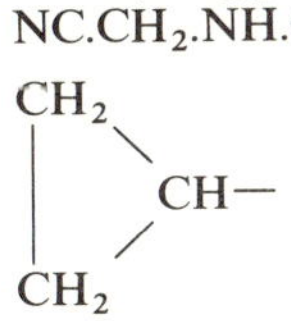	2·9297, 2·6211	3·7005	0·1667	35
C_6H_5—	3·0183, 2·8098	3·8854	0·1073	35
C_5H_4N-(α pyridyl)	3·0396, 2·8978	3·9583	0·0716	37
C_5H_4N-(β pyridyl)	3·0107, 2·8168	3·8850	0·0988	35
C_5H_4N-(γ pyridyl)	2·9353, 2·9073	3·8951	0·0144	36

* Average values.

TABLE 11.7. ^{14}N quadrupole coupling constants (Mc/s) of triply-bonded nitrogen compounds other than organic intriles

Compound	Formula	e^2Qq_z	η	Method	Temp. (°K)	Ref.
Nitrogen	N_2	4·648	0	NQR	4·2	11
		4·781	0	NQR	4·2	12
Hydrogen cyanide	HC≡N	−4·58	0	Micro	—	38
Hydrogen cyanide	HC≡N	4·0183	0·0085	NQR	77	33
Cyanogen fluoride	FCN	−2·67	0	Micro	—	
Cyanogen chloride	ClCN	−3·63	—	Micro	—	
Cyanogen chloride	ClCN	3·219	0·0157	NQR	77	39
Cyanogen bromide	BrCN	−3·83 (−3·85)	—	Micro	—	
Cyanogen bromide	BrCN	3·35	0·0056	NQR	77	27
Cyanogen iodide	ICN	−3·40	—	Micro	—	
Cyanogen iodide	ICN	3·80	0·000	NQR	273	
Silyl cyanide	H_3SiCN	−4·7	0	Micro	—	40
Germyl cyanide	H_3GeCN	−5·0 ± 0·1	0	Micro	—	41
Cyanoacetylene	HC≡C.CN	−4·28	—	Micro	—	
Methylcyanoacetylene	H_3C.C≡C.CN	−4·4	0	Micro	—	42
Cyanogen	NC.CN	4·269	0·0022	NQR	77	26
Sulphurdicyanide	$S(CN)_2$	−3·41	0·3	Micro	—	43
Methylisocyanide	CH_3.NC	10·51	0	Micro	—	
Thionitrosyl fluoride	F_3SN	+1·19	0	Micro	—	13
Methylthiocyanate	CH_3SCN	35154	0·4732	NQR	77	44
Ethylthiocyanate	CH_3CH_2.SCN	35903	0·4735	NQR	77	44
Ethylene dithiocyanate	NCS.CH_2CH_2.SCN	35240	0·4706	NQR	77	44
Potassium thiocyanate	KSCN	2431	0·0281	NQR	78	45
Potassium selenocyanate	KSeCN	2·8449	0·0497	NQR	78	45
Potassium azide	KN_3	1·79*	0·04	NMR	Room	46
		1·028†	0·03	NMR	Room	46

* Terminal nitrogen atoms; † Central nitogen atom.

quite different situation prevails for the nitrile group. In the gas phase compounds such as acetonitrile have a molecular symmetry which ensures cylindrically symmetric field-gradient. The asymmetry parameters of such molecules measured in the solid state are small (see Table 10.6), never more than 0·01. For compounds which do not possess such symmetry, such as propionitrile or monochloroacetonitrile, the asymmetry parameters are often quite considerable, as much as 0·1825 for methoxyacetonitrile. This may be due either to hyperconjugation or, when the substituent possesses lone-pair electrons, to intramolecular partial donor-acceptor bonds (V). Solid state intermolecular effects may of course also be partially responsible

(V)

for these large asymmetry parameters although the low asymmetry parameters of the symmetrical compounds, even such potentially strongly interacting ones as trichloroacetonitrile, is perhaps some evidence against this. However microwave measurements on some of these compounds would be most welcome. Some of these interactions could no doubt result in non-linear R—C—N bonds.

The above difficulties make it rather hazardous to discuss the asymmetry parameters in conjugated nitriles, such as benzonitrile, and indeed oddly enough few data are at present available for this series of compounds. A systematic study of *meta* and *para* substituted benzonitriles would certainly enable one to discover if they were behaving in a reasonable way but such a study has not yet been reported.

Although the situation at present regarding organic nitriles is somewhat confused the obviously very considerable polarizability of the nitrile group makes it very suitable for the detection of very subtle electronic effects. When sufficient data have been collected, it is very likely that the situation will become much more clear and the coupling constants may provide direct evidence of some of the interactions often postulated by the chemists who study reaction mechanisms.

3. The Isocyanides

The only representative of this class whose coupling constant has been measured is methyl isocyanide and its low value shows that the electronic structure of this compound is well represented by the formula

$$CH_3{-}\overset{+}{N}{\equiv}C^-$$

(VI)

4. The Azide Ion

The coupling constants of this ion have been already discussed briefly in Chapter 6. It is however interesting to note that if the sigma orbitals of the terminal nitrogen atoms are supposed to be given by the orbitals ϕ_3 and ϕ_4 of equation (8) where $\alpha^2 = 0{\cdot}5$ (the molecular symmetry ensures this for the sigma orbitals of the central atom) then Hückel molecular orbital theory predicts that the coupling constants of all three nitrogen atoms are zero. The low coupling constants observed now reflect mainly the terms neglected in the Townes–Dailey theory and particularly interesting in this respect would be a measurement of the *signs* of the coupling constants.

5. Thionitrosyl Fluoride

The microwave studies of Kirchhoff and Bright-Wilson[13] have established that the geometrical structure of this molecule is such that the sulphur atom is bonded to the three fluorine atoms and the nitrogen atom. The molecule has C_{3v} symmetry and of course the sulphur-nitrogen bond lies along the three-fold axis. Two extreme electronic structures can therefore be written for this molecule; (VII) and (VIII).

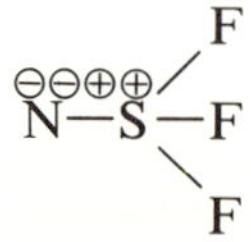
(VII)

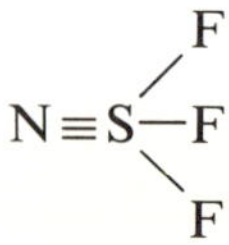
(VIII)

The purely covalent structure (VIII) naturally involves the outer 3*d* orbitals of the sulphur atom. This microwave investigation also established the value of the ^{14}N coupling constant as $+1{\cdot}19$ Mc/s and it is interesting to enquire whether we can decide on the basis of this evidence between the two extreme structures (VII) and (VIII), or more exactly what are the relative contributions of each.

In formula (VII) the occupation number of the nitrogen $2p_\pi$ orbitals is by definition 2, the sigma bond population is unspecified but we might expect it to be less than 1 owing to the large formal positive charge on the sulphur atom. Equation (10) then yields

$$\frac{e^2Qq}{e^2Qq_0} = 2\alpha^2 + b(1 - \alpha^2) - 2. \tag{23}$$

If $\alpha^2 = 0{\cdot}5$ and $b = 1{\cdot}0$ this predicts a positive coupling constant of about $+4{\cdot}5$Mc/s. If b is less than unity the predicted coupling constant is even greater than this while as α^2 approaches unity the predicted coupling constant approaches zero. For formula (VIII) we might reasonably expect a coupling constant similar to that of the cyanides, i.e. around $-3{\cdot}0$ Mc/s. The quadrupole coupling evidence seems thus to favour an electronic structure which involves an approximately equal contribution from both configurations.

D. NITROGEN ATOMS IN CYCLIC CONJUGATED SYSTEMS

1. Pyridine and the Azabenzenes

For the benefit of those readers who are not familiar with the trivial names of the azabenzenes the formulae of some of these conjugated planar molecules are set out below.

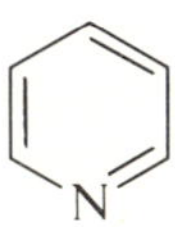

pyridine

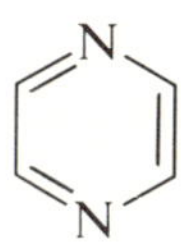

pyrazine

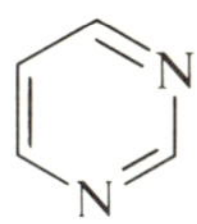

pyrimidine

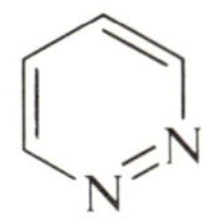

pyridazine

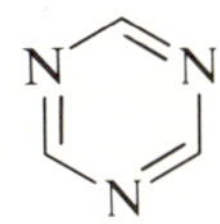

s-triazine

Their N^{14} coupling constants are shown in Table 8.

TABLE 11.8. N^{14} coupling constants (Mc/s at 77°K) of azabenzenes

	e^2Qq (Mc/s)	η	Ref.
Pyridine	4·584	0·396	36
Pyrazine	4·857	0·536	47, 15
Pyrimidine	4·436	0·386	48, 49
Pyridazine	5·188	0·085	48, 49
s-Triazine	4·572	0·443	47, 15

In a manner similar to that employed for the planar amides, an analysis of the coupling constants of compounds such as pyridine is fairly straightforward as long as the molecule has a two-fold symmetry axis going through the nitrogen atom. In this case the molecular symmetry ensures that one field-gradient axis is perpendicular to the molecular plane, a second lies along the two fold symmetry axis and the third is of course perpendicular to both of these. Unlike the amides the lone-pair electrons do not form part of the conjugated system and the population of the non-bonding orbital is thus two. In these circumstances for all reasonable values of the occupation numbers of the other orbitals the field-gradient axis lies in the molecular plane along the two-fold symmetry axis. The orbitals of equation (16) apply equally to the present case but in view of the above we now relabel them to give

$$\phi_1 = \psi_y;$$
$$\phi_2 = \cot\gamma \,.\, \psi_s + (1 - \cot^2\gamma)^{\frac{1}{2}}\psi_z$$
$$\phi_3 = \frac{1}{\sqrt{2}}\{(1 - \cot^2\gamma)^{\frac{1}{2}}\psi_s - \cot\gamma\psi_z + \psi_x\}$$
$$\phi_4 = \frac{1}{\sqrt{2}}\{(1 - \cot^2\gamma)^{\frac{1}{2}}\psi_s - \cot\gamma\psi_z - \psi_x\}. \quad (24)$$

The lone-pair of electrons thus occupy ϕ_2, and 2γ is the $\widehat{CNC}$ angle.

This relabelling also assumes that the field-gradient Y-axis lies perpendicular to the molecular plane and hence the X-axis in the molecular plane. This assignment was originally given on the grounds that it was the only one which made sense of the results for the azabenzenes[14,15] and has recently been confirmed experimentally by microwave measurements on pyridine[16] and pyridazine[17] (Table 9). These measurements also confirm the assumption that the N^{14} coupling constant in the azabenzenes is negative.

TABLE 11.9. Gas-phase N^{14} quadrupole coupling constants of azabenzenes

	e^2Qq (Mc/s)	η	Ref.
Pyridine	$-4{\cdot}88 \pm 0{\cdot}04$	$0{\cdot}405 \pm 0{\cdot}006$	16
Pyridazine	$-5{\cdot}65 \pm 0{\cdot}1$	$0{\cdot}16 \pm 0{\cdot}02$	17

If the populations of the pi(ϕ_1) and bonding sigma (ϕ_3 and ϕ_4) orbitals are set equal to a and b respectively then the Townes and Dailey analysis based on the above orbitals yields

$$(b - a) = \frac{2}{3}\eta \,.\, \frac{e^2Qq}{e^2Qq_0} \quad (25)$$

$$(2 - b) = \frac{1}{1 - \cot^2\gamma}\left(1 - \frac{\eta}{3}\right) \,.\, \frac{e^2Qq}{e^2Qq_0}. \quad (26)$$

Equation (26) is quite sensitive to the angular term $(1 - \cot^2\gamma)$. The bond angles in many of the azabenzenes have been measured experimentally and in all cases they are somewhat less than the regular hexagonal angle of 120° (Table 10). It is a matter of some uncertainty, however, whether in using such a simple theory as the Townes and Dailey theory it is reasonable to introduce such a refinement as an explicit consideration of the angular term. It is

TABLE 11.10. Bond angles in six-membered nitrogen containing heterocycles

Compound	CNC angle	Ref.
Pyridine	117° 34′	50
Pyrazine	115° 8′	51
Tetramethylpyrazine	117° 12′	52
Phenazine	116° 6′	53
s-triazine	113°	54
Pyrimidine	115·2°	55
Pyridazine	120·1°*	56

* Theoretically calculated value for the $\widehat{NNC}$ angle.

quite possible indeed that such a procedure would make the Townes and Dailey theory much less suitable for correlating the data. This point is illustrated for three symmetrical azabenzenes, pyridine, pyrazine and *s*-triazine in Table 11 where the value of e^2Qq_0 was taken as $-8{\cdot}0$ Mc/s in order to achieve a closer numerical agreement with the self-consistent field calculations on the pi-system. When γ is given the constant value of 60° the theoretical pi-electron densities correlate much better with those calculated with 2γ equal to the experimental interbond angle. Furthermore the sigma population should be roughly constant or at least much less variable than a.

TABLE 11.11. Orbital populations in azabenzenes from solid state N^{14} nuclear quadrupole coupling constants. The theoretical S.C.F. values of a are taken from Ref. 57

	θ = experimental value				$\theta = 60°$		
	b_1	b_2	a	a(S.C.F.)	a	b_2	b_1
Pyridine	1·210	—	1·060	1·100	1·100	—	1·250
Pyrazine	1·164	—	0·948	1·080	1·033	—	1·251
s-Triazine	1·155	—	1·022	1·116	1·116	—	1·290
Pyridazine	1·270	0·935	1·045	1·055	1·045	0·935	1·270
Pyrimidine	1·143	—	1·048	1·112	1·132	—	1·275

Again it is only if a constant value of γ is employed that such an expectation is realized. In addition to the above, in many of the cases we will be discussing the interbond angle is unknown so in general it seems to be a much more satisfactory procedure to use the constant value 60° throughout.

Equations (25) and (26) above apply, as already stated, to molecules such as pyridine where the nitrogen atom lies on a two-fold symmetry axis. If this is not the case (e.g. pyrimidine, pyridazine) then on the one hand the two sigma orbital populations will no longer be the same and so the electron distribution must be described in terms of three parameters, while on the other hand the Z axis no longer points along the bisector of the $\widehat{CNC}$ bond but lies at an angle θ to it. The molecular symmetry ensures only that one axis—the Y axis—remains perpendicular to the molecular plane. However an experimental measurement of the direction, θ, of the field-gradient Z-axis will provide a third experimental parameter which will enable the three unknown orbital populations to be determined from the data. From the point of view of the Townes–Dailey analysis the orbitals of equation (24) are now no longer in the principal axis system. The best way of dealing with the experimental data is however to calculate the field-gradients along these axes and equations (25) and (26) then become, if b_1 and b_2 are the separate values of the sigma orbital populations,

$$\tfrac{1}{2}\{b_1 + b_2 - a\} = \frac{2}{3}\eta' \frac{e^2Qq_{ZZ}}{e^2Qq_0} \tag{27}$$

$$-\tfrac{1}{2}\{4 - (b_1 + b_2)\} = \frac{1}{1 - \cot^2 \gamma}\left(1 - \frac{\eta'}{3}\right)\frac{e^2Qq_{ZZ}}{e^2Qq_0} \tag{28}$$

where

$$e^2Qq_{ZZ} = e^2Qq_{zz} \cos^2 \theta + e^2Qq_{xx} \sin^2 \theta$$

$$e^2Qq_{XX} = e^2Qq_{zz} \sin^2 \theta + e^2Qq_{xx} \cos^2 \theta$$

$$e^2Qq_{YY} = e^2Qq_{yy}$$

$$\eta' = \frac{e^2Qq_{XX} - e^2Qq_{YY}}{e^2Qq_{ZZ}}.$$

As before (in equations 13, 14, 15) we use upper case symbols to designate orbital axis systems and lower case symbols to indicate principal field-gradient axes. It may be readily shown that the off-diagonal element (given by equation 14) is related to the orbital populations by the equation

$$\frac{b_1 - b_2}{4 - (b_1 + b_2)} = \frac{\tan 2\theta}{\tan 2\gamma}. \tag{29}$$

In only one of the asymmetric azabenzenes, namely pyridazine, has the direction of the field-gradient axis been determined experimentally ($\theta = 9° \pm 1°$, to the adjacent N-atom). There is one difficulty, however, namely that the solid state coupling and the gas phase coupling are, as usual, not identical while pyridine is the only other azabenzene for which gas-phase data are available. One may therefore either compare pyridine and pyridazine in the gas-phase, which is done in Table 12, or assume that θ in the solid

TABLE 11.12. Orbital populations in pyridine and pyridazine from gas-phase coupling constants. The value of e^2Qq_0 was chosen to obtain agreement with the theoretical value of a in pyridine (Ref. 57)

	θ pyridine = 58°, e^2Qq = 8·87 Mc/s			
	b_1	b_2	a	a(S.C.F.)
Pyridine	1·249	—	(1.000)	1·100
Pyridazine	1·301	0·979	1·058	1·045
	θ pyridine = 60°, e^2Qq = 8·5 Mc/s			
	a	b_2	b_1	
Pyridine	(1·100)	—	1·255	
Pyridazine	1·018	0·935	1·271	

state has the same value as in the gas-phase. This has been done in the calculation of b_1 and b_2 shown previously in Table 11. Unlike the case of the unsymmetrical amines we have here just the same number of unknown orbital populations as we have experimental quantities and hence have no independent check on the off-diagonal term. We must therefore expect that the quantitative numerical predictions of equations (27), (28) and (29) leave something to be desired. If we start with its successes it is evident that the orbital population in the N—N bond of pyridazine will be less than that of the N—C bond. This implies, as is observed, that the field-gradient Z-axis should be rotated towards the adjacent nitrogen atom. From the localized-bond point of view, however, the N—N orbital population should be unity—although of course we may legitimately question this—whereas the values in the tables above are all markedly less than this. This difference could be almost entirely surmounted by taking the value of θ to be 8° instead of 9° (for example the sigma orbital populations for pyridazine in Table 11 become $b_1 = 1{\cdot}25$, $b_2 = 0{\cdot}98$) and we must conclude that the Townes and Dailey method in this case is extremely satisfactory. Whether or not this is purely a coincidence however must await the experimental results on other compounds in this series.

As far as the diazabenzenes are concerned pyridazine is the molecule in which the field-gradient axis is most different from the bisector of the internal angle of the hexagon. Since in this compound it is only 9° it is unlikely to be more than one or two degrees in pyrimidine. In this case the principal values of the coupling constants are little different from the components along the

X, Y, Z axes, and we can use equations (27 and 28) to obtain the average value of the sigma orbital population. The values given for pyrimidine in Table 11 have been calculated in this way. It is fair to say that on the whole the Townes and Dailey theory is here performing quite well. The reasons for this are fairly clear: the environment of the nitrogen atom is very similar from one molecule to another, the molecules are rigid and there are no problems due to hydrogen-bonds. The situation is an ideal one for correlating—as distinct from calculating exactly—the N^{14} coupling constants by means of a simple book-keeping theory.

The coupling constant in these molecules is unique in that it allows calculation of separate values of both pi and sigma electron populations. The absolute values of such numbers should not be stressed too greatly but the relative values of each of them in different molecules should be very significant. For this reason the N^{14} coupling constants of, say, a substituted pyridine should be an excellent method for investigating the properties of substituent groups and for distinguishing their inductive properties from their conjugative ones. In order to use equations (25) and (26) the pyridine derivatives should be symmetrically substituted, i.e. mono substituted in position 4 or disubstituted in the 2,6 and 3,5 positions. However, particularly for compounds mono-substituted in position 3, the field-gradient principal axis should be very close to the CNC bisector, i.e. the two sigma orbital populations should be almost identical, so that little error should arise by using equations (25) and (26) for these compounds also. Even in 2-substituted compounds the error should not be too bad if the substituent has not an extremely large inductive effect. Table 13 shows the coupling constants and the resultant sigma and pi populations for various symmetrically substituted pyridines, while Table 14 shows the same thing for unsymmetric derivatives. To those familiar with organic chemistry the results are just what would have been expected. Thus methyl groups in *ortho* or *para* positions increase the pi-electron population at the nitrogen atom without affecting the sigma population much; in the *meta* position however the substituent affects neither the sigma nor pi population. A similar situation prevails for the halogeno substituted derivatives. The acetyl group, however, withdraws pi-electrons from the nitrogen atom in *ortho* or *para* positions and does nothing in the *meta* position. This information is collected together in Table 14 which shows the changes in the pi and sigma populations produced by the various substituents. Where more than one compound has been studied an average value for the substituent is given. This table brings out in a very striking manner the negligible change in pi-population brought about by substituents in the *meta* positions. The anticipated regular alternation of the effect on the sigma populations in the sense $o > m > p$ is also very remarkable.

TABLE 11.13. N^{14} quadrupole coupling constants (Mc/s) and orbital populations in symmetrically substituted pyridines

Compound	Resonance frequencies	Coupling constant	Asymmetry parameter	Pi orbital population (*a*)	Sigma orbital population (*b*)	Ref.
Pyridine	—	4·60	0·396	1·201	1·335	36
4-Methylpyridine	3·6883, 2·9326	4·4140	0·3424	1·245	1·349	36
3,5-dimethylpyridine	3·9513, 3·0089	4·6401	0·4062	1·192	1·331	49
2,4,6-trimethyl-pyridine	3·4448, 2·9353	4·2500	0·2414	1·274	1·349	49
	3·4485, 2·9215					
4-chloropyridine	3·8409, 3·0105	4·562	0·363	1·209·	1·332	49
	3·8297, 3·0043					
3,5-dichloropyridine	3·8974, 3·0474	4·6299	0·3672	1·199	1·324	49
2,6-dichloropyridine	3·3052, 3·0793	4·2563	0·1061	1·285	1·318	49
2,6-dibromopyridine	3·3661, 3·1285	4·3297	0·1190	1·270	1·307	49
2,6-dimethoxy-pyridine	3·0022, 2·7822	3·8563	0·114	1·351	1·383	49
4-cyanopyridine	4·0997, 3·0576	4·7715	0·4368	1·168	1·321	49
4-acetylpyridine	4·0888, 3·0538	4·7617	0·4347	1·169	1·326	58
2,6-diacetylpyridine	4·1333, 3·0887	4·8147	0·4339	1·159	1·314	58

TABLE 11.14. N^{14} quadrupole coupling constants (Mc/s) and orbital populations in asymmetrically substituted pyridines

Compound	Resonance frequencies	Coupling constant	Asymmetry parameter	Pi orbital population (*a*)	Sigma orbital population (*b*)	Ref.
2-methylpyridine	3·7135, 2·9757	4·4594	0·3309	1·232	1·340	49
2,4-dimethylpyridine	3·5492, 2·9325	4·3211	0·2854	1·258	1·349	49
2-chloropyridine	3·6268, 3·0508	4·4517	0·259	1·239	1·323	49
3-chloropyridine	3·9126, 3·0365 3·9054, 3·0286	4·628	0·379	1·198	1·327	49
2-cyanopyridine	3·9476, 3·1298	4·7182	0·3466	1·185	1·305	49
3-cyanopyridine	3·9007, 3·0504	4·6338	0·3668	1·199	1·324	49
2-acetylpyridine	3·9961, 3·0379	4·6894	0·4087	1·184	1·325	58
3-acetylpyridine	3·9000, 3·0379	4·6253	0·3728	1·196	1·326	58

The electronic effects of substituents have been estimated in various ways such as the Hammett sigma parameter described in Chapter 10. Taft has attempted to separate the usual sigma parameter into two terms, one due to inductive effects and the other due to conjugative effects. Each substituent is thus characterized by two parameters σ_I (inductive) and σ_R (conjugative) (Table 15) and the *ortho*, *meta* or *para* Hammett sigma values are given by various combinations of σ_R and σ_I. Figure 2 shows the correlation between σ_I and σ_R for the change in sigma and pi orbital populations, respectively, produced by a substituent in the *ortho* position. The correlation with σ_R and $\Delta\pi$ is excellent, that between the sigma terms is less convincing but the anomalous electron-releasing effect of the methoxy group has already been discussed in Section **B4** of Chapter 10. It is interesting to notice how much more rapidly $\Delta\pi$ varies with σ_R than does $\Delta\sigma$ with σ_I but it is not clear why this should be so.

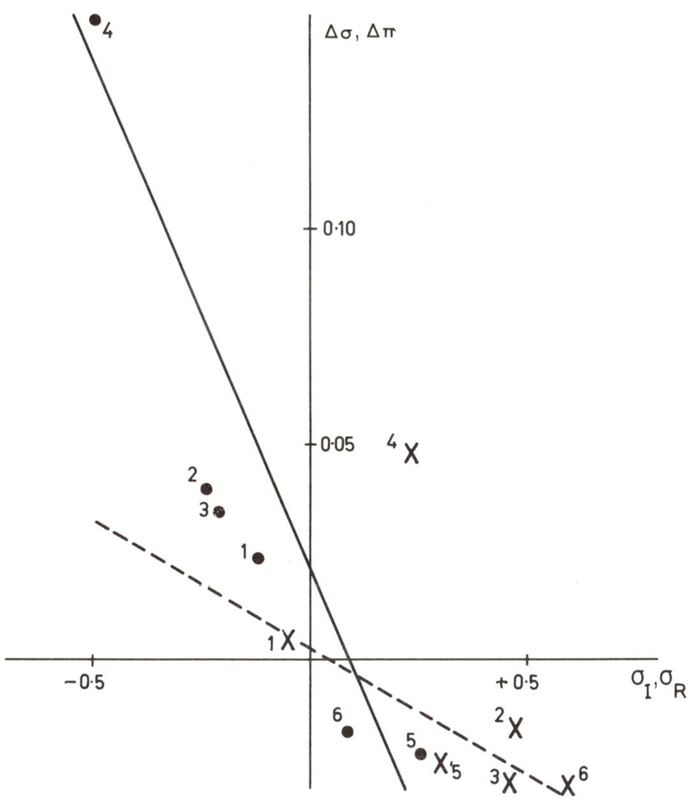

FIG. 11.2. Correlation between the changes in sigma and pi populations at the nitrogen atom in pyridine produced by the substituents listed in Table 2 in the ortho position and the corresponding Taft σ_R and σ_I parameters.

TABLE 11.15. Summary of substituent effects on sigma and pi populations in substituted pyridines

Substituent	Number in Fig. 2	*Ortho* $\Delta\pi$	*Ortho* $\Delta\sigma$	*Meta* $\Delta\pi$	*Meta* $\Delta\sigma$	*Para* $\Delta\pi$	*Para* $\Delta\sigma$
CH_3	1	+0·024	0·0	−0·004	0·0	+0·041	0·0
Cl	2	+0·040	−0·020	−0·002	−0·013	+0·008	−0·003
Br	3	+0·035	−0·028	—	—	—	—
CH_3O	4	+0·150	−0·048	—	—	—	—
CH_3CO	5	−0·019	−0·020	−0·005	−0·009	−0·032	−0·009
CN	6	−0·016	−0·030	−0·001	−0·011	−0·033	−0·014

The conjugative effects in the chloropyridines had also been studied from the point of view of the ^{35}Cl asymmetry parameters in Section **B2** of Chapter 10 where it was seen that the partial double-bond character of the carbon halogen bond was notably higher for 2,6-dichloropyridine than for 3,5-dichloropyridine. This is of course in complete qualitative agreement with the N^{14} coupling constants above. From the quantitative point of view, however, the agreement is not so good. If it is assumed that the difference between the asymmetry parameters in 3,5 and 2,6-dichloropyridine is entirely due to the conjugative effect in the latter compound giving rise to an increased pi-population at the nitrogen atom, then $\Delta\pi$ (nitrogen) may be calculated from Bersohn's formula (equation 2, Chapter 10) as 0·026, whereas the difference in N-atom pi-populations between these two compounds from the N^{14} quadrupole coupling constants is 0·086. Another compound which has been studied by both its Cl^{35} and N^{14} quadrupole coupling constants is cyanuric chloride (2,4,6-trichloro *s*-triazine). Here the ^{35}Cl asymmetry parameter is 0·23 and 0·26 (two crystallographically inequivalent chlorine atoms), approximately 0·13 greater than in the corresponding benzene compound. Again assuming that the difference between the two compounds is entirely due to an increased pi-population on the nitrogen atom we have $\Delta\pi = 0{\cdot}055$. The N^{14} quadrupole coupling data for this compound[18] are $e^2Qq = 4{\cdot}083$ Mc/s, $\eta = 0{\cdot}017$ which yield a sigma and pi population of 1·325 and 1·320 respectively, i.e. $\Delta\pi = 0{\cdot}204$ (cf. Table 11). Again the change in pi-population estimated from the N^{14} coupling constants is much higher than that estimated from the Cl^{35} couplings. These differences could to some extent be surmounted by a lower value for e^2Qq_0. We have furthermore remarked in Chapter 7 that Bersohn's equation underestimates the partial double-bond character of a halogen bond.

It is clear that on the whole the Townes and Dailey analysis is remarkably successful in correlating the N^{14} coupling constants of the azabenzenes. Improvements could possibly be obtained by using different values of e^2Qq_0

for sigma-p-orbitals than for pi-p-orbitals (see Chapter 7) but it would probably be unwise to attempt to do this until an even greater body of data becomes available.

Finally in connection with the N^{14} coupling constants of the azabenzenes complete S.C.F.-calculations on molecules as large as pyridine, including all valence electrons on all atoms, are slowly becoming available.[19] It will be most interesting to see how well these correlate with the N^{14} coupling constants. However to do this the field-gradients must be calculated exactly from the wave-function. It would be very unsatisfactory to try to use such wave-functions in conjunction with the Townes and Dailey method.

2. Five-membered Heterocyclic Systems

The formulae of some of these planar molecules are shown below

pyrrole pyrazole imidazole 1,2,4-triazole

Two types of nitrogen atom will be encountered: a three coordinate nitrogen atom whose coupling constants will be analysed by formulae similar to those used for the planar amides, and a two-coordinate nitrogen atom whose coupling constants will be analysed by formulae analogous to those used for pyridine. The data in this series are sparse and are assembled in Table 16. Only

TABLE 11.16. Pure quadrupole resonance frequencies (Mc/s at 77°K) for N^{14} in five-membered heterocycles (Refs. 59 and 60)

Compound	Frequencies (Mc/s) ν_1	ν_2	Coupling constant (Mc/s) e^2Qq	Asymmetry η
Pyrrole	1·6850	1·4078	2·060	0·269
Imidazole	2·559	2·348	3·271	0·129
	1·417	—	—	—
1,2,4-Triazole	3·8759	2·4690	4·2301	0·665
	2·5510	2·2790	3·220	0·169
	2·1648	—	—	—
Pyrazole	3·6649	2·3653 }	3·9949	0·657
	3·6397	2·3148 }		
	2·1182	—	—	—
	2·0737	—	—	—
Tetrazole	3·3023	—	—	—
	2·9782	—	—	—

for pyrrole have both the resonance frequencies for the tertiary nitrogen atoms been observed, in the other cases only the higher frequency resonance has been detected. In what follows we assume that, as usual, these coupling constants are negative.*

In pyrrole the molecular symmetry ensures that one field-gradient axis is perpendicular to the molecular plane, that another lies along the NH bond and the third, lying in the molecular plane, is perpendicular to these two. Since the p_π orbital of the nitrogen atom contains the lone-pair of electrons, even if these are in fact delocalized over the ring, it is almost certain that the Z-axis lies perpendicular to molecular plane (cf. the amides)*. If the N—C sigma orbital populations are less than the N—H sigma orbital populations then the X-axis will lie along the NH bond. The bonding orbitals of the nitrogen atom in pyrrole are then the same as those given in equation (16) for the amides but with the X and Y axes interchanged. Equations (17) and (18) then become, with the same symbols for the populations of orbitals ϕ_1, ϕ_2, ϕ_3 and ϕ_4,

$$\frac{e^2Qq}{e^2Qq_0}\left(1 + \frac{\eta}{3}\right) = (a - c) \tag{30}$$

$$\frac{e^2Qq}{e^2Qq_0} = \tfrac{3}{2}(b - c)(1 - \cot^2 \gamma). \tag{31}$$

Thus a is the pi population, b is the N—H sigma population and c the N—C sigma populations. 2γ is the $\widehat{CNC}$ angle.

We again have to worry about the angular term in equation (31) which as usual turns out to have a very profound influence. For the reasons previously given it is probably better to maintain a constant value of γ. For a regular pentagon $2\gamma = 180°$, which is very close to the tetrahedral angle 109° 28′. For the latter angle $(1 - \cot^2 \gamma) = 0{\cdot}5$ so for convenience in what follows we allow the angular term to maintain this constant value. Even with this simplification equations (30) and (31) contain three unknowns for only two pieces of experimental information, so that one of the values a, b or c must be given an assumed value. For the amides we assumed that the N—H population was 1·33, thus obtaining values of about 1·20 for the N—C population. This is a little low in comparison with what we have obtained in the pyridine series so in order to give a better comparison with these molecules we adopt a higher value, 1·45, for the NH sigma population. (This assumption in the amides would yield populations each higher by 0·12 units.) Application of

* A recent microwave measurement (B. Bak and L. Nygaard, private communication cited by D. W. Davies and W. C. Mackrodt, *Chem. Comm.* 1226 (1967)) shows that the principal value of the coupling constant (−2·67 Mc/s) lies along the axis perpendicular to the molecular plane. The asymmetry parameter is small (0·09), but contrary to the conclusions above the y-axis lies along the N—H bond.

equations (30) and (31) to the coupling in pyrrole then yields $a = 1{\cdot}62$ ($b = 1{\cdot}45$), $c = 1{\cdot}37$. The value of the pi population a should be compared with the predictions of Hückel theory 1·69[20] or 1·52.[21] The use of the same parameters in the amides yields $a = 1{\cdot}80$ (urea), $a = 1{\cdot}75$ (thiourea). The delocalization of the lone-pair is thus, as expected, greater in pyrrole than in the amides.

In all the other compounds in this series we have the additional problem that the loss of symmetry means that we no longer know the directions of the field-gradient axes, beyond the fact that one of them is perpendicular to the molecular plane. This problem is particularly serious in pyrazole (cf. pyridazine). In addition it is known that these molecules have hydrogen bonds between the NH group and the two coordinated nitrogen atom in the solid state.[22] We must therefore be content with a rather qualitative discussion.

In all these compounds the coupling constant of the tricoordinated nitrogen atom is apparently very low. According to equation (30) this implies that $(a - c)$ be very small (unless η is extremely large) and since c is presumably fairly constant a must be smaller in pyrazole and imidazole than in pyrrole. This would be the expected effect of the additional electron-withdrawing nitrogen atom in these molecules.

Application of equations (25) and (26) to the data for the bicoordinate nitrogen atom with the value 0·5 for $(1 - \cot^2 \gamma)$ yields either $a = 1{\cdot}27$, $b = 1{\cdot}30$ or $a = 1{\cdot}27$, $b = 1{\cdot}24$ for the pi and sigma populations, depending on whether it is assumed that it is the y or the x axis which is perpendicular to the molecular plane. The former choice yields a N—C sigma population close to that obtained in the pyridine series while the pi population, independently of the axis assignment, is noticeably higher than in the six-membered ring compounds. This again is to be expected since in the five-membered hydrocarbon (the cyclopentadienyl anion) which is the homonuclear parent of these compounds, each atom has a pi-population of 1·20 compared with 1·0 for the six-membered ring.

Finally, the most notable difference between the two-coordinated nitrogen atoms in pyrazole and imidazole is the very high asymmetry parameter of the former. Inspection of equation (27) shows that this is a consequence of the decrease in value of the N—N sigma bond population compared with the N—C sigma population. In this compound the X-axis is almost certainly now perpendicular to the molecular plane.

To conclude, Table 17 lists miscellaneous compounds measured in the gas-phase by microwave spectroscopy and for which only the values in the inertial axis system are available. To quote these values without showing the inertial axes in each case would be pointless so the reader is referred to the original publication. In any case there is little point in discussing these results at the Townes and Dailey level unless the principal values of the coupling constants are available.

TABLE 11.17. Compounds for which only the coupling constants in the inertial axis system are available

Compound	Formula	Ref.
Hydrazoic acid	HN_3	61, 62
Nitryl fluoride	ONOF	63
Cyanic acid	HNCO	64
Thiocyanic acid	HNCS	64
Methyl nitrite	CH_3ONO	65
1,2,5-Thiadiazole	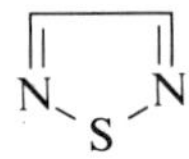	66
1,2,5-Oxadiazole	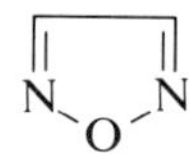	67
Ethylene imine	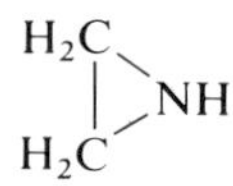	68
Methyl isocyanate	CH_3NCO	69
N-methyl methyleneimine	$CH_3N{=}CH_2$	70
Thionitryl fluoride	N≡S—F	71
Difluoramine	NF_2H	4
Nitrous oxide	N_2O	
Nitric acid	$HONO_2$	72
Methylamine	CH_3NH_2	73
	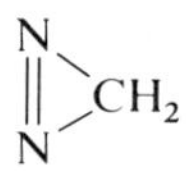	74
Oxazole	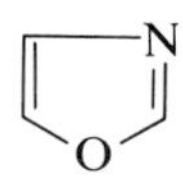	75
Isoazole	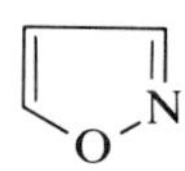	

REFERENCES

1. Y. Abe and S. Kojima, *J. Phys. Soc. Japan* **18**, 1843 (1963); also, Y. Abe, Y. Kamishina and S. Kojima, *J. Phys. Soc. Japan* **21**, 2083 (1966).
2. J. E. Wollrab, Thesis, Stanford University 1964. *Diss. Abs.* **26**, 65–6361 (1965). Reported in B. Starck *Molecular Constants from Microwave Spectroscopy*, Landolt–Börnstein, New Series, Group II, Vol. IV, Springer-Verlag, 1967.
3. L. Pierce, R. G. Hayes and J. F. Beecher, *J. Chem. Phys.* **46**, 4352 (1967).

4. D. R. Lide Jr., *J. Chem. Phys.* **38**, 456 (1963).
5. D. H. Smith and R. M. Cotts, *J. Chem. Phys.* **41**, 2403 (1964).
6. R. C. Neuman, D. N. Roark and V. Jonas, *J. Am. Chem. Soc.* **89**, 3412 (1967).
7. C. A. Coulson, *Valence*, Oxford University Press, 1952.
8. W. B. Moniz and H. S. Gutowsky, *J. Chem. Phys.* **38**, 1155 (1963).
9. M. Gourdji and L. Guibé, *C.r. rebd. Acad. Sci. Paris* **260**, 1131 (1965).
11. T. A. Scott, *J. Chem. Phys.* **36**, 1459 (1962).
12. H. Meyer and T. A. Scott, *J. Phys. Chem. Solids* **11**, 215 (1959).
13. W. H. Kirchhoff and E. Bright Wilson Jr., *J. Am. Chem. Soc.* **84**, 334 (1962).
14. E. A. C. Lucken, *Trans. Faraday Soc.* **57**, 729 (1961).
15. L. Guibé and E. A. C. Lucken, *Mol. Physics* **10**, 273 (1966).
16. G. Sørensen, *J. Mol. Spec.* **22**, 325 (1967).
17. W. Werner, H. Dreizler and H. D. Rudolph, *Z. Naturforschung* **22**a, 531 (1967).
18. S. Kojima and M. Minematsu, *J. Phys. Soc. Japan* **15**, 355 (1960).
19. E. Clementi, *J. Chem. Phys.* **46**, 4731 (1967).
20. C. A. Coulson and H. C. Longuet-Higgins, *Rev. Sci. Paris* **85**, 929 (1947).
21. R. D. Brown, *Australian J. Chem.* **12**, 152 (1959).
22. S. Martinez-Carrera, *Acta Cryst.* **20**, 783 (1966).
23. M. T. Weiss and M. W. P. Strandberg, *Phys. Rev.* **83**, 567 (1951).
24. J. Sheridan and W. Gordy, *Phys. Rev.* **79**, 513 (1950).
25. D. R. Lide and D. E. Mann, *J. Chem. Phys.* **28**, 572 (1958).
26. P. J. Haigh and L. Guibé, *C.r. hebd. Acad. Sci. Paris* **261**, 2328 (1965).
27. G. D. Watkins and R. V. Pound, *Phys. Rev.* **85**, 1062 (1952).
28. S. S. Lehrer and C. T. O'Konski, *J. Chem. Phys.* **43**, 1941 (1965).
29. P. J. Haigh, Private communication.
30. C. T. O'Konski and T. J. Flautt, *J. Chem. Phys.* **27**, 815 (1957).
31. M. Minematsu, *J. Phys. Soc. Japan* **14**, 1030 (1959).
32. L. Guibé, Private communication.
33. H. Negita, P. A. Casabella and P. J. Bray, *J. Chem. Phys.* **32**, 314 (1960).
34. P. A. Casabella and P. J. Bray, *J. Chem. Phys.* **29**, 1105 (1958).
35. A. Colligiani, L. Guibé, P. J. Haigh and E. A. C. Lucken, *Mol. Phys.* **14**, 89 (1968).
36. L. Guibé, *Ann. Phys.* **7**, 177 (1962).
37. H. Negita and P. J. Bray, *J. Chem. Phys.* **33**, 1876 (1960).
38. J. K. Tyler and J. Sheridan, *Trans. Faraday Soc.*, **59**, 2661 (1963).
39. P. A. Casabella and P. J. Bray, *J. Chem. Phys.* **28**, 1182 (1958).
40. J. Sheridan and A. Turner, *Proc. Chem. Soc.* **21** (1960).
41. R. Varma and K. S. Buckton, *J. Chem. Phys.* **46**, 1565 (1967).
42. J. Sheridan and L. F. Thomas, *Nature, Lond.* **174**, 798 (1954).
43. L. Pierce, R. Nelson and C. Thomas, *J. Chem. Phys.* **43**, 3423 (1965).
44. R. Ikeda, D. Nakamura and M. Kubo, *J. Phys. Chem.* **70**, 3626 (1966).
45. R. Ikeda, D. Nakamura and M. Kubo, *Bull. Chem. Soc. Japan* **40**, 701 (1967).
46. R. A. Forman, *J. Chem. Phys.* **45**, 1118 (1966).
47. L. Guibé and E. A. C. Lucken, Proceedings of the XIII Colloque Ampère, Lewven 1964, North Holland Publishing Co. Amsterdam 1965.
48. E. Schempp and P. J. Bray, *J. Chem. Phys.* **46**, 1186 (1967).
49. L. Guibé and E. A. C. Lucken, *Mol. Phys.* **14**, 79 (1968).
50. A. M. Liquori and A. Vaciago, *Ric. Sci.* **26**, 1848 (1956).
51. P. J. Wheatley, *Acta Cryst.* **10**, 182 (1957).
52. Calculated by the author from the data given by D. J. Cromer, A. J. Ihde and H. L. Ritter, *J. Am. Chem. Soc.* **73**, 5587 (1951).

53. F. H. Herbstein and G. M. J. Schmidt, *Acta Cryst.* **8**, 399 (1955).
54. L. Lancaster and B. P. Stoicheff, *Can. J. Phys.* **34**, 1016 (1956).
55. P. J. Wheatley, *Acta Cryst.* **13**, 80 (1960).
56. C. A. Coulson, *J. Chem. Soc.* 5893 (1963).
57. T. E. Peacock, *J. Chem. Soc.* 1946 (1960).
58. P. J. Bray and E. Schempp, Personal Communication.
59. E. Schempp and P. J. Bray, *Phys. Lett.* **25**A, 414 (1967).
60. L. Guibé and E. A. C. Lucken, *Mol. Phys.* **12**, 73 (1968).
61. R. A. Forman and D. R. Lide, *J. Chem. Phys.* **39**, 1133 (1963).
62. R. Kewley, K. V. L. N. Sastry and M. Winnewisser, *J. Mol. Spec.* **12**, 387 (1964).
63. D. F. Smith and D. W. Magnuson, *Phys. Rev.* **87**, 226 (1952).
64. J. N. Shoolery, R. G. Shulman and D. M. Yost, *J. Chem. Phys.* **19**, 250 (1951).
65. D. Stelman, *Diss. Abs.* XXV, 1609 (1963).
66. V. Dobyns and L. Pierce, *J. Am. Chem. Soc.* **85**, 3553 (1963).
67. E. Saegebarth and A. P. Cox, *J. Chem. Phys.* **43**, 166 (1965).
68. W. M. Tolles and W. D. Gwinn, *J. Chem. Phys.* **42**, 2253 (1965).
69. R. F. Curl, V. M. Rao, K. V. L. N. Sastry and J. A. Hodgeson, *J. Chem. Phys.* **39**, 3335 (1963).
70. K. V. L. N. Sastry and R. F. Curl, *J. Chem. Phys.* **41**, 77 (1964).
71. W. H. Kirchhoff and E. Bright Wilson, *J. Am. Chem. Soc.* **85**, 1726 (1963).
72. D. J. Millen and J. R. Morton, *J. Chem. Soc.* 1523 (1960).
73. D. R. Lide, *J. Chem. Phys.* **27**, 343 (1957).
74. L. Pierce and V. Dobyns, *J. Am. Chem. Soc.* **84**, 2651 (1962).
75. W. C. Mackrodt, A. Wardley, P. A. Curnuck, N. L. Owen and J. Sheridan, *Chem. Comm.* 692 (1966).

CHAPTER 12

QUADRUPOLE COUPLING CONSTANTS OF COMPOUNDS OF TYPICAL ELEMENTS OF THE PERIODIC TABLE

A. INTRODUCTION

In this chapter the quadrupole coupling constants for compounds of the typical elements of the periodic table will be surveyed. Most of the coupling constants are from halogen atoms but the data available for other nuclei will also be discussed, with of course the exception of N^{14} which has been covered in the previous chapter. Likewise the material of Chapter 10 will not be repeated here. In Chapter 10, however, we were more concerned with the effect of a substituent on the coupling constant of a halogen atom bonded to a particular multivalent centre, where as here we will be obliged to compare the coupling constants of halogen atoms connected to different centres. As a framework for discussion we will continue to use the Townes and Dailey theory as a basis but we must remember that this will be less reliable here than in Chapter 10 where, since the atom bonded to the halogen remains the same, many of the terms which are neglected in the Townes and Dailey theory are roughly constant. As in Chapter 10 and 12 we will usually use formal *s-p* hybrid orbitals to describe the bond framework and most of the formulae which are required for this purpose have been set out in these two chapters.

B. GROUP IV. CARBON, SILICON, GERMANIUM, TIN AND LEAD

1. Halogen Resonances

Since the results for organic halides have been discussed in Chapter 10, Group IV is an obvious choice for starting this survey. Furthermore in this

Group there arise few of the structural problems which are so prevalent in other groups of the periodic table.

The tetrahalides of these elements all have regular tetrahedral symmetry in the vapour phase and most of them appear to form molecular crystals where the identity of the MX_4 unit is preserved. The halogen pure quadrupole resonance frequencies are set out in Table 1; of course since these molecules do not have dipole moments they cannot be studied by microwave spectroscopy and thus gas-phase data are not available.

TABLE 12.1. The halogen pure quadrupole resonance frequencies (Mc/s at 77°K) for symmetrical tetrahalides of Group IV (References are superior numbers)

	^{35}Cl	^{81}Br	^{127}I $\pm\frac{1}{2} \leftrightarrow \pm\frac{3}{2}$	^{127}I $\pm\frac{3}{2} \leftrightarrow \pm\frac{5}{2}$
C	40·6*[63]	268·0†[75]	319·549[60]	
Si	20·273[63]	147·511[75]	198·736[60]	397·430
	20·408	147·576	199·999	399·993
	20·415			
	20·464			
Ge	25·451[63]	175·549[75]	222·672[60]	445·294
	25·714	175·584	225·088	450·173
	25·736			
	25·746			
Sn	23·719[63]	162·032[75]	207·683[60]	415·320
	24·140	165·208	209.133	418.257
	24·226	165·270		
	24·294	165·401		
Pb	22·68[38]	—	—	—

* Average of 15 resonances; † Average of four closely-spaced groups of resonances.

The electronegativity of carbon is 2·5–2·6 while the electronegativities of silicon, germanium, tin and lead are all about the same, namely 1·8–1·9 (Table 2). In Fig. 1 the sigma orbital population of the halogen atom in these compounds, calculated from equation (22), Chapter 7 with $s = 0$, is plotted as a function of composition (arbitrary units). As expected the sigma orbital populations of the carbon compounds are uniformly much lower than those of the corresponding compounds of the remaining elements and the populations of the halogen atoms for a particular central atom are in the expected order Cl < Br < I. Not in accord with the relative electronegativities, however, is the uniformly much higher sigma orbital population of the silicon compounds. A possible explanation of this behaviour springs at once to the mind of any chemist. This is that the outer 3*d* orbitals of silicon are available for bond-formation with the lone-pair electrons of the halogen atoms. Such

TABLE 12.2. The electronegativity scale (Pauling)

Li 1·0	Be 1·5	B 2·0											C 2·5	N 3·0	O 3·5	F 4·0
Na 0·9	Mg 1·2	Al 1·5											Si 1·8	P 2·1	S 2·5	Cl 3·0
K 0·8	Ca 1·0	Sc 1·3	Ti 1·5	V 1·6	Cr 1·6	Mn 1·5	Fe 1·8	Co 1·8	Ni 1·8	Cu 1·9	Zn 1·6	Ga 1·6	Ge 1·8	As 2·0	Se 2·4	Br 2·8
Rb 0·8	Sr 1·0	Y 1·2	Zr 1·4	Nb 1·6	Mo 1·8	Tc 1·9	Ru 2·2	Rh 2·2	Pd 2·2	Ag 1·9	Cd 1·7	In 1·7	Sn 1·8	Sb 1·9	Te 2·1	I 2·5
Cs 0·7	Ba 0·9	La–Lu 1·1–1·2	Hf 1·3	Ta 1·5	W 1·7	Re 1·9	Os 2·2	Ir 2·2	Pt 2·2	Au 2·4	Hg 1·9	Tl 1·8	Pb 1·8	Bi 1·9	Pa 2·0	At 2·2
Fr 0·7	Ra 0·9	Ac 1·1	Th 1·3	Pa 1·5	U 1·7	Np-No 1·3										

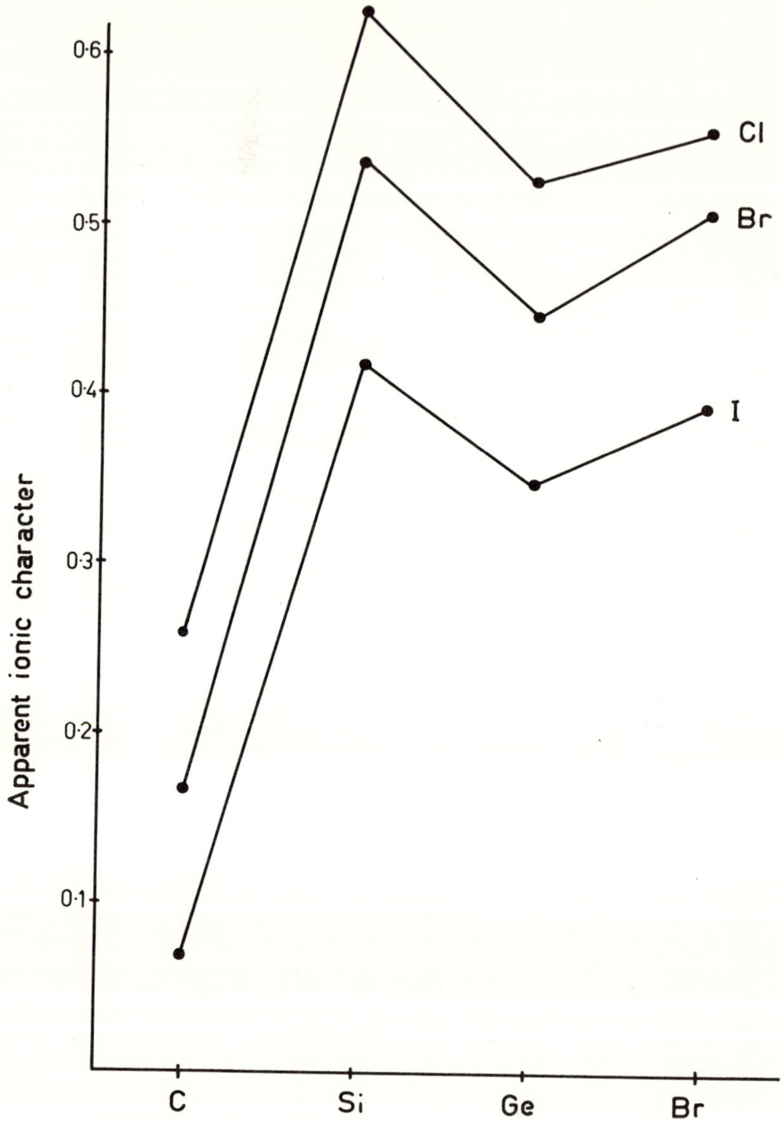

FIG. 12.1. Apparent ionic characters (equation 11, Chapter 7) in the Group IV halides.

π-bonding will naturally lead to a decrease in coupling constant according to the formula

$$\frac{e^2Qq}{e^2Qq_0} = \pi - \sigma \tag{1}$$

where π and σ are respectively the pi and sigma orbital populations of the halogen atom. Much of the chemistry of the elements of the typical groups can be rationalized with the assumption that this double bonding is non-existent for the first short period, is a maximum with the element of the second short period and again vanishes or considerably decreases its importance with the elements of the long periods. It is clear that such behaviour would also explain the nuclear quadrupole coupling constants.

Support for this hypothesis comes from a consideration of the effect of substituents on the halogen quadrupole resonance frequencies of the chlorosilanes, germanes and stannanes described in Section C, Chapter 10. Here it was seen that whereas the frequencies of the carbon germanium and their derivatives were quite strongly affected by the nature of the substituent, the silicon compounds were hardly affected at all. This phenomenon is readily explained with the aid of the back-bonding hypothesis and equation (1). Thus in a compound R—$SiCl_3$, where halogen pi-bonding is supposed to be appreciable, an electron-withdrawing substituent will remove electrons both from the π and σ orbitals leaving the difference between them approximately constant, whereas if the π population is fixed at 2, as in the corresponding carbon compound, the whole of the effect is immediately seen in the quadrupole coupling constant. One can even envisage situations in which the substituent is such a strong pi-electron acceptor that the quadrupole resonance frequency actually decreases with increasing electron-withdrawing power of the substituent.

We may polarize a bond in yet another way, namely by the external application of an electrostatic field (see Chapter 9). If the field lies along the MX bond with the negative end directed towards M the main effect, if the lone-pair π-orbitals of X do not participate in the bonding, is to pull sigma electrons towards M and thus decrease the ionic character of the MX bond and hence *increase* the coupling constant. If, however, the pi-electrons are also mobile the electric field will also increase the transfer of pi-electrons to the atom M and hence, according to equation (1), *decrease* the coupling constant. Thus once again if our basic assumption is correct the silicon halides should be less affected by an electrostatic field than the corresponding carbon compounds. Table 3 gives the relevant experimental data;[1a] the calculated values of the electrostatic field shift were made assuming no M—X multiple bonding. The results are again clearly compatible with the hypothesis of Si—Cl multiple bonding. However it should be stressed that none of these observations distinguish between transfer of the halogen pi-electrons to a metal 3*d* orbital and transfer to some other orbital, say 4*p*. They only show that some sort of transfer probably does occur.

In these symmetrical compounds it is not of course possible to detect partial double bonding by measurements of the asymmetry parameter which, crystal field effects apart, should be identically zero. This is certainly

TABLE 12.3. Electrostatic shifts of quadrupole resonance frequencies (Ref. 1a)

Bond	Shift [c.p.s. $(kV/cm)^{-1}$] calculated	experimental
$C—Cl^{35}$	95	40–70
$C—Br^{81}$	640	610
$C—I^{127}$	1430	930
$Si—Cl^{35}$	130	23
$Sn—Cl^{35}$	200	128

the case for the tetraiodides of silicon, germanium and tin where both resonances have been observed.

2. Central Atom Resonances

Of the central atoms in this group only Germanium (Ge^{73}, $I = -\frac{9}{2}$, abundance 7·61 per cent) has a stable isotope with a quadrupole moment but quadrupole coupling constants are available for the Sn^{119m} nucleus from Mössbauer spectroscopy. The central atoms in the tetrahalides are of course in a cubic environment and thus have no quadrupole coupling constant, but for Ge^{73} a few coupling constants for unsymmetrical compounds are available from microwave spectroscopy. They are listed in Table 4, but the data are so few that there is really very little one can say about them.

TABLE 12.4. Ge^{73} nuclear quadrupole coupling constants (Megacycles/sec) from microwave measurements

Compound	Coupling constant (Mc/s)	Ref.
H_3GeF	-92 ± 3	76
H_3GeCl*	-95 ± 3	77
H_3GeCH_3	3	78
GeS	$187{\cdot}38 \pm 0{\cdot}31$	79

* $(e^2Qq)_{Cl^{35}} = -46{\cdot}95 \pm 0{\cdot}015$ Mc/s.

Mössbauer spectroscopy of the Sn^{119m} nucleus presents few problems and data are abundant. A large number of organic derivatives of tetravalent tin have been studied and a representative list of results for such compounds is shown in Table 5. With the exception on SnF_4, which has an ionic structure in which SnF_6 octahedra share two edges, the tetrahalides and organic compounds of formula SnR_4 have the zero quadrupole splitting expected for a tetrahedral molecule. However, most of the compounds of formula R_3SnX,

TABLE 12.5. Sn^{119m} Mössbauer spectral data for tetravalent tin compounds, measured at 77°K. Isomer shifts are given with respect to SnO_2.

Absorber	Isomer shift (mm sec^{-1})	Quadrupole splitting (mm sec^{-1})	Ref.
$(CH_3)_2SnL_2$			
$(CH_3)_2Sn(OCH_3)_2$	0·990	2·31	80
$(CH_3)_2Sn(CH_3COCH_2CH_2COCH_3)$	1·18	3·93	80
$(CH_3)_2Sn(OCOC_5H_4N)_2$	1·28	4·43	80
$(CH_3)_2SnCl_2$(1,10 phenan)	1·32	4·03	80
$(CH_3)_2Sn(OCOC_6H_5)_2$	1·40	3·96	80
$(CH_3)_2Sn(OCHO)_2$	1·45	4·72	80
$(CH_3)_2SnCl_2$	1·53	3·41	81
$(CH_3)_3SnL$			
$(CH_3)_3SnOH$	1·07	2·71	81
$(CH_3)_3Sn$(imid)	1·16	2·76	81
$(CH_3)_3SnF$	1·26	3·77	80
$(CH_3)_3Sn$(Bz.imid)	1·20	2·89	80
$(CH_3)_3SnN_3$	1·24	3·23	81
$(CH_3)_3Sn$(1,2,4 triazole)	1·27	2·96	80
$(CH_3)_3Sn$(1,2,3 triazole)	1·31	3·18	80
$(CH_3)_3SnOCOCH_3$	1·35	3·52	80
$(CH_3)_3SnCl$	1·40	3·09	81
$(CH_3)_3Sn[OCOC{=}CCO_2](CH_3)_3Sn$	1·40	3·81	80
$(C_2H_5)_2SnL_2$			
$(C_2H_5)_2SnCl_2$	1·6	3·4	82
$(C_2H_5)_3SnL$			
$(C_2H_5)_3SnN(NO_2)CH_3$	1·49	3·54	83
$(C_2H_5)_3Sn[OCOC(CH_3){=}CH_2]$	1·5	3·0	84
$(C_2H_5)_3Sn(OCOCH_3)$	1·6	3·2	82
$(C_3H_7)_2SnL$			
$(C_3H_7)_2SnCl_2$	1·7	3·6	82
$(C_4H_9)_2SnL_2$			
$(C_4H_9)_2SnS$	0·9	1·9	85
$(C_4H_9)_2SnSO_3$	1·3	4·0	85
$(nC_4H_9)_2Sn(OCOCH_3)_2$	1·34	3·50	86
$(nC_4H_9)_2Sn(OCOC_{11}H_{23})_2$	1·34	3·35	86
$(nC_4H_9)_2Sn(OCOC_7H_{15})_2$	1·35	3·45	86
$(nC_4H_9)_2Sn(OCOC_{17}H_{35})_2$	1·36	3·56	86
$(C_4H_9)_2Sn(Cl)OCHO.H_2O$	1·44	3·47	80
$(C_4H_9)_2SnF_2$	1·45	3·9	85
$(C_4H_9)_2SnCl_2$	1·5	3·4	82
$(C_4H_9)_2SnCl_2$	1·6	3·25	85
$(C_4H_9)_2Sn(OCOC_6H_5)_2$	1·62	3.44	80

TABLE 12.5. *continued*

Absorber	Isomer shift (mm sec^{-1})	Quadrupole splitting (mm sec^{-1})	Ref.
$(C_4H_9)_2SnBr_2$	1·7	3·15	85
$(C_4H_9)_2SnI_2$	1·8	2·9	85
$(C_4H_9)_2SnSO_4$	1·8	4·8	85, 87
$(C_4H_{93}SnL$			
$(C_4H_9)_3Sn(OCOCH{=}CHCO_2)$	1·25	3·62	80
$(C_4H_9)_3SnOCOCH_3$	1·38	3·67	80
$(C_6H_5)SnL_3$			
$(C_6H_5)SnCl_3$	2·8	4·8	82
$(C_6H_5)_2SnL_2$			
$(C_6H_5)_2Sn$(8-oxyquin)$_2$	0·637	1·53	80
$(C_6H_5)_2SnOC(C_6H_5)(CH_2)_2{-}C(CH_3)CO$	0·716	2·03	80
$(C_6H_5)_2SnCl_2$	1·31	—	81
$(C_6H_5)_3SnL$			
$(C_6F_5)_3SnCl$	1·01	1·55	88
$(C_6H_5)_3Sn[OCOC(CH_3){=}CH_2]$	1·15	2·1	82
$(C_6H_5)_3SnF$	1·17	3·34	81
$(C_6H_5)_3Sn(C_6F_5)$	1·18	1·10	88
$(C_6H_5)_3SnOH$	1·18	2·68	81
$(C_6H_5)_3Sn$(Bz.Imid)	1·19	2·59	80
$(C_6H_5)_3SnBr$	1·20	2·51	88
$[mCF_3(C_6H_4)]_3SnBr$	1·22	1·94	81
$[pF(C_6H_5)]_3SnI$	1·23	1·92	81
$(C_6H_5)_3Sn$(1,2,4 triazole)	1·29	2·76	80
$(C_6H_5)_3Sn$(1,2,3 Bz.triazole)	1·31	2·98	81
$[pCl(C_6H_4)]_3SnCl$	1·37	2·49	81
$(C_6H_5)_3SnCl$	1·37	2·45	81
$(C_5H_5)_3SnN_3$	1·40	3·19	81
(neophyl)$_3$SnL*			
$[(CH_3)_2C(C_6H_5)CH_2]_3SnOH$	1·13	1·08	81
$[(CH_3)_2C(C_6H_5)CH]_3SnN_3$	1·33	2·48	81
$[(CH_3)_2C(C_6H_5)CH]_3SnF$	1·33	2·79	81
$[(CH_3)_2C(C_6H_5)CH]_3SnOCOCH_3$	1·35	2·45	81
$[(CH_3)_2C(C_6H_5)CH]_3SnCl$	1·39	2·65	81
$[(CH_3)_2C(C_6H_5)CH]_3SnNO_3$	1·40	3·18	81
$[(CH_3)_2C(C_6H_5)CH]_3SnI$	1·41	2·40	81
$[(CH_3)_2C(C_6H_5)CH]_3SnBr$	1·42	2·65	81
$[(CH_3)_2C(C_6H_4)CH]_3SnClO_4$	1·57	3·83	81
Oxides and sulphides			
$[pI(C_6H_4)]_2SnO$	0·84	1·73	86
$[pCl(C_6H_4)]_2SnO$	0·88	1·73	86

TABLE 12.5. *continued*

Absorber	Isomer shift (mm sec^{-1})	Quadrupole splitting (mm sec^{-1})	Ref.
$[pBr(C_6H_4)]_2SnO$	0·92	1·83	86
$[(CH_3)_2SnO]_x$	0·924	1·92	81, 80
$[(C_6H_5)_2SnO]_x$	0·951	1·87	81, 80
$n(C_4H_9)_2SnO$	0·95	2·2	85
$[(C_6H_5)_3Sn]_2O$	1·08	2·15	81
$n(C_5H_{11})_2SnO$	1·1	2·11	86
$[(C_3H_7)_2SnO]_x$	1·1	2·4	82
$[(C_6H_5)_3Sn]_2S$	1·22	1·17	81
$[(nC_4H_9)_2Sn(OCOCH_3)]_2O$	1·30	3·24	86
$[(CH_3)_2SnOCOCH_3]_2O$	1·38	3·57	80
<u>Hydrides</u>			
SnH_4	1·27	0	89
$[(CH_3)_2CH]SnH_3$	1·40	0	89
$(CH_3)_2SnH_2$	1·23	0	89
$(C_4H_9)_2SnH_2$	1·45	0	90
$(CH_3)_3SnH$	1·24	0	89
$(C_6H_5)_3SnH$	1·45	0	90
$[pF(C_6H_4)]_3SnH$	1·37	0	81
<u>Compounds with metal–metal bonds</u>			
$[(C_6H_5)_3Sn]_2$	1·41	0	81
$[(C_2H_5)_3Sn]_2$	1·25	0	91
$[(pC_6H_4F)_3Sn]_2$	1·33	0	81
$[(mC_6H_4F)_3Sn]_2$	1·40	0	81
$[(pC_6H_4Cl)_3Sn]_2$	1·44	0	81
$[(C_6H_5)_3Sn]_4Sn$	1·33	0	90
$[(C_6H_5)_3Sn]_4Ge$	1·13	0	90
$[(C_6H_5)_3Sn]_4Pb$	1·39	0	90
$[(C_6H_5)_3Sn]Li$	1·40	0	90
<u>SnL_4</u>			
SnF_4	−0·4	1·8	92
$SnCl_4$	0·8	0	92
$SnBr_4$	1·1	0	92
SnI_4	1·8	0	92
$(C_6H_5)_4Sn$	1·208	0	81
$(C_6F_5)_4Sn$	1·02	0	88
$(CH_3)_4Sn$	1·22	0	89
$(CH_3CO.O)_4Sn$	0·176	0	83

* Neophyl = $(C_6H_5)C(CH_3)_2.CH_2$—.

where X is an element or group more electronegative than tin, have a very marked quadrupole coupling constant. The structures of relatively few of these compounds are known but $(CH_3)_3SnF$[1b] and $(CH_3)_3SnOH$[2] have been shown by X-ray crystallography to have structures in which the tin atom has a five-coordinate structure in which planar or near-planar $(CH_3)_3Sn^+$ ions are joined together by two anion bridges. Furthermore infrared spectroscopic studies of compounds where X is an oxyanion such as $CH_3CO.O^-$, NO_3^-, ClO_4^- indicate that similar structures prevail here.[3] In these circumstances the marked quadrupole coupling constant is not surprising. The quadrupole moment of the 23·9 keV level of Sn^{119} has been estimated[4] as $-0{\cdot}07$ barns while the value of $\langle 1/r^3 \rangle$ for the $5p$ electron in neutral tin has been calculated as 6·75 a.u.[5] This implies that the quadrupole splitting produced by one valence $5p$ electron is about 2·3 mm/sec, which would be the expected coupling for a planar R_3Sn^+ ion in which the R—Sn bonds are exactly homopolar. Qualitatively at least therefore the coupling constants are of the right order of magnitude but in the absence of detailed structural information it is not possible to say much more. Even if this were available we have seen in Chapter 11 that for polycoordinated nuclei quite small changes in geometry can have profound effects on nuclear quadrupole coupling constants and this of course renders the use of the Townes and Dailey method very hazardous as a basis for the discussion.

There is one compound in Table 5 which most probably retains the regular tetrahedral structure coupling constant, namely $[C_6H_5]_3Sn[C_6F_5]$. Our estimate of $(e^2Qq_0)_{5p,Sn}$ implies therefore that the difference in population between the C_6H_5—Sn and C_6F_5—Sn sigma bonds is around 0·4. This figure is rather too high and suggests in turn that $(e^2Qq_0)_{5p,Sn}$ is too low and indeed a higher value would seem to be more satisfactory for the five-coordinate compounds.

In contrast to the results for the compounds R_3SnX discussed above, the quadrupole splittings of compounds R_3SnM where M is another metal atom, or hydrogen, are below the limit of detection (0·2–0·3 mm/sec). If all these compounds retain a regular tetrahedral configuration the small quadrupole splitting is somewhat surprising in view of the marked splitting observed for $(C_6H_5)_3(C_6F_5)Sn$, and although it is very probable that in all cases the four-coordinated structure is retained considerable distortion may well be taking place. Once again the absence of structural information makes discussion difficult.

Both tin and lead exhibit an extensive chemistry in which they are divalent and essentially ionic. The halogen coupling constants of the divalent halides have not been determined but would presumably be small. The Sn^{119m} quadrupole coupling constants of most of the common Sn(II) compounds have been determined and a representative list is given in Table 6. Surprisingly

TABLE 12.6. Quadrupole splittings (mm/sec) and isomer shifts (mm/sec from SnO_2) for divalent tin compounds. Data from Ref. 93 with isomer shifts recalculated from $SnO_2 = +2{\cdot}1$ mm/sec relative to grey tin

Compound	Isomer shift	Quadrupole splitting
SnO	+2·8	1·5
SnS	+3·5	1·0
SnF_2	+3·7	1·7
$SnCl_2$	+4·5	0·0
$SnCl_2.2H_2O$	+3·7	1·0
$SnBr_2$	+4·2	0·0
$SnSe_4$	+3·9	1·0
SnC_2O_4	+3·8	1·4
$Sn_3(PO_4)_2$	+3·2	1·6

enough the crystal structures of the simple tin halides appear to be unknown, either of the anhydrous or the hydrated compounds, and again detailed comment is impossible. In accordance with their formulation as ionic compounds the quadrupole splitting is on the whole small, and the formal presence of the $5s^2$ electrons is responsible for the markedly larger values of the isomer shifts.

C. GROUP III. BORON, ALUMINIUM, GALLIUM, INDIUM AND THALLIUM

1. Halogen Resonances of Boron Compounds

The halides of Group III of the periodic table are interesting since not only have the coupling constants of the halogen atoms been determined for most of these compounds but often the coupling constant of the central atom is also available. However, despite the common empirical formula, MX_3, these halides, particularly in the solid state, are not iso-structural and a straightforward comparison of the coupling constants is not possible.

The trihalides of boron in the gas-phase have the trigonal planar configuration and form molecular crystals in the solid phase. The crystals of chloride and bromide have been shown by X-ray diffraction[6] to be isomorphous while the quadrupole resonance data suggest that the triiodide has a similar structure. It is usually supposed that the reason for the unique trigonal planar structure for the boron trihalides is the possibility of back-bonding from the halogen atom to the vacant $2p_\pi$ orbital on the boron atom. As will be seen below the remaining halides satisfy the vacant orbital of the metal atom by forming dimers consisting of two MX_3 fragments joined together by a double halogen bridge and it is thought that in these cases the increased metal-halogen bond length makes back-bonding much less efficient.

A direct measure of back-bonding is the halogen asymmetry parameter and although the bromide is liquid melting at -46°C, Chiba[7] has succeeded in carrying out Zeeman studies on single crystals of this compound. For the iodide, the asymmetry parameter is given directly from the two separate resonance frequencies of the polycrystalline solid. The melting point of the chloride is -107°C, thus making the preparation and handling of a single crystal of this compound very difficult. The halogen quadrupole resonance data are summarized in Table 7, the coupling constants for the boron nucleus do not seem to have been determined.

TABLE 12.7. Halogen pure quadrupole resonance frequencies and coupling constants (Mc/s) at 77°K of the boron trihalides (References in parentheses)

$B^{35}Cl_3$ (7)	$B^{79}Br_3$ (7)			$B^{127}I_3$ (94)			
ν	ν	e^2Qq	η	ν_1	ν_2	e^2Qq	η
21·582	175·293	340	0·45 ± 0·05	212·6*	340·1*	1176	0·456
21·578	175·264	—	—	—	—	—	—

* Measured at an unspecified temperature. Each line is a narrow (~25 kc/s) doublet.

The high asymmetry parameters of the bromide and iodide are clearly compatible with the hypothesis of back-bonding. Analysis of the data by equation (2), Chapter 10, yields 13·3 per cent and 15·6 per cent respectively for the partial double bond character of the boron–halogen bond in these compounds while Equation (3), Chapter 10, yields 1·49 and 1/41 for the corresponding sigma-bond populations. These double-bond characters seem at first sight rather small but for the isoelectronic fragment

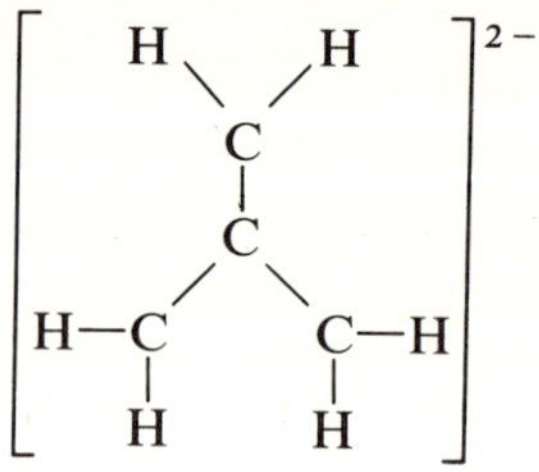

(I)

Hückel molecular orbital theory predicts the value 33·3 per cent and in this light the results for the above compounds are very reasonable. A recent S.C.F.–M.O. calculation yielded values in the range 10–17 per cent, depending on the values given to certain of the parameters used in the calculation.[8]

The interest and significance of these results cannot be too highly stressed. This type of back-bonding has been taken for granted for many years, certainly ever since Pauling first wrote "The Nature of the Chemical Bond" in 1938, but the value of the asymmetry parameter of the nuclear quadrupole coupling tensor is the first direct demonstration that it is indeed taking place.

2. Boron Resonances

It is unfortunate that no data are available for the B^{10} or B^{11} coupling constants in these compounds but P. Love[9] has measured the B^{10} and B^{11} coupling constants of a series of trialkyl boranes which, at least in the gas-phase, have the trigonal planar structure of the trihalides. Although the crystal structures are unknown it is unlikely that they are other than molecular crystals and in accordance with this a near zero asymmetry parameter (0·09 in $(C_2H_5)_3B$) is observed. The most interesting thing about these results (Table 8) is the steady increase in coupling constant in the series $(CH_3)_3B < (CH_3CH_2)_3B < [(CH_3)_2CH]_3B$. Although the effect is small the constancy of the couplings for $(CH_3CH_2)_3B$, $(CH_3CH_2CH_2)_3B$ and $(CH_3.CH_2.CH_2.CH_2)_3B$ encourage one to interpret it as being intramolecular in origin and to associate it with hyperconjugation.

TABLE 12.8. B^{11} pure quadrupole resonance frequencies of trialkyl boranes

Compounds	Resonance frequency (Mc/s at 77°K)
$(CH_3)_3B$	2·438
$(C_2H_5)_3B$	2·502
$({}^iC_3H_7)_3B$	2·625
$({}^nC_3H_7)_3B$	2·500
$({}^nC_7H_9)_3B$	2·500

Several systems containing planar trigonal BR_3 fragments have had their B^{11} quadrupole coupling constants determined (Table 9). In all these compounds the sigma orbital framework is given by the orbitals of equation (16), Chapter 11, where the populations of ϕ_2, ϕ_3 and ϕ_4 are all the same ($\equiv b$) while that of ϕ_1, the pi orbital, contains a electrons. The trigonal symmetry of these molecules ensures that $\gamma = 60°$ and the asymmetry parameter zero. The usual Townes and Dailey analysis leads straightforwardly to the equation:

$$\frac{e^2Qq}{e^2Qq_0} = (a - b). \qquad (2)$$

The experimental value[10] of e^2Qq_0 for B^{11} is $-5{\cdot}39$ Mc/s and with this the values of $|a - b|$ for the trigonal compounds so far discussed have been calculated and shown in Table 9. In all these compounds b is almost certainly greater than a and in the trialkyl boranes b is probably a little less than 1·0 (c.f. Table 2 for the relative electronegativities of boron and carbon). This implies that only few electrons are released into the vacant $2p_z$ orbital of the

TABLE 12.9. Sigma and pi population differences from the B^{11} quadrupole coupling constants of trigonal planar BR_3 molecules.

Compound	e^2Qq	$\|a - b\|$	Ref.
$B(CH_3)_3$	4·876	0·904	9
$B(C_2H_5)_3$	5·003	0·928	9
$B(^iC_3H_7)_3$	5·250	0·974	9
$B(OH)_3$	2·56	0·475	14
$(BN)_n$	2·96	0·549	95
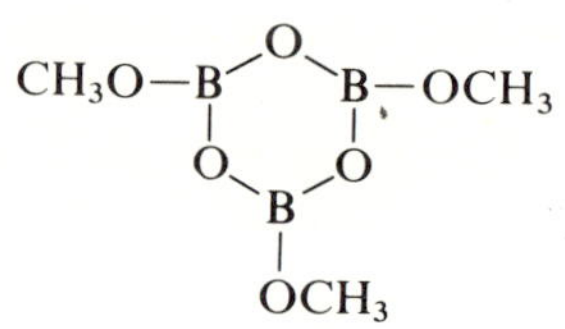	2·14	0·397	96
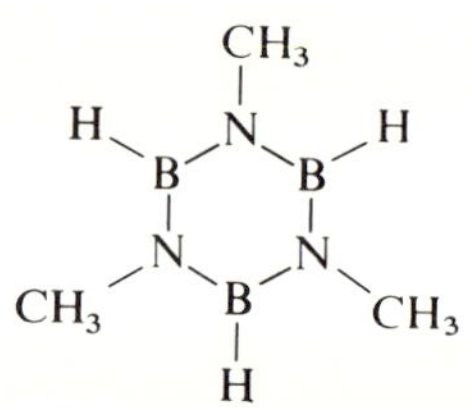	1·71	0·317	96

boron atom by a hyperconjugative mechanism, certainly not more than 0·1 electrons in all. This is in agreement with the relative values for the series $B(CH_3)_3$, $B(C_2H_5)_3$ and $B(i\text{-}C_3H_7)_3$ where the hyperconjugative power of the substituent would be expected to progressively decrease. In orthoboric acid, $B(OH)_3$, however the π-delocalization should be comparable with that observed in the trihalides (i.e. $a \approx 0{\cdot}4$). Since b should here be significantly less than unity this is in good agreement with the value of $|a - b|$ reported in the table. Similar comments apply to boron nitride and the heterocyclic compounds. Bassompierre has calculated that the B^{11} coupling constant in the $[BO_3]^{3-}$ ion should be 2·6 Mc/s[11].

The above analyses imply that the coupling constants of all these compounds are positive but this awaits experimental verification.

Apart from this information relating to electronic structure, B^{11} nuclear quadrupole coupling constants have proved useful in problems of molecular structure. Thus a boron carbide purported to have the empirical formula B_4C and the crystal structure shown[12] in Fig. 2, was shown by Bray[13] to exhibit three separate B^{11} quadrupole coupling constants (~0·2 Mc/s, 1·3 Mc/s and 5·58 Mc/s) instead of the two expected on the basis of the crystal structure. Bray[13] was able to show that the highest coupling constant, 5·58 Mc/s, was due to a B^{11} atom in the centre of the previously supposed C_3 chain at the centre of the unit cell, i.e. this chain is in fact C—B—C and the empirical formula of this carbide should be $B_{13}C_2$.

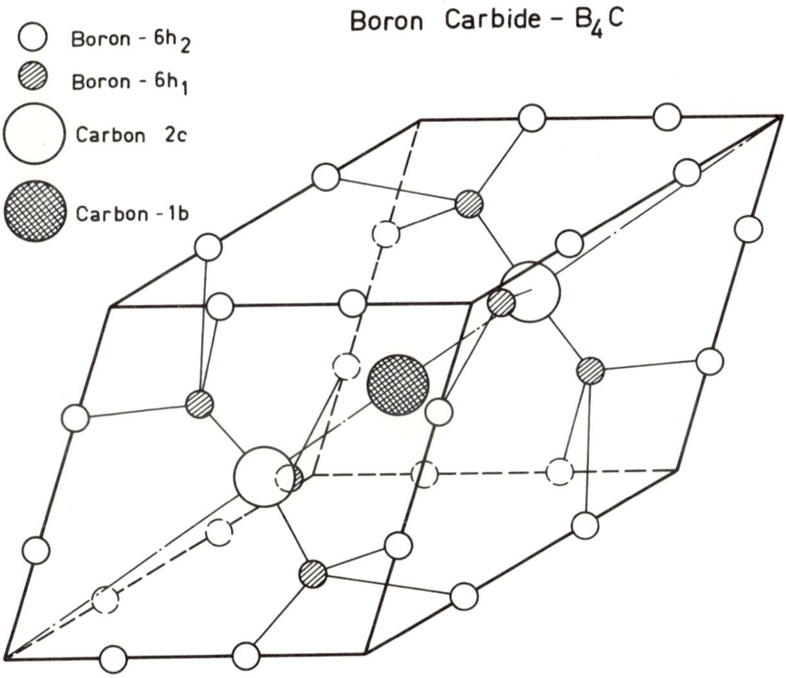

FIG. 12.2. The crystal structure of boron carbide, B_4C.

In an investigation of borate glasses Bray *et al.*[14] have shown that they consist of trigonal (BO_3) groups linked together by random sharing of oxygen atoms, thus confirming the conclusions from a previous X-ray investigation.[15] Bray and co-workers showed that the tetrahedral Borate ion has the expectedly small value (0·3–0·6 Mc/s) of the B^{11} coupling constant and the absence of such couplings in borate glasses shows that a model based on linked tetrahedra must be rejected.

3. Halogen Resonances of Trivalent Al, Ga, In and Tl Compounds

In the vapour phase the trihalides of many of the chlorides, bromides and iodides of Group III, aluminium, gallium, indium and thallium, exhibit a common dimeric structure in which two trihalide molecules are joined by halide bridges[16] (see Fig. 3). The environment of the metal atom is approximately tetrahedral. In the crystal this dimer is known to persist for aluminium

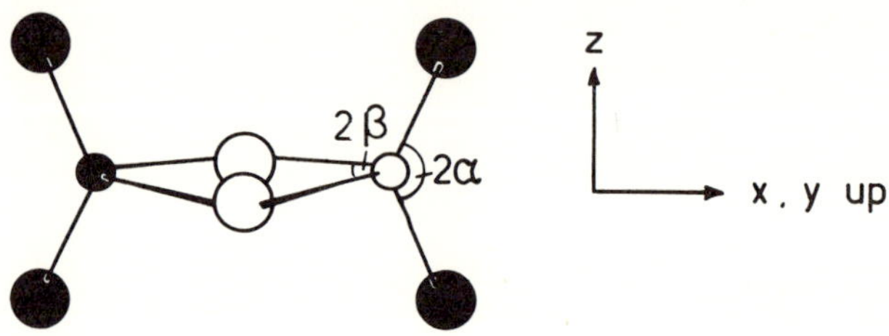

FIG. 12.3. The structure of Al_2Br_6 and the axes to which the orbitals of equation (3) are referred.

bromide while aluminium chloride adopts an ionic layer structure as do also indium and thallium trichlorides.[16] The crystal structures of the remaining compounds are unknown, but as will be seen below the pure quadrupole resonance results cast some light on this problem.

The halogen pure quadrupole resonance data for the chlorides, bromides and iodides of aluminium, gallium, indium and thallium are collected in Table 10. For aluminium bromide, in accordance with the crystal structure, three

TABLE 12.10. Halogen coupling constants (Mc/s at 77°K) of Group III halides (Ref. 97)

	Cl^{35}	Br^{79}	I^{127} ($\pm\frac{1}{2} \leftrightarrow \pm\frac{3}{2}$)	I^{127} ($\pm\frac{3}{2} \leftrightarrow \pm\frac{5}{2}$)
Al	—	97·497	112·314	215·614
		113·790	131·371	263·228
		115·450	131·844	263·920
Ga	14·667*	120·602	—	—
		120·635	135·719	253·347
	19·084	168·375	176·496	352·948
		168·430		
	20·255	168·560†	177·438	354·519
In	—	104·112	122·728‡	229·190‡
		128·110	173·177	346·289
		128·608	173·633	347·234
Tl	14·925	125·7**	—	—
	15·080	125·9	—	—

* At 33°C: † At 24°C: ‡ At 24°C. On going to 77°K a phase change takes place and the resonance frequencies are 176·763, 177·146: 353·466, 354·580: ** At room temperature this last line is also seen to be a narrow doublet.

Br^{79} resonances are observed, two closely spaced at around 114 Mc/s and a third at a decidedly lower frequency. The same pattern is observed in gallium chloride, indium bromide and the iodides of aluminium, gallium and indium. The set of three narrow doublets in gallium bromide also conform to this scheme and the doublet splitting is probably due to chlorine atoms bonded to the different isotopic species, Ga^{69} (60·2 per cent) and Ga^{71} (39·8 per cent). It is therefore extremely probable that all these compounds are isomorphous.

No single-crystal studies have been reported, therefore asymmetry parameters are only known for the iodides. It is almost certain that the bridging halogen atoms are those exhibiting the lowest resonance frequency, and in accordance with this hypothesis the asymmetry parameters corresponding to the high frequency lines are essentially zero, while those of the low frequency lines are considerable (Table 10). For the bridging halogens the hybrid atomic orbitals are given by the expressions

$$\begin{aligned}
\phi_1 &= 3\psi_z \\
\phi_2 &= \cot\gamma \, . \, 3\psi_s + (1 - \cot^2\gamma)^{\frac{1}{2}} 3\psi_y \\
\phi_3 &= \frac{1}{\sqrt{2}}\{(1 - \cot^2\gamma)^{\frac{1}{2}} 3\psi_s - \cot\gamma 3\psi_y + 3\psi_x\} \\
\phi_4 &= \frac{1}{\sqrt{2}}\{(1 - \cot^2\gamma)^{\frac{1}{2}} 3\psi_s - \cot\gamma 2\psi_y - 3\psi_x\}
\end{aligned} \tag{3}$$

where 2γ is the Al$\widehat{\text{X}}$Al angle. The axes to which the atomic orbitals are referred are those given in Fig. 3. Assuming that no pi-bonding effects are present the population of orbitals ϕ_1 and ϕ_2 are each two while for the bridging orbitals ϕ_3 and ϕ_4 we let the population be represented by the symbol a. Then the usual Townes–Dailey analysis leads to the equation

$$\frac{e^2Qq}{e^2Qq_0}\left(1 + \frac{\eta}{3}\right) = 2 - a. \tag{4}$$

This equation is apparently independent of the interbond angle but in fact the asymmetry parameter is entirely determined by γ, according to the relationship

$$\eta = -3\cos 2\gamma. \tag{5}$$

The interorbital angles for three iodides, calculated from equation (5) are shown in Table 11. The measured[17] interbond angle in aluminium tribromide is 82°. It is of course impossible to form sp hybrid orbitals at an angle of less than 90° to each other so that the above analysis can likewise never give a value of less than 90°. Nevertheless it is very satisfactory that the quadrupole resonance data suggest an interorbital angle in the neighbourhood of 90° —as is observed experimentally—rather than in the neighbourhood of either

TABLE 12.11. $\widehat{MIM}$ interorbital angles in Group III triiodides

Compound	Interorbital angle
AlI_3	93° 26′
GaI_3	94° 32′
InI_3	94° 32′

109° or 120° which are other possible alternatives. This result incidentally shows that the participation of the *s*-orbitals of chlorine in the bonding is small, even when the chlorine atom is dicoordinated. When the chlorine atom is singly coordinated the *s*-participation is expected to be even less.

TABLE 12.12. Al^{27} and In^{115} quadrupole coupling constants in $AlBr_3$ and InI_3 measured at room temperature

	e^2Qq (Mc/s)	η	Reference
$AlCl_3$	0·422	0·0	98
$AlBr_3$	13·57	0·747	18
InI_3	320·41	0·690	99

The quadrupole coupling constants for Al^{27} in $AlBr_3$ and $AlCl_3$ and In^{115} in InI_3 are given in Table 12. The low value for Al^{27} in $AlCl_3$ reflects the ionic nature of the crystal. The sp^3 hybrid orbitals for a system having C_{2v} symmetry such as the metal atoms in these complexes are given by

$$\phi_1 = \frac{1}{\sqrt{2}}\{(1 - \cot^2\beta)^{\frac{1}{2}}\psi_s + \cot\beta\,.\,\psi_x + \psi_y\}$$

$$\phi_2 = \frac{1}{\sqrt{2}}\{(1 - \cot^2\beta)^{\frac{1}{2}}\psi_s + \cot\beta\,.\,\psi_x - \psi_y\}$$

$$\phi_3 = \frac{1}{\sqrt{2}}\{(1 - \cot^2\alpha)^{\frac{1}{2}}\psi_s - \cot\alpha\,.\,\psi_x + \psi_z\}$$

$$\phi_4 = \frac{1}{\sqrt{2}}\{(1 - \cot^2\alpha)^{\frac{1}{2}}\psi_s - \cot\alpha\,.\,\psi_x - \psi_z\} \qquad (6)$$

where the axes and angles are those shown in Fig. 3 for the $AlBr_3$ dimer. Note that α and β are not independent but are related by the equation

$$\cot^2\alpha + \cot^2\beta = 1. \qquad (7)$$

It follows from these equations that, as before, the asymmetry parameter is given by the analogue of equation (5), namely

$$\eta = -3 \cos \beta. \tag{8}$$

The asymmetry parameters then yield 104° and 103° for the interorbital angles in $AlBr_3$ and InI_3 respectively. The interbond angle in $AlBr_3$ is 98°.

The above results have been proposed[18] as supporting evidence for "bent-bonds" in these bridged dimers. However it should be clearly realized that if the metal and halogen bonds are considered to be made up only of *s* and *p* atomic orbitals then any departure from a square configuration of the

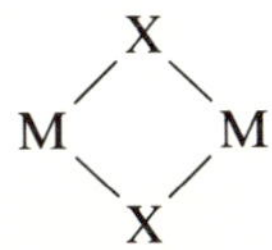

fragment necessarily implies, within the framework of this hypothesis, the existence of "bent-bonds". It is interesting to note however that in all cases where the interbond angles are known the interorbital angles derived above are greater. The above sp^3 orbitals also show that, according to the Townes and Dailey theory, the coupling constant is given by the simple relationship

$$\left|\frac{e^2Qq}{e^2Qq_0}\right|\left(1 + \frac{\eta}{3}\right) = |b - a| \tag{9}$$

where b is the population of the terminal AX sigma orbitals (ϕ_3 and ϕ_4) while a is the population of the bridging orbitals (ϕ_1 and ϕ_2). For Al^{27} the value of e^2Qq_0 is −37·5 Mc/s[19] while for In^{115} it is −899·10 Mc/s.[20] The values of $|b - a|$ thus afforded by equation (9) for $AlBr_3$ and InI_3 are 0·452 and 0·438 respectively.

Table 13 shows a complete analysis for bond populations of the two different types of halogen atom in the three iodides, according to equation (4) for the bridging atoms and according to the usual Townes–Dailey equation for the non-bridging groups. As fairly elementary considerations of resonance structures would suggest the total negative charge on the bridging atoms is

TABLE 12.13. Orbital populations and net charges for the iodine atoms in Group III triiodides

Compound	Orbital populations		Net charge	
	Bridging	Non-bridging	Bridging	Non-bridging
AlI_3	1·666	1·617	−0·333	−0·617
GaI_3	1·598	1·486	−0·196	−0·486
InI_3	1·637	1·496	−0·274	−0·496

less than that on the non-bridging atoms. In accordance with the lower electronegativity of aluminium, gallium and indium compared with boron, the bond-populations, despite the different structures, are all greater than in boron triiodide. Furthermore, as for the halides of the analogous Group IV atoms the apparent sigma bond populations have the order B <<< Al >> Ga < In. This is of course exactly the behaviour observed for the corresponding halides of Group IV and the reason for it is perhaps the same, namely back-bonding by the p-π electrons of the halogen atoms.

For InI_3 the In^{115} coupling constants analysed above permit a check on the self-consistency of the populations calculated in the various orbitals. According to the I^{127} results shown in Table 13 $|b - a|$ in equation (9) should be 0·14, considerably less than the 0·44 obtained from the In^{115} data. This is yet another example of the difficulties inherent in using the Townes–Dailey method on nuclei of high coordination number.

Table 14 shows the halogen quadrupole coupling constants of a few miscellaneous trivalent compounds of Group IV. The $(AlCl_4)^-$ anion is isoelectronic with silicon tetrachloride. For this latter compound the apparent chlorine sigma orbital population is 1·628. The average resonance frequency of $(AlCl_4)^-$ is 10·61 Mc/s, yielding an apparent chlorine sigma orbital population of 1·807 and implying that the negative charge in this compound resides mostly on the halogen atoms.

TABLE 12.14. Chlorine pure quadrupole resonance frequencies (Mc/s at 77°K) for miscellaneous Group III compounds

Compound	Resonance frequency	Ref.
$B_3N_3Cl_3$ (NN′N″ trichloroborazole)	19·937	100
$AlCl_3.O(C_6H_5)_2$	11·30*	101
$Ga[AlCl_4]$	10·296†	102
	10·382	
	10·460	
	11·312	

* Average of two resonances; † At 24°C.

4. Diatomic Halides

There remains to discuss in this section the interesting results which have been obtained from the microwave spectroscopy of the gaseous diatomic halides of Group III. Many of these molecules are either unstable in the solid phase or form ionic-type lattices, e.g. thallium chloride, where the symmetry of the environment of the halogen atom precludes any possibilities of a quadrupole coupling constant, but have been made accessible to study by the techniques of high-temperature microwave spectroscopy. The interest of these results is considerably increased by the fact that with the exception of thallium, the metal nuclei also possess quadrupole moments and produce hyperfine structure in the microwave spectrum. Thus data are available for both ends of the molecule and the results are summarized in Table 15.

TABLE 12.15. Quadrupole coupling constants (Megacycles/sec) of halogen (^{35}Cl, ^{79}Br, ^{127}I) and metal nuclei in gaseous diatomic halides of Group III

		Quadrupole Couplings (Mc/s) Halogen	Metal	Ref.
Al^{27}	F	—	$-37{\cdot}6$	21b
Al^{27}	Cl^{35}	$-8{\cdot}8$	$-29{\cdot}2$	21b
Ga^{69}	Cl^{35}	-20 ± 2	$-84.7 \pm 1{\cdot}0$	21a
Ga^{69}	Br^{79}	-134 ± 3	-74 ± 5	21a
Ga^{69}	I^{127}	-549 ± 31	-66 ± 8	21a
In^{115}	Cl^{35}	-18 ± 2	-655 ± 6	21a
In^{115}	Br^{79}	138 ± 11	-642 ± 12	21a
Tl	Cl^{35}	$-15{\cdot}8 \pm 0{\cdot}5$	—	21a
Tl	Br^{79}	$13{\cdot}0 \pm 10$	—	21a
Tl	I^{127}	-537 ± 25	—	21a

As far as the halogen coupling constants are concerned their universally rather low values imply, at least at first sight, that the bonding is highly ionic in all these compounds. The details of the bonding however demand a rather closer inspection than this. The electronic configuration of the valence shell of all these metals is ns^2np, ^2P and in consequence there arises the strong possibility of back-donation of the lone-pair electrons of the halogen atoms to the vacant p_π valency orbitals of the metal. This of course reduces the coupling constant and increases the apparent ionic character and if we consider the halogen couplings in the series AlCl, GaCl, InCl, TlCl there is a clear indication that pi-bonding is taking place. The ^{35}Cl coupling constants of the chlorides of gallium, indium and thallium are roughly what would be expected from the electronegativity difference between the metal and halogen atoms but the coupling constant of AlCl is much too low. However, about 10 per cent triple bond character for the AlCl bond would yield a much more reasonable value for its ionic character.

The results for the quadrupole coupling constants of the metal atoms are in many ways even more interesting than those of the halogens. The coupling constants of a valency p-electron have been determined by atomic beam methods for the ^{27}Al ($-37{\cdot}5$ Mc/s), ^{69}Ga (-125 Mc/s) and ^{115}In (-899 Mc/s) nuclei and it is clear that the molecular coupling constants are a significant fraction of these values. Even allowing for pi-bonding effects the halogen resonances indicate a considerable ionic character for the metal halogen bond; this would at first sight tend to imply that the metal coupling constants would likewise be small. The explanation for the high coupling constants of the metal atoms cannot lie in pi-bonding effects since these would also decrease the coupling constant and the only reasonable explanation seems to lie in assuming a considerable extent of *s*-character in the metal atom sigma bond orbital. Thus formally speaking if the bonding sigma orbital—and hence also the lone-pair orbital—were exactly *sp* hybrid, but the bond entirely ionic, the metal coupling would be exactly that of the free atom. By assuming that the ionic character could be deduced entirely from the halogen coupling constant, Barrett and Mandel[21a] and, following them, Lide[21b] have calculated the characteristics of those metal-halogen bonds as set out in Table 16 below. The halogen sigma orbitals were supposed to be 15 per cent hybridized and no account was taken of possible pi-bonding. The precise numerical values should certainly not be taken too seriously but there is no doubt that considerable hybridization of the metal orbitals does occur.

TABLE 12.16. Characteristics of bonds in Group III diatomic halides

Compound	Ionic character	*s*-hybridization of metal atom
AlCl	0·91	0·27
GaCl	0·79	0·21
GaBr	0·80	0·18
GaI	0·72	0·13
InCl	0·81	0·24
InBr	0·79	0·23

To the present author at least it seems a little odd to talk about the hybridization of an atomic orbital which, being situated at the positive end of a highly ionic bond, is largely unoccupied. It seems much more reasonable to think of the phenomenon as being simply the distortion of the lone-pair electrons. The result of course is the same but it is now clear that it is nothing to do with the requirements of the bonding electrons but rather with the behaviour of the s^2 inert pair under the influence of the neighbouring negative charge. The probable extent of this distortion may be estimated quite simply as follows. Under the influence of a potential gradient, dV/dr, the distorted *s*-orbital, s^*, may be expressed in the way familiar to perturbation theory

$$s^* = as_0 + \sum_i b_i p_i \tag{10}$$

where s_0 is the unperturbed atomic *s* orbital and p_i the various atomic *p*-orbitals of higher energy. It is clear that the most important term in the expansion in the present case will be just that due to the valence *p*-orbital and if we truncate the expansion at this point we may calculate the coefficients from the standard expression

$$\frac{b}{a} = \frac{\int s_0 | r(\mathrm{d}V/\mathrm{d}r) | p \, \mathrm{d}\tau}{\Delta E} \tag{11}$$

where ΔE is the energy difference between the valency state *s*- and *p*-orbitals. If these orbitals are expressed in terms of the product of the (same) radial function multiplied by the appropriate spherical harmonic angular function then it is easily shown that in a *uniform* potential gradient

$$\frac{b}{a} = \frac{r \, \mathrm{d}E/\mathrm{d}r}{\Delta E}. \tag{12}$$

The energy difference between the *s* and *p* levels can be estimated from the experimentally observed energy difference between the atomic ground state ns^2np, ^2P and the excited state $nsnp^2$, ^2P which for all these metal atoms is around 30,000–40,000 cm^{-1} or 4–5 eV and since the potential energy due to a unit charge at a distance of 2·1 Å, a typical metal-halogen bond-length in these compounds (Table 17), is 6·8 eV it is quite clear that by a purely electrostatic effect alone the *s*-orbital will be distorted to a complete *s-p* hybrid. This explanation also predicts that the metal coupling constant will increase as the metal-halogen bond-length decreases and reference to Tables 15 and 17 shows that this trend is indeed observed.

TABLE 12.17. Bond-lengths in Group III diatomic halides

Molecule	Bond-length (Å)
AlF	1·65437 ± 0·00001
AlCl	2·12983 ± 0·00001
GaCl	2·2017 ± 0·0001
GaBr	2·3525 ± 0·0001
GaI	2·5747 ± 0·0001
InCl	2·4011 ± 0·0001
InBr	2·5432 ± 0·0001

5. Central Atom Resonances in Trivalent Compounds

Apart from the values already given in Sections 3 and 4 there are few other data of interest for the Group III metal atom quadrupole coupling constants. Al^{27} has been fairly widely studied in aluminosilicate minerals but the coupling constants are small since the aluminium atoms are in general at the centre of an AlO_4 tetrahedron. Quadrupole effects have been observed in the spectra of metallic aluminium,[22] gallium[23] and indium,[24] and In^{115} resonances have been observed[25] in $(NH_4)_4(InCl_5.H_2O)$.

D. GROUP V. NITROGEN, PHOSPHORUS, ARSENIC, ANTIMONY AND BISMUTH

1. Trivalent Compounds, Halogen Resonances

The chlorides, bromides and iodides of nitrogen are unstable explosive substances and their halogen quadrupole coupling constants are unknown. The quadrupole coupling constants of *N*-chloro and *N*-bromo amines and amides are however very high (10.C, 10.D) and indicate that as expected from electronegativity considerations the *N*-halogen bond is almost purely covalent. Microwave spectroscopy studies of nitrosyl chloride[26] (NOCl) and nitryl chloride[27] (NO_2Cl) have yielded values for the ^{35}Cl coupling constant and since the latter compound has the planar symmetrical structure the principal axes of the halogen field-gradient tensor coincide with the inertial axes. In general agreement with the results for the *N*-chloroimides the ^{35}Cl coupling constant is $-94{\cdot}3$ Mc/s. The presence of a small asymmetry parameter ($\eta = 0{\cdot}09$) indicates a small amount of double bond character in the nitrogen-chlorine bond corresponding to contributions of resonance structures such as:

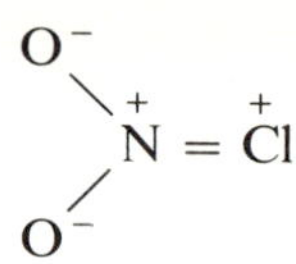

In the unsymmetrical, NOCl, however, it is of course not possible to be sure of the exact value of the coupling constant. If the principal field-gradient axis lies along the N—Cl bond the coupling constant is $-57{\cdot}2$ Mc/s and the asymmetry parameter is 0·325. The minimum value it can have is 0·208. This of course implies a much higher degree of partial double-bond character in the nitrogen chlorine bond in NOCl than in NO_2Cl. Even more surprising is the very low value of the coupling constant itself and a possible explanation lies in the contribution of resonance structures such as (II)

$N{\equiv}O^+$

Cl^-

(II)

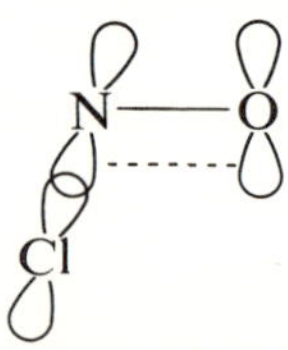

(III)

to the ground state of the molecule. In molecular orbital terms this, of course, implies overlap between the N—Cl sigma bond and the pi system of oxygen (III) and is exactly analogous to the suggested explanation of the low coupling constants in the fluorine- and methoxy-substituted chloromethanes. (10.B.4).

Similar conclusions have been reached for NOBr where the ^{79}Br coupling constants[28] along the N—Br bond are 439·0 Mc/s, $\eta = 0{\cdot}322$.

The data for the remaining elements in this group are set out in Table 18; the resonance frequencies for PI_3 do not appear to have been reported. Most of these trihalides are volatile and their molecular structures in the vapour phase have been determined by electron-diffraction to be trigonal pyramids. In the crystalline phase arsenic tribromide[29,30] and antimony trichloride[31] have been shown by X-ray diffraction to form molecular crystals. The tri-iodides of all three elements however form layer-lattices by the sharing of edges of MI_6 octahedra. Once again therefore the quadrupole resonance data are not directly comparable and we limit any detailed discussion of these results to cases in which structural information is available.

TABLE 12.18. Halogen pure quadrupole resonance frequencies (Megacycles/sec at 77°K) of the trihalides of Group V. (References are superior numbers)

	Cl^{35}	Br^{81}	I^{127} $\pm\frac{1}{2} \leftrightarrow \pm\frac{3}{2}$	I^{127} $\pm\frac{3}{2} \leftrightarrow \pm\frac{5}{2}$
P	26·107[63]	182·636*[60]	—	—
	26·202	184·257	—	—
As	24·960[63]	171·480[104]	207·011[104]	395·78
	25·058	172·728	—	—
	25·406	173·143	—	—
Sb	19·305[63]	133·449[60]	174·509[105]	254·637
	20·912	141·923	—	—
	—	142·622	—	—
Bi	15·955[103]	—	110·32[105]	201·38
	19·155	—	—	—

* At 83°K.

In the triiodides[32] each halogen atom forms a bridge between two metal atoms. In this case the principal axis of the field-gradient should bisect the $\widehat{MXM}$ angle and Zeeman studies[33] on single crystals of AsI_3 have confirmed this conclusion. As for the dimeric trihalides of Group III these bridging halogen atoms should also exhibit an appreciable asymmetry parameter and, again assuming that the halogen atom only makes use of its valence *s* and *p* orbitals, the interorbital angle is given by equation (5). The asymmetry parameters and coupling constants for the triiodides are listed in Table 19 together with the results of the application of equations (4) and (5).

TABLE 12.19. Coupling constants, asymmetry parameters and derived data of the triiodides of Group V

	e^2Qq/h (Mc/s)	η	Interorbital angle	Halogen sigma orbital population
AsI_3	1330·23	0·1891	94°	1·383
SbI_3	895·83	0·565	101°	1·536
BiI_3	682·18	0·274	95°	1·675

The $\widehat{AsIAs}$ interbond angle in AsI_3 has been determined by X-ray diffraction to be 88°.[32] The ideal close-packed structure from which the crystal structure of AsI_3 is derived would have an interbond angle of 90° and it is clear that the observed asymmetry parameters are in accord with small distortions from this.

The orbital populations of the iodine atoms follow the trend observed in previous groups. It is unfortunate that the results for phosphorus triiodide are not available.

In accordance with the structure determined by X-ray crystallography[29,30] $AsBr_3$ shows three closely-spaced resonances and it is tempting to conclude from the similarity of their resonance spectra that $AsCl_3$ and $SbBr_3$ are isomorphous with it. Moreover, above 90°C AsI_3 undergoes a phase-change yielding, at 110°C, the quadrupole resonance spectrum shown in Table 20. If indeed all these compounds are isomorphous, forming like $AsBr_3$ molecular crystals, then the large asymmetry parameter in AsI_3 implies considerable back-bonding of the lone-pair electrons on the iodine atoms to vacant atomic orbitals on the arsenic atom. In $AsBr_3$ however single crystal Zeeman[29] studies reveal an asymmetry parameter of the bromine atoms as only $0{\cdot}04 \pm 0{\cdot}02$.

TABLE 12.20. I^{127} Pure quadrupole resonance frequencies of AsI_3 (Megacycles at 110°C) (Ref. 104)

$\pm\frac{1}{2} \leftrightarrow \pm\frac{3}{2}$	$\pm\frac{3}{2} \leftrightarrow \pm\frac{5}{2}$	e^2Qq/h	η
206·20	403·16	1348·53	0·131
208·45	406·96	1361·48	0·135
211·74	412·46	1380·50	0·141

Antimony trichloride also forms molecular crystals but is not isomorphous with $AsBr_3$. In accordance with its crystal structure[31] there are two distinct chlorine resonances. It is very likely that PCl_3 and PBr_3 are isomorphous with $SbCl_3$. Finally although $BiCl_3$ also shows two chlorine resonances, the considerable difference between them makes it unlikely either that it is isomorphous with $SbCl_3$ or that it forms molecular crystals.

Assuming that the three compounds PCl_3, PCl_2F and $PClF_2$ form molecular crystals, their chlorine resonance frequencies (Table 21) exhibit a similar trend to that observed in the carbon tetrahalide series CCl_4, CCl_3F, CCl_2F_2, $CClF_3$, namely a decrease of ^{35}Cl frequency on increasing substitution by fluorine instead of the increase which would have been expected on the grounds of the high electronegativity of the fluorine atom. It is possible that the reason for this behaviour is the same suggested for the carbon compounds, namely contributions of resonance structures such as (IV)

$$\overset{+}{F} = P—Cl$$

$$Cl^-$$

(IV)

to the ground state electronic structure. In view of the unwillingness of phosphorus to form p_π-p_π double bonds in the trivalent state this is perhaps unlikely. An alternative explanation could be p_π-d_π double-bonding as in the silicon halides.

TABLE 12.21. ^{35}Cl Resonance frequencies (Mc/s at 77°K) of trivalent phosphorus halides

Compound	Resonance frequency		Ref.
PCl_3	26·107,	26·202	63
$FPCl_2$	24·4,	24·7	106
F_2PCl	22·2,	22·6	106

2. Trivalent Compounds. Central Atom Resonances

Apart from phosphorus all the elements in this Group have stable isotopes possessing quadrupole moments. In addition to solid-state data the coupling constants of a few simple derivatives of arsenic and antimony have been determined by microwave spectroscopy and the results are set out in Table 22 together with the structural information which will be required for their analysis. All these compounds have the formula MR_3 and the appropriate Townes and Dailey formula for the analysis of their coupling constants was given in Chapter 11 for the amines (equation 3, Chapter 11). Unfortunately neither for As^{75} nor $Sb^{121,123}$ are there available direct experimental values of the atomic coupling constant e^2Qq_0 and we must fall back on the present estimates[34] of Q and the Hartree–Fock value[5] for $\langle 1/r^3 \rangle$. These yield

TABLE 12.22. As^{75} and $Sb^{121,123}$ quadrupole coupling constants (Megacycles/sec) for simple MX_3 compounds measured by microwave spectroscopy. The structural data were taken from Ref. 48

Compound	Nucleus	e^2Qq (Mc/s)	Ref.	$X\widehat{M}X$
AsH_3	As^{75}	$-160{\cdot}1 \pm 0{\cdot}4$	107	$91{\cdot}83 \pm 0{\cdot}33°$
$As(CH_3)_3$	As^{75}	$-203{\cdot}15 \pm 0{\cdot}10$	108	$96 \pm 5°$
$AsCl_3$	As^{75}	-173 ± 20	109	$98{\cdot}4 \pm 0{\cdot}5°$
AsF_3	As^{75}	$-236{\cdot}23 \pm 0{\cdot}05$	110	$102 \pm 2°$
SbH_3	Sb^{121}	$458{\cdot}7 \pm 0{\cdot}8$	111	$91{\cdot}30 \pm 0{\cdot}33°$
	Sb^{123}	$586{\cdot}0 \pm 0{\cdot}8$		

-412 Mc/s and $+954$ Mc/s for As^{75} and Sb^{121}, respectively. As seen in Chapter 11, equation (3) is rather sensitive to the value of the interbond angle and this is particularly the case when this is in the region of 90° (where the angular term is 0·0). It makes it extremely doubtful whether an analysis of the data is worthwhile for these compounds but at least qualitatively it is clear that the As—H sigma orbital must have, in accordance with the relative electronegativities of arsenic and hydrogen, a population of about unity, while the similarity of the coupling constants for AsF_3 and AsH_3 is mainly due to the very different interbond angles in these compounds. One might indeed be tempted to use the quadrupole coupling data to at least restrict the limits of uncertainty of the bond angle in $As(CH_3)_3$. In view of the relative electronegativities of Cl and CH_3 the coupling constants for these two molecules are only compatible with our ideas about electronic structure if the bond angle in trimethyl arsine is greater than that in arsenic trichloride, so that a value close to 100° might be the appropriate one. However perhaps this is pushing the Townes and Dailey method too far.

The solid state data for these nuclei are more abundant but here the structural problems are even more acute. A selection of data is shown in Table 23. For those compounds such as the trichlorides and tribromides and triphenyl derivatives which form molecular crystals in which the isolated trigonal pyramid present in the gas-phase is retained, the coupling constants are similar to those measured by microwave spectroscopy. Two totally different resonance frequencies have been reported for AsI_3; in view of the high co-ordination number (6) of arsenic in this compound the smaller value is probably the correct one. In SbI_3 only one (25·402 Mc/s) of the two possible resonance frequencies (Sb^{121}, $I = \frac{5}{2}$) was observed and this was claimed to be the low-frequency component. The coupling constant has been calculated on this basis assuming $\eta = 0$. The value so obtained is again in accordance with the polymeric structure of this iodide. The values shown for Bi^{209} are the only ones which appear to have ever been measured for this nucleus.

TABLE 12.23. Central atom nuclear quadrupole coupling constants (Mc/s at 77°K) of trivalent Group V compounds. For As^{75} (spin = $\frac{3}{2}$) the coupling constants were obtained by doubling the resonance frequency

Compound	Nucleus	$\|e^2Qq\|$	η	Ref.
$AsCl_3$	As^{75}	157·90	(0)	112
$AsBr_3$	As^{75}	122·36	(0)	112
AsI_3	As^{75}	58·676	(0)	113
AsI_3	As^{75}	233·720	(0)	112
$As(C_6H_5)_3$	As^{75}	197·4	(0)	112
$SbCl_3$	Sb^{121}	383·69	0·184	113
$SbBr_3$	Sb^{121}	343·94	0·100	113
SbI_3	Sb^{121}	169·4	(0)	113
$Sb(C_6H_5)_3$	Sb^{121}	509·0 520·84	0·080 0·048	105
$BiCl_3$*	Bi^{209}	318·76	0·555	35
$BiBr_3$	Bi^{209}	340·5	0·553	114
$Bi(C_6H_5)_3$†	Bi^{209}	669·06	0·09	35

* At 21°C. † At an unspecified temperature.

Robinson *et al.*[35] have estimated that e^2Qq_0 for this nucleus is around +1500 Mc/s, the quadrupole moment obtained from atomic beam measurements ($-0{\cdot}34b$) (Table 7.8) together with the Hartree–Fock value[5] for $\langle 1/r^3 \rangle$ (14·5709 a.u.) yields +930 Mc/s. This however neglects any relativistic correction which for atoms as heavy as this can be 1·5 or more. This would bring us close to Robinson's estimate and indeed the value of e^2Qq_0 for Bi^{209} is almost certainly at least 20 per cent greater than that of Sb^{121}. It must be stressed however that the evaluation of e^2Qq_0 for such heavy atoms is a very hazardous affair since the states of the atoms for which the hyperfine measurements are available are all intermediate between L-S and *jj* coupling and the relativistic corrections are large. One is better off using the results for the coupling constants of a closely related series of molecules of known structure.

Arsenic and antimony trihalides form a variety of molecular complexes whose electronic structure has been studied by means of the halogen and metal atom nuclear quadrupole coupling constants. These data are discussed in Chapter 13.

3. Pentavalent Compounds

Phosphorus, arsenic, antimony and bismuth all exhibit a valency of five and all form, for example, the pentafluorides MF_5. Of the remaining halides, however, PCl_5, PBr_5 and $SbCl_5$ are the only known examples of unmixed pentahalides although the mixed chlorofluorides of phosphorus have also been prepared. Although all these unmixed halides exhibit the same trigonal bipyramidal structure in the vapour phase, none of them have the same structure in the crystalline state to which the quadrupole resonance results refer. Thus PCl_5 forms an ionic lattice of $[PCl_4]^+$ cations and $[PCl_6]^-$ anions, PBr_5 an ionic lattice of $[PBr_4]^+$ cations and Br^- anions, while $SbCl_5$ forms molecular crystals in which the trigonal bipyramidal structure of the vapour phase is retained.

For phosphorus pentachloride one would expect two widely-differing sets of ^{35}Cl resonance frequencies corresponding to the complex anion and cation. McCall[36] reported, without comment, three resonance frequencies (32·282, 32·384, 32·63 Mc/s at 77°K) differing very little from each other. The crystal structure determined at room temperature[37] shows that all the chlorine atoms in the PCl_4 fragment are crystallographically equivalent and thus will give only one pure quadrupole resonance frequency while for the $[PCl_6]^-$ anion the chlorine atoms fall into three crystallographically equivalent sets containing two, two and eight atoms respectively and thus should give rise to three distinct quadrupole resonances. DiLorenzo and Schneider[38] recently reinvestigated this problem and showed that a phase change occurs. At 77°K the frequencies of Cl^{35} are 29·61, 30·07, 30·45, 30·62, 32·28 and 32·61 Mc/s while at 195°K the frequencies are 29·27, 29·92 and 31·96 Mc/s. These are in accordance with the room temperature crystal structure and by a study of various compounds containing the PCl_4^+ or PCl_6^- ions (Table 24) they showed that the lower frequencies were due to the PCl_6^- ion.

TABLE 12.24. ^{35}Cl Pure quadrupole resonance frequencies of complex halide ions of Group V elements (Ref. 38)

Complex	Ion	Resonance frequencies (Mc/s)
$[PCl_4][PF_6]$	PCl_4^+	32·46
$[PCl_4][SbCl_6]$	PCl_4^+	31·87, 32·35, 32·51
$[AsCl_4][AlCl_4]$	$AsCl_4^+$	36·82, 37·03
$[(C_2H_5)_4N][PCl_6]$	PCl_6^-	29·32, 30·06, 30·34
$[PCl_4][SbCl_6]$	$SbCl_6^-$	22·80, 23·02
$[(C_2H_5)_4N][SbCl_6]$	$SbCl_6^-$	24·01, 24·21, 24·67, 24·86
$[NO][SbCl_6]$	$SbCl_6^-$	22·35, 22·97, 23·15, 23·37, 25·03, 25·48

The first measurements[36] made on $SbCl_5$ presented a similar problem in that three frequencies were observed at 77°K (27·88, 28·30 and 30·40 Mc/s) whereas the crystal structure at 243°K[39] showed that $SbCl_5$ forms a molecular crystal containing the trigonal bipyramidal structure present in the gas phase and that only two distinct chlorine atoms were present, i.e. only two ^{35}Cl resonances would be expected. DiLorenzo and Schneider[38] have also re-investigated this problem. They were unable to detect any ^{35}Cl resonances at temperatures greater than 228°K, although at lower temperatures a large number of lines were visible. The situation is complicated by the presence of Sb^{121} and Sb^{123} resonances in the same frequency region and indeed they were unable to assign some of the resonances with certainty. At 249°K they did however succeed in obtaining a complete Sb^{121} and Sb^{123} spectrum which was in accord with the crystal structure. Their results are shown in Table 25; the large range in values for $SbCl_5$ at 77°K indicates a complex polymeric structure probably containing chlorine bridges.

TABLE 12.25. ^{35}Cl and $Sb^{121,123}$ data for $SbCl_5$ (Ref. 115)

Resonance frequencies at 77°K		Resonance frequencies at 249°K	
Cl^{35}	Sb	Sb^{121}	Sb^{123}
18·76	27·93	12·67	7·72
27·42	28·69	25·39	15·43
27·85	31·09	—	23·11
28·01	38·48		
29·80	39·63		
29·98	42·89		
30·42	54·22		
30·50	—		

The low temperature resonances in the column headed Cl^{35} have corresponding Cl^{37} resonances and must definitely be assigned to this isotope. The figures in the column headed Sb cannot with certainty be assigned to Sb^{121} or Sb^{123} and indeed some of them may be chlorine resonances.

The complex ion PCl_4^+ is isoelectronic with $SiCl_4$ and with the complex anion $AlCl_4^-$. It is interesting to note that the relative frequencies of these three molecules indicate that the charge on the central atom remains approximately constant throughout the series. It is not clear whether this charge is maintained constant through the sigma-bond framework or by a greater or lesser extent of d_π-p_π bonding since of course the ^{35}Cl quadrupole coupling constant is given (see equation 1) by the difference of these two terms. As in Group IV the compound of the third element in the Group ($AsCl_4^+$) has a higher coupling constant than that of the second element, the absolute magnitude of the difference being about the same. If the d_π-p_π bonding explanation for this phenomenon was correct for the Group IV halides it is presumably also correct in this case.

The mixed chlorofluorides, PCl_4F, PCl_3F_2, PCl_2F_3 and $PClF_4$ condense from the gas phase to form molecular liquids which at room temperature slowly rearrange to form ionic solids such as $[PCl_4]^+F^-$. The quadrupole coupling constants for the frozen and presumably polycrystalline molecular liquids are shown in Table 26. Infrared and Raman studies, together with

TABLE 12.26. ^{35}Cl resonance frequencies (Mc/s at 77°K) of pentavalent phosphorus chlorofluorides

	Axial chlorine	Equatorial chlorine	Ref.
PCl_4F	28·99	32·54	106
PCl_3F_2	—	31·26, 31·89	106
PCl_2F_3	—	31·49	106
$PClF_4$	—	32·14	116

dipole moment studies show that PCl_3F_2 has the non-polar structure in which both fluorine atoms are in axial positions while PCl_2F_3 has the polar structure where the chlorine atoms are both in equatorial positions. These facts together with spectroscopic studies of PCl_4F indicate that the fluorine in this compound is in an axial position. The high ^{35}Cl frequencies in PCl_3F_2 and PCl_2F_3 together with the relative intensities of the high and low ^{35}Cl frequencies in PCl_4F show clearly that in this last compound the low frequency (28·99 Mc/s) is to be ascribed to the axial chlorine atom. A further example of the low frequency of chlorine atoms in an axial position is yielded by $F_3C.PCl_4$ where infrared and vibrational studies support the more symmetrical C_{3v} structure and where the ^{35}Cl resonance frequencies are 31·1 and 28·8 Mc/s at 77°K.[40]

There are two extreme descriptions of the electronic structure of such trigonal bipyramidal molecules of Group V elements. In the first the molecule forms a planar trigonal framework PCl_3 and the remaining *p*-orbital is used to form a four-electron three-centre bond with the remaining two axial groups. In resonance terms this may be written as resonance between the two structures

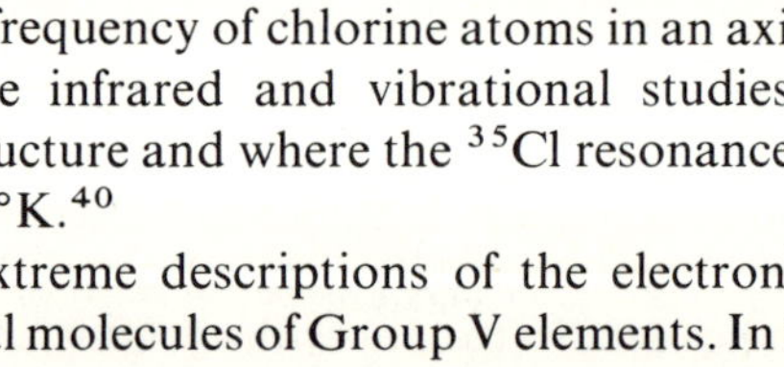

In the second extreme description the first part of the skeleton is formed as above but now an unoccupied $3d_{z^2}$ orbital hybridizes with the axial $3p_z$ orbital permitting the formation of two, two-centre, two-electron axial bonds.

Because *d*-orbitals are more diffuse than *s*- or *p*-orbitals they screen the nucleus much less efficiently. The effective electronegativity of the *p-d* hybrid axial orbitals is thus less than those of the *s-p* hybrid equatorial orbitals and the halogen atoms in the axial positions would be expected to carry a slightly larger negative charge, as indeed is observed. Of course in the previous model with no *d*-hybridization, the axial halogen atoms are also expected to carry a negative charge but from a rather naive standpoint at least one would have expected here a much larger difference between the frequencies of the axial and the equatorial chlorine atoms. The ^{35}Cl pure quadrupole resonance spectrum may thus be taken to provide some evidence of *d*-orbital participation in the bonding in these molecules. Both of these structures of course provide a ready rationalization of the preference of electronegative groups such as F or CF_3 for axial positions. It is also possible that partial double-bond character of the phosphorus-halogen bond, involving donation of halogen lone-pair electrons to vacant phosphorus orbitals (4*p*, 3*d*) may play a part in determining the structure and ^{35}Cl quadrupole coupling constant of these compounds. However the problem of the bonding in PCl_5 is very similar to that of the bonding in the trihalide anions and for these entities quadrupole coupling constants are available for all three nuclei and there are no uncertainties introduced by the equatorial bonds. They are therefore much more suitable subjects for a detailed analysis of the data and we return to this problem in Section F.

With the MX_6^- ions a bonding problem similar to that in the pentahalides presents itself. Are the 3*d* orbitals used to form a d^2sp^3 sigma-bond framework or not? Again the question is difficult to answer with such a large molecule as this but on the whole it seems likely that there must be some *d*-orbital participation. If we use a very simple-minded model for the ionic-resonance form of PCl_6^-, namely that we start by forming a distorted PCl_4^+ ion and assume that the coupling constants of the chlorine atoms remain the same as in the tetrahedral molecule (32 Mc/s), that we then add two Cl^- ions to form PCl_6^- and finally redistribute the negative charge among the six chlorine atoms, then the calculated ^{35}Cl resonance frequency is around 11·0 Mc/s. (This calculation assumes that the total charge on the phosphorus atom in PCl_6^- is the same as in PCl_4^+.) Even admitting the extremely crude nature of this estimate the difference between it and the observed resonance frequency (30 Mc/s) makes it hard to believe that PCl_6^- is, in resonance terms, a hybrid of $PCl_4^+/2Cl^-$ ionic structures. Of course the problem is complicated by the possibility or even probability of d_π-p_π bonding in PCl_4^+ but we may well ask that if the 3*d* orbitals are used there where there is no apparent "necessity" then why not for the sigma-bonds in PCl_6^- where they really are needed?

The ^{35}Cl quadrupole coupling constants of the phosphoryl halides are discussed in (Section C, Chapter 10). A microwave study[41] of the related

$PSCl_3$ indicates a ^{35}Cl quadrupole coupling constant in the neighbourhood of -60 Mc/s. The ^{35}Cl pure quadrupole resonances of the interesting cyclic polymers, the phosphonitrilic chlorides, $(PNCl_2)_n$ are shown in Table 27. They fail to provide any evidence for the increasing aromatic character with ring size predicted on the basis of molecular orbital calculations by Craig and co-workers. It must be admitted however that the ^{35}Cl resonance frequency would not be expected to be very sensitive to this. A single crystal Zeeman study[4] of $(PNCl_2)_3$ showed no evidence for "bent bonds" in this molecule and sł ɔwed that the asymmetry parameter of the chlorine quadrupole coupling tensor was small ($0{\cdot}0 \leqslant \eta \leqslant 0{\cdot}02$).

TABLE 12.27. ^{35}Cl pure quadrupole resonance frequencies (Mc/s at 77°K) of phosphonitrile chloride cyclic polymers (Ref. 117)

n	Resonance frequencies
$n = 3$	28·3158, 28·3275, 28·5471, 28·6523
$n = 4$ (α phase)	28·6371, 28·6595
$n = 4$ (β phase)	28·6380, 28·6587
$n = 5$	27·6876, 27·8415, 27·9270, 27·9423, 28·2220, 28·2474, 28·4400, 28·4850
$n = 6$ (α phase)	28·4
$n = 6$ (β phase)	27·6822, 27·7470, 27·7938, 27·8262, 27·8730, 27·9216, 27·9720, 28·0260, 28·0395, 28·0890, 28·3095, 28·4157

E. GROUP VI. OXYGEN, SULPHUR, SELENIUM AND TELLURIUM

1. Halogen Resonances

The halogens form a variety of oxides and oxyanions and the quadrupole coupling constant of these, particularly for the chlorine compounds, have been well studied. Compounds of the type RSO_2Cl, having a sulphur chlorine bond, have been discussed in Section C. Chapter 10; the parent compound in this series SO_2Cl_2 and the related $SOCl_2$ will be discussed here for the sake of completeness, as will a few other miscellaneous compounds of Group IV.

The microwave spectrum of the lowest oxide of chlorine, Cl_2O shows not only that it has the expected triangular configuration (interbond angle $110{\cdot}8° \pm 1°$) but it yields also the quadrupole coupling constant.[43] These are of course in the inertial axis system, but if the usual assumption is made that the principal axis coincides with the OCl bond the values $e^2Qq = -140$ Mc/s,

$\eta = 0{\cdot}06$ are obtained. Such a high value of the coupling constant (formal sigma bond population, 0·72) is completely in accord with the higher electronegativity of oxygen compared to chlorine and the low asymmetry parameter with the closed-shell nature of the pi-system. The microwave spectrum of the related methyl hypochlorite[44] has also been determined and using the same assumption for the direction of the principal field-gradient axis the coupling constants are $e^2Qq = -114{\cdot}0 \pm 5{\cdot}2$ Mc/s, $\eta = 0{\cdot}04 \pm 0{\cdot}04$. The lower value of the coupling constant is the expected consequence of the lower electronegativity of the methyl group compared to chlorine.

FIG. 12.4. Molecular geometry and inertial axes of chlorine dioxide.

Chlorine dioxide is one of the few stable gaseous free-radicals and its microwave spectrum[45] has yielded values of the quadrupole coupling constant. The molecular structure of this molecule is illustrated in Fig. 4 and the quadrupole coupling constants along the inertial axes shown in Fig. 4 are:

$$e^2Qq_{aa} = -51{\cdot}90 \text{ Mc/s}$$
$$e^2Qq_{bb} = +2{\cdot}28 \text{ Mc/s}$$
$$e^2Qq_{cc} = +49{\cdot}62 \text{ Mc/s}.$$

(The unwary reader should not be mislead by the values given in the original paper where the quadrupole coupling constants are given as $e^2Qq/2\text{I}(2\text{I} - 1) = \frac{1}{6}e^2Qq$.)

Both the theoretical analysis of the electronic structure of this molecule as well as the hyperfine coupling constants show that the unpaired electron occupies an antibonding π-orbital. The familiar C_{2v} sp^2 hybridized sigma orbitals then yield by a population analysis:

$$\frac{q_a}{q_0} = \left(\frac{a}{2} - 1\right)(1 - \cot^2\gamma) - \tfrac{1}{2}(b - a)$$

$$\frac{q_b}{q_0} = (2 - a)(1 - \cot^2\gamma) - \tfrac{1}{2}(b - a)$$

$$\frac{q_c}{q_0} = \left(\frac{a}{2} - 1\right)(1 + \cot^2\gamma) + (b - a) \qquad (13)$$

where a is the population of each chlorine sigma orbital and b is the population of the π-orbital. The remaining sigma orbital is presumed to be occupied by two electrons. From the known bond angle (2γ) and coupling constants we have

$$a = 1{\cdot}45; \qquad b = 2{\cdot}10.$$

These numbers appear to be rather high, in part no doubt due to the use of the same value (109·7 Mc/s) for the coupling constant of a valence p-electron as is appropriate for singly-bonded chlorine atoms, and a somewhat higher value would probably be more appropriate.

Related to the chlorine dioxide molecule by the addition of an electron is the chlorite anion. The resonance frequency[46] of $AgClO_2$ at room temperature is 54·08 Mc/s while that of the sodium salt is 51·82 Mc/s. This anion like ClO_2 has a bent symmetrical configuration and Curl[45] has used the coupling constants for ClO_2 to "predict" a resonance frequency of 55 Mc/s for the chlorite anion. It would be expected of course that the coupling tensor for ClO_2^- would exhibit a considerable anisotropy; this has not however been demonstrated experimentally.

The chlorate ion, ClO_3^-, has been investigated in a large number of its salts and the frequencies shown to be fairly independent of the nature of the counterion. The related bromate and iodate have likewise been well studied and Table 28 lists a few typical values. The ion has the pyramidal symmetrical C_{3v} structure, and the asymmetry parameter is consequently expected to be

TABLE 12.28. Pure quadrupole resonance frequencies (Mc/s at 77°K) for chlorates, bromates and iodates. (References are superior numbers)

Anion / Cation	$X = Cl^{35}$	$X = Br^{81}$	$X = I^{127}$ $\pm\frac{1}{2} \leftrightarrow \pm\frac{3}{2}$	$X = I^{127}$ $\pm\frac{3}{2} \leftrightarrow \pm\frac{5}{2}$
Na	30·630[118]	151·909[121]	—	—
K	28·952[119]	149·361[121]	144·37	288·82[122]
			144·91	289·85
			145·35	290·71
Ag	29·419[120]	141·7*[121]	—	—
Ca	—	149·3†[121]	150·75	297·470[122]
			150·97	298·709
				307·612

zero. This may be seen to be the case for potassium iodate and extensive Zeeman[47] measurements have shown that this is also the case for the chlorate. Of course in an ionic crystal if the anion is not in a site of at least trigonal symmetry, crystal field effects would be expected to induce a slight asymmetry. The sigma bonding in these ions may be described in terms of sp^3 hybrids, the fourth orbital containing a lone-pair of electrons. As previously noted the coupling constant of the central ion in this situation is given by equation (3), Chapter 11, and is unfortunately very sensitive to the value of α, the interorbital angle. Moreover in these three ions the experimental values of the interbond angle[48] vary widely (ClO_3', 106·7°; BrO_3', 111·8°; IO_3', 97·3°). If the interorbital angle is assumed to be the tetrahedral angle in all three cases the central atom has apparent charges of Cl, +1·21; Br, +0·87; I, +0·69. On the other hand if the experimental interbond angle is used these charges become Cl, +1·49; Br, +0·72 and I, +2·73. Thus it is extremely difficult if not impossible to draw any valid conclusions about the nature of the bonding in these ions.

In iodic acid the hydrogen atom is covalently bonded to one of the oxygen atoms so that the molecule loses its threefold symmetry axis. This should of course be reflected in the asymmetry parameter for the ^{127}I nucleus and indeed the pure quadrupole resonance data for this compound show e^2Qq = 1126·9 Mc/s, η = 0·4505. In addition there have been two independent Zeeman[49,50] studies of single crystals of iodic acid which have yielded results in agreement with the crystal structure. In one of these[49] a doublet splitting in the Zeeman hyperfine structure was shown to be due to interaction with the magnetic moment of the hydrogen atom and the magnitude of the effect was used to confirm the location of the hydrogen atom in the crystal derived from neutron diffraction.

TABLE 12.29. ^{35}Cl pure quadrupole resonance frequencies (Mc/s at 77°K) of sulphur–chlorine and selenium–chlorine compounds)

	ν_Q (Mc/s at 77°K)	Ref.
$SOCl_2$	31·887, 32·091	63
SO_2Cl_2	37·613, 37·822	67
S_2Cl_2	35·605, 35·994	123
$CCl_3C.SCl$	39·166, 39·827, 40·117	123
$SeCl_4$	34·534, 34·685, 35·041, 35·696, 36·171, 36·794	67
$TeCl_4$	27·304, 27·370, 27·471, 27·926, 28·110, 28·461	124

The coupling constants[51] of the anhydride of iodic acid, I_2O_5 are e^2Qq = 1068 Mc/s at 77°K, η = 0·33, similar as might be expected to those of the acid itself. No crystal determination of this compound appears to have been made.

To turn now to compounds having a sulphur-halogen bond the resonance frequencies of four such chlorine compounds are shown in Table 29.

Inasmuch as the extreme ionic structures of thionyl chloride and sulphuryl chloride are (V) and (VI)

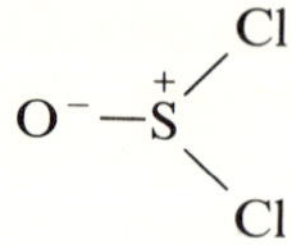

(V)

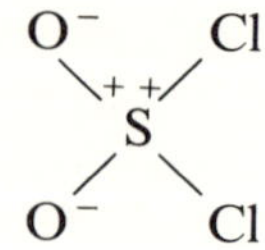

(VI)

the higher frequency of sulphuryl chloride is perfectly reasonable. Of course some back-bonding from oxygen undoubtedly occurs and the actual formal charges are probably not more than half those shown in the above formulae. Moreover the high value of the coupling constant of S_2Cl_2, having the molecular structure shown in (VII), is completely in accordance with the electronegativity of the sulphur atom and suggests that in the oxygen-containing compounds the considerable formal positive charge on the

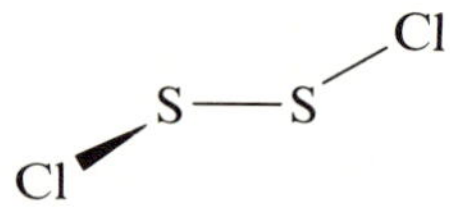

(VII)

sulphur-atom gives rise to extensive back π-bonding by the chlorine atom leading to a lowering in frequency. No measurements of asymmetry parameters in these compounds appear however to have been made.

The coupling constants of the ^{35}Cl nucleus in $ClSF_5$[52] is $-81{\cdot}5 \pm 5$ Mc/s and of ^{79}Br in the corresponding $BrSF_5$[53] is 800 ± 5 Mc/s, both determinations having been made by microwave spectroscopy. They correspond to formal sigma orbital populations on the halogen atom of 1·257 in $ClSF_5$ and 0·96 in $BrSF_5$ and the unusually large difference between the two halogen atoms cannot pass without comment. The problem of the electronic structure of the sulphur hexahalides is of course identical with that of the structure of the hexahalide anions of phosphorus. It may well be that the precise structure, intermediate between the ionic resonance form and the *d*-orbital form, depends very critically on the nature of the halogen. The present result may indicate that ionic forms are more important with fluorine and the *d*-orbital form with bromine and iodine, chlorine occupying an intermediate position.

The only other halogen compounds of Group VI for which data are available are $SeCl_4$ and $TeCl_4$ whose resonance frequencies are also listed in Table 29. The structure of these molecules are probably not tetrahedral but more closely correspond to that of a trigonal bipyramid with one of the equatorial positions occupied by a lone-pair. The large spread in quadrupole resonance frequencies certainly supports this suggestion but no detailed crystal structure seems to be available. The magnitude of the coupling constant is much as would be expected.

2. Central Atom Resonances

The Group VI nuclei are not very well suited to nuclear quadrupole coupling studies since the only two members which possess stable quadrupolar isotopes do so in such low abundance that for most purposes they can only be measured by microwave spectroscopy (O^{17}, $I = \frac{5}{2}$, 0·037 per cent; S^{33}, $I = \frac{3}{2}$, 0·74 per cent). In addition Se^{79} ($I = \frac{7}{2}$), a radioactive isotope of selenium has been incorporated into O=C=Se and its quadrupole coupling constant has been similarly determined. The results of such measurements are shown in Table 30.

TABLE 12.30. O^{17}, S^{33} and Se^{79} quadrupole coupling constants (Mc/s) from microwave data. The principal field-gradient axes are shown in the Table for the asymmetric tops.

Molecule	Nucleus	e^2Qq	η	Reference
O=C=S	S^{33}	$-29·07 \pm 0·01$	0	125, 126
C≡S	S^{33}	$12·835 \pm 0·026$	0	127
H–N=C=S	S^{33}	$-27·5 \pm 0·8$	0	128
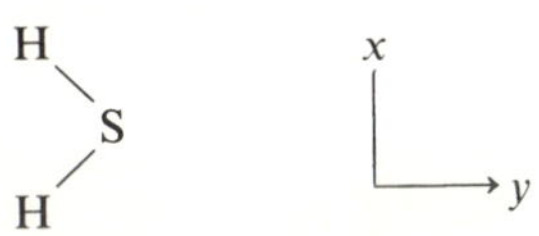	S^{33}	40	0·6	129, 130
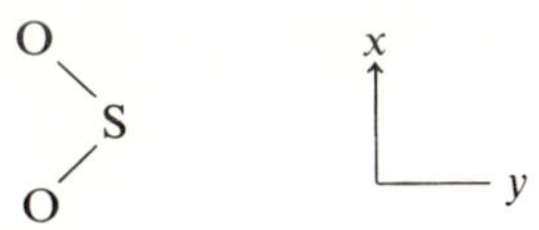	S^{33}	$25·71 \pm 0·03$	0·87	130, 131
S_8	S^{33}	$\pm 45·8$	—	132
SO	S^{33}	$-15·9$	0	133
S=C=O	O^{17}	$-1·32$	0	134
C≡O	O^{17}	$4·43 \pm 0·4$	0	135
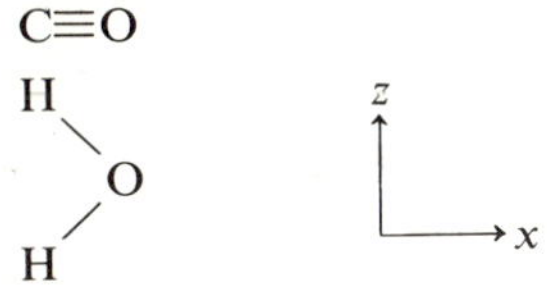	O^{17}	$-9·83$*	$0·4 \pm 0·1$*	136
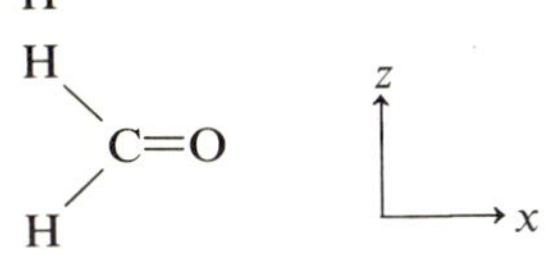	O^{17}	12·37	0·695	137
O=C=Se	Se^{79}	$752·09 \pm 0·05$	0	138

* Recalculated by the author in the principal axis system.

The results of exact calculations of the field-gradient at the oxygen atom in carbon monoxide and formaldehyde have been discussed in Chapter 6. We will content ourselves here with a brief and much more qualitative discussion. The quadrupole moments of O^{17} and S^{33} are both negative ($-0{\cdot}03$b and $-0{\cdot}11$b respectively). The S^{33} coupling are therefore in general greater than the O^{17} couplings in analogous molecules. In view of this and the similarity between the structures of H_2O and H_2S it is at first sight surprising that the signs of the coupling constants and the directions of the field-gradient axes are quite different in the two molecules. In fact this is readily explicable by the simple Townes and Dailey theory and is due to the fact that the interbond angle in water (104·5°) is considerably greater than that in hydrogen sulphide (92·27°). The reader may perhaps like to show for himself that if one forms a pair of sp^3 hybrid orbitals for the two sigma bonds and put two electrons in each of the remaining orbitals (equation 6) then when the interorbital angle is 90° (i.e. pure *p*-orbitals for bonding) the field-gradient is perpendicular to the molecular plane, whereas by the time the interorbital angle is 120° the field-gradient is in the molecular plane and perpendicular to the bisector. This assumes that the population of the bonding orbitals is around unity and the cross-over angle depends very much on the actual value of this population. For these compounds we can calculate the interorbital angle directly from the asymmetry parameter, according to equation (8) for H_2O.

Equation (8) yields the angle between the non-bonding orbitals, and the angle required may be then calculated from equation (7). One thus obtains 125° for H_2O and 102° for H_2S in both cases, notably greater than the interbond angle. This of course was just what was found on previous occasions when equations (5) or (8) were applied.

F. GROUP VII. FLUORINE, CHLORINE, BROMINE AND IODINE

The coupling constants of the simple diatomic halogen and interhalogen molecules were discussed briefly in Chapter 7 where the basis of the Townes and Dailey method was described. For convenience they are collected together here in Table 31. As seen previously the coupling constants fit the predictions fairly well and indeed the success with the Townes and Dailey method for such simple compounds was a strong factor in its more widespread application.

From the point of view of a more fundamental understanding of the coupling constants in heavy molecules it is unfortunate that for the homonuclear molecules they are all (necessarily) derived from solid state data. Julg[54] has attempted to estimate the "gas-phase" coupling constant of molecular chlorine by interpolation from the value of gas phase interhalogen couplings,

TABLE 12.31. The coupling constants (Mc/s) of the halogen and interhalogen diatomic molecules

Molecule	Temperature °K	Quadrupole coupling constants (Mc/s)			Ref.
		Cl^{35}	Br^{79}	I^{127}	
Cl_2*	20	±108·95**	—	—	139
Br_2*	4	—	±765·86†	—	140
I_2*	253	—	—	±2156‡	141
FCl	—	−146·0	—	—	142
BrCl	—	−103·6	876·8	—	143
ICl	—	−82·5	—	−2944	144
FBr	—	—	1089·0	—	145
IBr	—	—	722 ± 15	−2731 ± 60	146

* Solid state data; † $\eta = 0·20$; ‡ $\eta = 0·175$; ** $\eta \leqslant 0·03$ (Ref. 147).

and the value he thus obtains (116 Mc/s) is significantly higher than that predicted by the Townes and Dailey theory and Julg[54] has suggested that *d*-orbital contributions to the bonding are responsible for this. Similarly Whitehead and Jaffé[55] have suggested that the slight incompatibility between the Townes and Dailey analysis for the sigma populations at either end of the bond in BrCl is due to *d*-orbital participation in sigma bonding. In both cases however the magnitudes of the discrepancies are too small and the Townes and Dailey theory too simple for any valid conclusion to be drawn. These last authors have attempted[56] to separate *s*-hybridization from ionicity in the Townes and Dailey equation by calculating the ionicity from their "orbital electronegativities".[57] They calculate values in the range 5–15 per cent for the *s*-character.

Bersohn[58] has discussed the quadrupole coupling constants in the solid halogens from the point of view of the intermolecular bonding in the solid state. These molecules crystallize in a layer-lattice and in principle the molecules could interact to form infinite graphite-like sheets with pi-electrons delocalized over the whole of the lattice plane (Fig. 5). He naturally employed the *d*-orbitals in this formulation but his theory implies that a value of the coupling constant lower than that of the atomic halogen goes hand-in-hand with a non-zero asymmetry parameter. Thus although it fits the data for solid iodine reasonably well (indeed such suggestions for this compound had been made on at least two separate occasions[59,60]) it is unable to account for the simultaneous presence in bromine of a large value for η and a value of e^2Qq close to that of atomic bromine.

Some of the most interesting compounds which can be studied by means of their nuclear quadrupole coupling constants are the trihalide ions, X_3^- or X_2Y^-, and the related polyiodide ions I_5^- and I_7^-. X-ray crystallographic

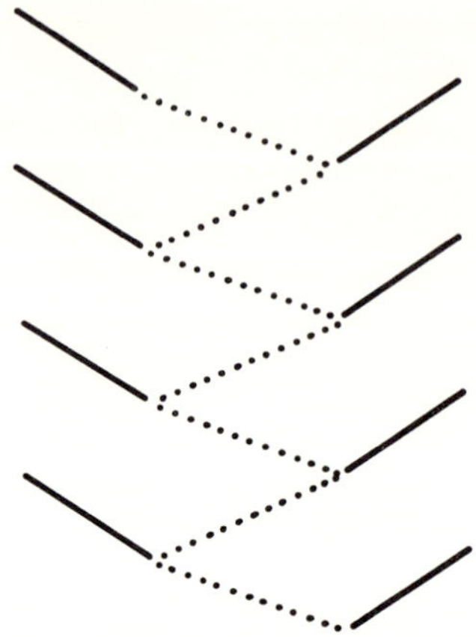

FIG. 12.5. Schematic drawing of the crystal structure of solid I_2 and Br_2. One layer of the structure is shown, the solid lines represent intramolecular bonds, the dotted lines intermolecular bonds.

studies on the triiodides show that in general an almost linear but more or less unsymmetrical structure is adopted, the degree of asymmetry depending on the nature of the cation. A similar situation prevails for the Br_3^- ion while the ICl_2^- ion is usually symmetric and linear, the iodine atom occupying the central position. The electronic structure of these ions may be described in two extreme forms. If we start with the idealized symmetrical structure then either we consider the bonding as due to a four-electron three-centre bond using only the valence *p*-orbitals (this is equivalent in resonance terms to a resonance hybrid $X - X X^- \leftrightarrow X^- X - X$) or we introduce an appropriate 3*d* orbital, on the central atom, in this case d_{z^2}, to yield two hybrid *dp* orbitals (Fig. 6). The quadrupole coupling constants should in principal

3 centre, four electron   *dp* hybrids

FIG. 12.6. Bonding in the trihalide ions. (a) 3 centre, four electron; (b) *dp* hybrids.

be able to decide what mixture of these two extremes constitutes the real ground-state structure of the system since any use of *d*-orbitals will result in a deficiency of *p*-electrons as calculated from the coupling constants. This test can of course only be made if the coupling data for all three sites are available, and in these circumstances it will not matter even if the ion is unsymmetrical since the same rule holds, namely that the total p-σ electron population should add up to 4.0.

For a number of triiodides and dichloroiodides coupling data for all three centres are available. For the tribromides only two of the three different bromine atoms have had their coupling constants determined, and all these data are collected in Table 32. Table 33 shows the results of the Townes and Dailey analyses of the data and it is gratifying to see how close the sum of the p-orbital populations is to 4·0 in all cases. At first sight this would

TABLE 12.32. Quadrupole coupling constants (Mc/s) of trihalide anions

Molecule	Nucleus	e^2Qq (Mc/s)	η	Quadrupole reference	Structure reference
NH_4I_3	I^{127}	466·8	0·130	148	153
		1725·0	0·151		
		2458·7	0·018		
RbI_3	I^{127}	774·2	0·056	148	
		1449·8	0·058		
		2465·7	—		
CsI_3	I^{127}	819·0	0·02	148	152
		1436·6	0·04		
		2477·5	—		
$KICl_2$	Cl^{35}	36·81	—	68	
		37·53	—		
		38·00	—		
		39·25	—		
	I^{127}	3081	—	149	
		3129	—		
$RbICl_2$	Cl^{35}	—	—	68	
	I^{127}	3094	—	149	
$CsICl_2$	Cl^{35}	39·72	—	68	151
	I^{127}	3099	—	149	
$PBr_4.Br_3$	Br^{79}	573·2	—	150	
		806·0	—		
$CsBr_3$	Br^{79}	503·4	—	150	
		814·0	—		
NH_4Br_3	Br^{79}	816·4	—	150	

appear to dispose entirely of d-orbitals in these compounds but it should be noted that any s-p hybridization of the sigma orbitals of the terminal halogen atoms would result in a total p-electron population of greater than 4·0. It is thus conceivable that the observed total population is due to the cancellation of the effects of s-hybridization on the terminal atoms and d-hybridization on the central one. However if the Br_3^- ions, when the coupling data is finally complete, also yield a total population of 4·0, it

TABLE 12.33. p-σ populations in linear trihalide ions $[XYZ]^-$

Molecule	p-σ populations X	Y	Z	Total
NH_4I_3	1·80	0·95	1·25	4·00
RbI_3	1·66	0·94	1·37	3·97
CsI_3	1·64	0·94	1·37	3·95
$KICl_2$	1·65	0·65	1·65	3·95
$CsICl_2$	1·64	0·65	1·64	3·93

would be unlikely for three coincidences to have occurred and evidence for the three-centre four-electron structure would be very strong. Indeed, it is already very probable that this is the correct formulation. The actual magnitudes of the p-populations at the various sites are quite well in accord with the predictions of simple Hückel theory. For a linear symmetrical homonuclear molecule the three-centre four-electron bond should have a p-electron population of 1·5 on the terminal atoms and 1·0 on the central atom. If the two outer atoms are more electronegative than the centre one, as in ICl_2^-, then the population at the outer atoms should be greater than 1·5 while at the central atom it should be less than 1·0. If in a homonuclear molecule one bond is longer than the other then the terminal atom at the long bond should have a population more than 1·5 while at the terminal atom of the short bond it should be less than 1·5. The population at the central atom should be but little affected by asymmetry. All these predictions are observed in the data presented in Table 33.

This is a convenient point to place the results for the HCl_2^- ion which in some senses is related to the trihalide ions. Both symmetrical and non-symmetrical forms of this exist and the quadrupole coupling data are shown in Table 34. The much lower coupling constant of the symmetrical ion compared with the ICl_2^- ion is of course the consequence of the lower electronegativity of the hydrogen atom.

TABLE 12.34. ^{35}Cl quadrupole resonance frequencies (Mc/s) of $[H_2Cl]^-$ ion

$[(C_2H_5)_4N]H_2Cl$	$11{\cdot}89 \pm 0{\cdot}03$	154
$[(CH_3)_4N]H_2Cl$	20·22	155

Related to the triiodide is the pentaiodide ion. This is non-linear with a central bent I_3 section, $I\hat{I}I$ angle of 94°, and with the two terminal atoms each approximately collinear with its corresponding I-I fragment. Partial

I^{127} quadrupole coupling data are available[51] for this species but they cannot be discussed until a complete study has been made.

The ICl_4^- ion has the planar regular tetragonal structure with a central iodine atom. From the point of view of its electronic structure we can consider it as two non-interacting ICl_2^- fragments (Fig. 7). The coupling constants (Table 35) are in excellent agreement with this formulation, the σ-p populations for each ICl_2 fragment adding up to very nearly 4·0 as theory requires.

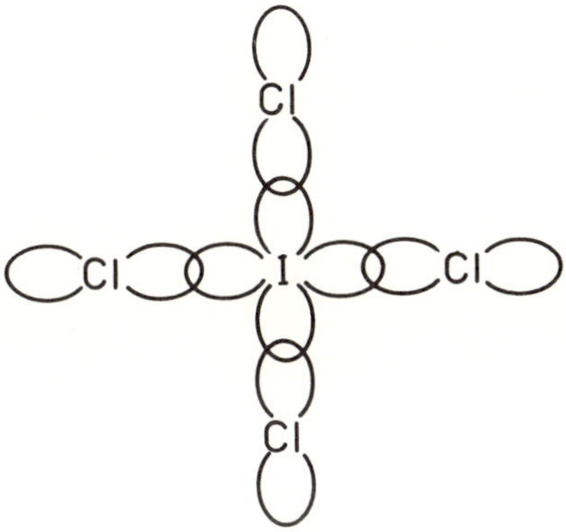

FIG. 12.7. Three-centre bonds in ICl_4^-.

TABLE 12.35. Quadrupole coupling constants (Mc/s) of tetrachloroiodides and related compounds

Molecule	Nucleus	e^2Qq(Mc/s)	p-σ population	Reference
$NaICl_4.2H_2O$	Cl^{35}	41·18	—	68
		44·35	—	
		45·77	—	
		46·52	—	
	I^{127}	3078	—	
$KICl_4$	Cl^{35}	44·74	1·59	68
	I^{127}	3184	0·61	149
$KICl_4.H_2O$	Cl^{35}	40·27	—	68
		49·71		
		56·34		
	I^{127}	3059	—	149
$RbICl_4$	Cl^{35}	44·78	1·59	68
	I^{127}	3060	0·65	149
		3137		
$CsICl_4$	Cl^{35}	44·40	—	—
		45·18	1·59	68
	I^{127}	3021	0·68	148
XeF_4	Xe^{129m}	2666	0·47	156
$XeCl_4$	Xe^{129m}	1635	1·06	156

Thus in this system also it does not appear necessary to introduce the *d*-orbitals. Related to the ICl_4^- cation is the rare gas fluoride XeF_4 for which Mössbauer spectroscopy has yielded the coupling constant of the Xe nucleus. The corresponding $XeCl_4$, which has not yet been prepared in the laboratory, has been studied by Mössbauer emission spectroscopy in the following very ingenious way.

$$[I^{129}Cl_4]^- \xrightarrow{B} Xe^{129m}Cl_4 \xrightarrow[\text{Mössbauer}]{\gamma} Xe^{129}Cl_4 \qquad (14)$$

We thus have a way of studying non-existent compounds! The values of the Xe^{129} coupling constants are in agreement with the relative electronegativities of fluorine and chlorine.

Microwave studies[61] of gaseous ClF_3 and X-ray diffraction studies[62] of crystalline ClF_3 show that in both phases the molecule has the planar T-shaped configuration shown in Fig. 8. The dimensions and angles are those from the crystalline phase; those in the gas-phase are only a little different from these. (In the gas phase the $F\widehat{Cl}F$ angle is 87·9°).

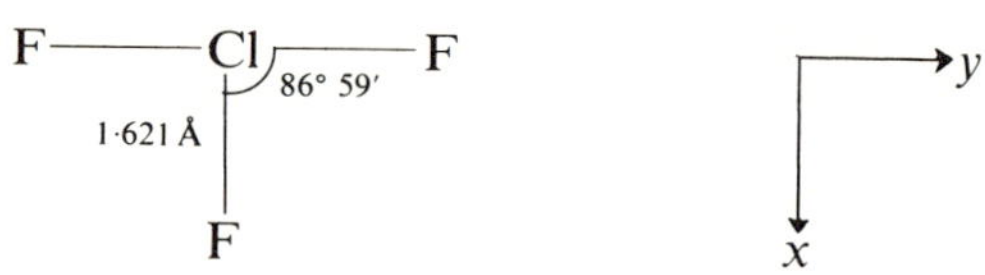

FIG. 12.8. The structure of ClF_3 and the axes to which the atomic orbitals in the text are referred.

In addition to giving the dimensions of the molecule the microwave investigations also yielded the values of the ^{35}Cl coupling tensor, and since ClF_3 has C_{2v} symmetry with the chlorine atom on the symmetry axis, the inertial axes and ^{35}Cl field-gradient axes coincide. The coupling constants are -65 Mc/s along the unique Cl—F bond, 146 Mc/s perpendicular to the molecular plane and -81 Mc/s along the third axis, these results being in very satisfactory agreement with the solid state results[63] ($2\nu_Q = 150{\cdot}26$ Mc/s). These results can be analysed to yield the populations of the p_x and p_y valence orbitals if it is assumed that the p_z orbital contains two electrons. In this way if $e^2Qq_0 = 109$ Mc/s we have:

$$a_x = 0{\cdot}71; \qquad a_y = 0{\cdot}61; \qquad a_z = (2{\cdot}0).$$

The central chlorine atom in this molecule bears a positive charge (+1·7 units from the above calculations) so that the value $e^2Qq_0 = 109$ Mc/s is probably too small. Application of Townes and Dailey's $(1 + \varepsilon)$ correction factor brings these values up to approximately $a_x = 0{\cdot}9$; $a_y = 0{\cdot}8$; $a_z = (2{\cdot}0)$.

It is interesting now to compare these values with those yielded by a recent S.C.F. calculation on the valence orbitals. Two calculations were made, one with just the valence 3*s* and 3*p* orbitals and another in which the 3*d* orbitals were included.[64] For the *s-p* set the values were $a_x = 1{\cdot}159$, $a_y = 1{\cdot}109$, $a_z = 2{\cdot}000$ while the *s-p-d* set yielded $a_x = 0{\cdot}788$, $a_y = 0{\cdot}861$, $a_z = 1{\cdot}966$. Thus while the absolute magnitudes of the occupation numbers are much better reproduced by the *s-p-d* set the relative values of a_x and a_y are given only by the *sp* set. Nevertheless in both calculations $a_x - a_y$ is small and on the whole the coupling constants favour the *d*-orbital model.

Related to ClF_3 is the molecule $C_6H_5ICl_2$ which has been shown by X-ray crystallography to have the structure

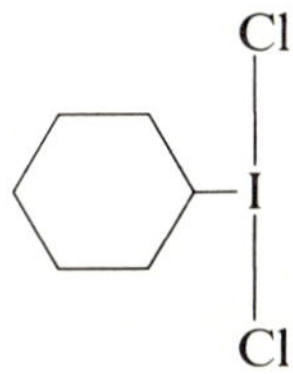

FIG. 12.9. The structure of $C_6H_5ICl_2$.

shown in Fig. 9 where the Cl—I—Cl fragment is perpendicular to the plane of the benzene ring. The ^{35}Cl resonance frequencies[65] at 27°C are 22·406 and 24·260 Mc/s. This implies a total *p*-σ population of almost 1·6 at each chlorine atom.

Although the S.C.F. calculations on ClF_3 appear to favour a *d*-orbital description of the bonding it is perhaps worthwhile to look at the predictions of a simple Hückel calculation for these systems. The bond forming the vertical arm of the "T" may be thought of as a normal single bond while the linear F—Cl—F or Cl—I—Cl makes up a three-centre four-electron bond. This is the picture we used above to describe the trihalide negative ions (Fig. 6) to which indeed the present molecules are related. As we have seen Hückel theory then predicts a central atom p_y-σ population in ClF_3 of less than 1·0—as observed—while in $(C_6H_5)ICl_2$ the terminal p_σ populations would be expected to be greater than 1·5, again as observed. This discussion serves to bring out the differences between the predictions of Hückel theory which is well-known to exaggerate charge separations and those of the S.C.F. calculations typified by those given above for ClF_3. All conclusions about, say, *d*-orbital participation in bonding must thus be treated with great caution.

There remains two other polyhalogen molecules for which coupling constants are available. The molecular structure of I_2Cl_6 determined by X-ray crystallography[66] is shown in Fig. 10. The departure from 90° inter-bond angles is small and once again we may consider to a first approximation that only valency σ-*p* orbitals are used in bond-formation. The low asymmetry parameter of the iodine nucleus ($\eta = 0{\cdot}0772$) is completely in accordance with this assumption. In this approximation the electronic structure may be considered to be made up of four non-interacting Cl—I—Cl four-electron three-centre bonds and one might expect the resonance frequencies (Table 36)

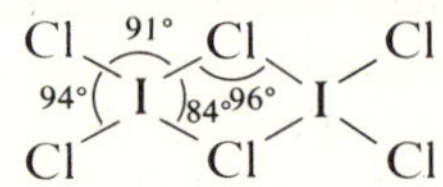

FIG. 12.10. The crystal structure of I_2Cl_6.

to be not too different from those observed in the ICl_2^- and ICl_4^- anions. This is certainly the case for the iodine atom, the ^{35}Cl frequencies have however risen significantly above those of the chloroiodide anions and in fact above the limit of 27 Mc/s which would have been expected for the limiting population of 1·5 for the terminal members of a homonuclear X_3

TABLE 12.36. Halogen quadrupole resonance frequencies in I_2Cl_6

Molecule	Nucleus	Resonance frequencies (Mc/s)	*p-σ* population	Ref.
I_2Cl_6	Cl^{35} (terminal)	33·413	—	67
		34·918	1·38	
	Cl^{35} (bridging)	13·740	1·75	69
	I^{127}	451·22, 452·87 (ν_1)	0·69	67
		895·76, 899·70 (ν_2)		

molecule. The original measurements on this compound only revealed the two higher ^{35}Cl frequencies and Bray[67] attributed the higher of these to the four external chlorine atoms and the lower to the bridging atoms; Cornwell and Yamasaki,[68] however, pointed out that the external chlorine atoms fall into two non-equivalent sets for which therefore slightly differing resonance frequencies would be expected. The two observed frequencies they therefore reassigned to the external halogen atoms and suggested that the bridging chlorine atoms were not observed and had a much lower frequency. This hypothesis was proved correct by the subsequent[69] detection of the low frequency resonances for the bridging chlorine atoms. The sum of the *p-σ* populations is not too far removed from 4·0 as the very simple description as four ICl_2 fragments requires. Evans and Lo[69] measured the Cl^{35} resonance frequencies in the bent ICl_2^+ ion[70] (38·690, 39·086 Mc/s).

G. GROUPS I AND II. THE ALKALI AND ALKALINE EARTH METALS

It is not proposed to discuss here the coupling constants of crystalline compounds of the alkali and alkaline earth elements. Almost all these compounds form essentially ionic crystals and the small coupling constants observed are completely consistent with this. As already discussed in Chapters 5 and 6, a satisfactory quantitative estimation of the expected values of the coupling constants in such cases is rendered very difficult by the problems of correctly evaluating the interionic field-gradient and the Sternheimer shielding parameter. The alkali metal halides, however, form, at sufficiently high temperatures, diatomic molecules which have been extensively studied in the gas-phase by molecular beam and microwave spectroscopy. As will be seen the coupling constants are almost all those to be expected for a molecule with an ion-pair structure but they present several points of interest and in addition provide useful standards with which the coupling constants of other diatomic metal halides such as those of the monovalent Group III elements can be compared.

Table 37 shows the coupling constants of the alkali metal halides, either measured in various vibrational states or as an average value over all populated vibrational states. If the molecules are to be considered as ion-pairs then the field-gradients should be easily calculable from the internuclear distance and the Sternheimer shielding factor.

$$q = \pm \frac{2}{R_{MX}^3}(1 - \gamma_\infty). \tag{15}$$

Irrespective of the actual value of the latter term q should be proportional to R_{MX}^{-3}, at least if the Sternheimer shielding is constant. Figure 11 shows the relationship between R_{MX}^{-3} (Table 38) and the metal atom coupling constant ($\mathbf{v} = 0$) and clearly, with the exception of LiF, the expected linear relationship obtains. The lithium halides are of course the members of this class of compounds in which the effects of covalency should be a maximum and the Li^7 coupling constant of LiF has already been discussed in Chapter 6.

Despite the good linear relationship between R_{MX}^{-3} and the coupling constants of the alkali metal ions, the lines in Fig. 11 do not extrapolate to zero as equation (15) implies. Several authors[71,72] have pointed out that the electrostatic dipole moment induced in the neighbouring ion, X (given by $\mu_X = \alpha_x/R_{MX}^2$, where α_x is the dipole polarizability of X) produces a field-gradient $6\mu/R_{MX}^4$ at the nucleus of M so that equation (15) should be modified to

$$q_M = \frac{2}{R_{MX}^3}(1 - \gamma_\alpha)\left(1 + \frac{3\alpha_x}{R_{MX}^3}\right). \tag{16}$$

TABLE 12.37. Nuclear quadrupole coupling constants of alkali-metal halide gases

Compound	Nucleus	Coupling Constant (Mc/s) $v = 0$	$v = 1$	$v = 2$	$v = 3$	$v = 4$	Ref.
LiF	Li^{7}	0·41602	—	—	—	—	157
LiCl	Li^{7}	0·192*	—	—	—	—	158
LiBr	Li^{7}	0·184*	—	—	—	—	158
LiI	Li^{7}	0·172*	—	—	—	—	158
NaF	Na^{23}	−8·4401	−8·3279	−8·2205	—	—	159, 160
NaCl	Na^{23}	−5·40*	—	—	—	—	158
NaBr	Na^{23}	−4·68*	—	—	—	—	158
NaI	Na^{23}	−3·88*	—	—	—	—	158
KF	K^{39}	−7·933878	−7·837739	−7·742809	−7·649539	−7·556619	161
KCl	K^{39}	−5·66843	−5·61579	−5·56469	−5·51639	−5·46502	161
KBr	K^{39}	−5·003	−4·984	−4·915	—	—	162
RbF	Rb^{85}	−70·342	−69·555	−68·779	−68·013	−67·259	163, 164, 165
RbCl	Rb^{85}	−52·675	−52·306	−51·903	—	—	—
CsF	Cs^{133}	1·2383	1·224	—	—	—	166
CsCl	Cs^{133}	<4	—	—	—	—	167
LiCl	Cl^{35}	−3·07172	−3·4793	−3·8730	−4·250	—	168
KCl	Cl^{35}	−0·0585	−0·1003	−0·2588	−0·4137	−0·5544	161
RbCl	Cl^{35}	0·744	0·612	0·470	—	—	—
LiBr	Br^{79}	38·463	41·304	44·067	46·751	49·355	169
NaBr	Br^{79}	58	—	—	—	—	167
KBr	Br^{79}	10·244	11·224	12·204	—	—	161
LiI	I^{127}	−198·780	−207·453	—	—	—	170
NaI	I^{127}	−259·87	−264·52	−267·59	−271·03	—	167
KI	I^{127}	−60	—	—	—	—	167

* Values averaged over vibrational states.

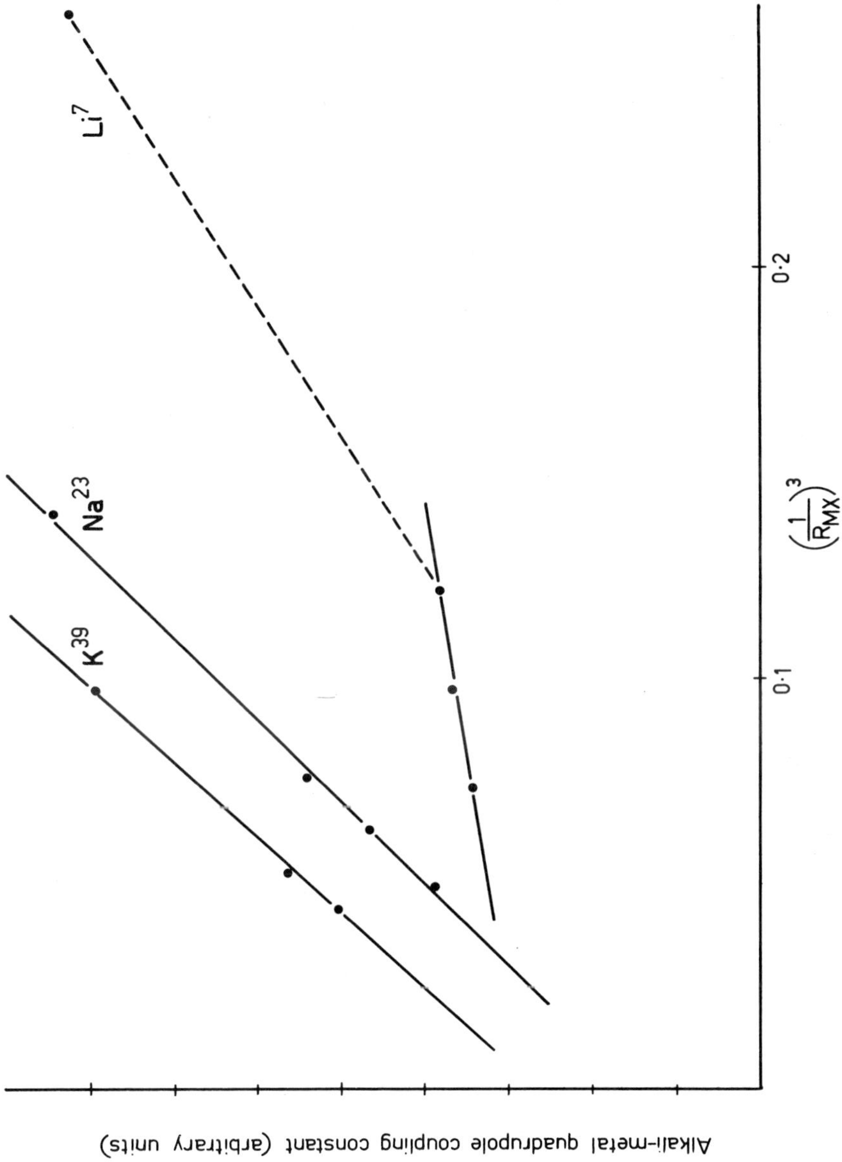

FIG. 12.11. Alkali-metal coupling constants and internuclear distances in the diatomic alkali-metal halides.

TABLE 12.38. Equilibrium internuclear distances (Å) in gaseous diatomic alkali-metal halides. Data taken from Ref. 171

	F	Cl	Br	I
Li	1·563892	2·02067	2·17042	2·39191
Na	1·92595	2·3606	2·50201	2·71143
K	2·17144	2·6666	2·82075	3·04781
Rb	2·26554	2·78670	2·94471	3·17684
Cs	2·3453	2·9062	3·07221	3·31515

According to this equation the quantity $e^2Qq_M.R_{MX}^3(1 + 3\alpha_x/R_{MX}^3)^{-1}$ ($\equiv \varkappa$) should be equal to the theoretical value of $(1 - \gamma_\infty)$. Table 39 shows a comparison of $\varkappa$ and $(1 - \gamma_\infty)$ taken from the paper by Burns.[71] The constancy of $\varkappa$ and the agreement between it and $(1 - \gamma_\infty)$ is excellent.

TABLE 12.39. Theoretical values of shielding factors $(1 - \gamma_\infty)$, see Table 5.2, and "experimental" values calculated according to equation 15 (Column headed $e^2Qq_M.R_{MX}^3$) or equation 16 (column headed $\varkappa$)

Molecule	Shielding factors		
	Theoretical	"Experimental"	
	$1 - \gamma_\infty$	$e^2Qq_M.R_{MX}^3$	$\varkappa$
LiF	0·74	1·30	0·788
LiCl	—	1·43	0·681
LiBr	—	1·69	0·759
LiI	—	2·11	0·875
NaF	5·1	7·26	5·18
NaCl	—	10·21	5·94
NaBr	—	10·53	5·75
NaI	—	11·52	5·64
KF	14	15·71	12·03
KCl	—	22·60	14·56
KBr	—	23·04	14·08
RbF	72	46·04	36·07
RbCl	—	66·02	43·91
CsF	145	76·57	59·90

The coupling constants depend to a greater or lesser extent on the vibrational state of the molecule, i.e. on the value of the vibrational quantum number **v**. For the alkali-metal nuclei this can be quite well correlated with the effect of the anharmonicity of the internuclear potential on the average value of R_{MX}^{-3}. The internuclear potential, $V(r)$, may be expressed as a power series

$$V(r) = a_0\xi^2[1 + a_1\xi + a_2\xi^2 \ldots] \tag{17}$$

where $\xi \equiv [(r - r_e)/r_e]$ is the displacement from the equilibrium internuclear distance r_e. The various constants $a_1, a_2 \cdots$ may be derived from vibrational and rotational spectroscopic data. It can be shown[73] that the average values of the various powers of the displacements are given by

$$\langle \xi \rangle_{\mathbf{v}} = -3a_1\left(\frac{B_e}{\omega_e}\right)(\mathbf{v} + \tfrac{1}{2}) \tag{18}$$

$$\langle \xi^2 \rangle_{\mathbf{v}} = 2\left(\frac{B_e}{\omega_e}\right)(\mathbf{v} + \tfrac{1}{2}) \tag{19}$$

where B_e is the equilibrium rotational constant and ω_e the fundamental vibrational frequency. Since

$$\left(\frac{r}{r_e}\right)^n = 1 + n\xi + \tfrac{1}{2}n(n - 1)\xi^2 \ldots \tag{20}$$

then

$$\langle r^n \rangle_{\mathbf{v}} = r_e^n\left[1 + \frac{B_e}{\omega_e}\{n(n - 1) - 3na_1\}(\mathbf{v} + \tfrac{1}{2})\right]. \tag{21}$$

Empirically, the field-gradient at the alkali metal nucleus can be fairly well expressed by an equation of the form

$$q_{\mathrm{M}} = q_0 + q_{\mathbf{v}}(\mathbf{v} + \tfrac{1}{2}). \tag{22}$$

Substitution of the expression for $\langle \mathrm{R}_{\mathrm{MX}}^3 \rangle$ given by equation (21) in equation (16) gives the theoretical value of $q_{\mathbf{v}}$

$$\frac{q_{\mathbf{v}}}{q_{\mathrm{M}}} = \frac{B_e}{\omega_e}\left[(12 + 9a_1) + \frac{3\alpha_x}{\mathrm{R}_{\mathrm{MX}}^3}(42 + 18a_1)\right]. \tag{23}$$

A comparison between theory and experiment for the values of $q_{\mathbf{v}}$, taken from a paper by de Wijn,[74] shows the excellent agreement obtainable (Table 40).

No such simple theory can be used to calculate the coupling constants of the halide ions in these molecules, nor can it be improved by the addition of small amounts of covalent character which, since the alkali metal bonding orbital would be essentially the valence *s*-orbital, would little affect the previous excellent agreement between the theoretical and experimental alkali-metal coupling constants. A further difficulty lies in the much greater sensitivity of the halogen coupling constant to the vibrational state of the molecule. In the most extreme example of this phenomenon, KCl, the ^{35}Cl coupling constant even changes sign on going from $\mathbf{v} = 0$ to $\mathbf{v} = 1$. Within the framework of the ionic model for the structure of these compounds various attempts have been made to obtain better theoretical values for the field-gradient at the halogen nuclei. De Wijn[74] has suggested that partial quenching of the shielding takes place, essentially owing to the anion-cation

TABLE 12.40. The effect of vibrational state on field-gradient at the alkali metal in the diatomic alkali metal halides. The theoretical values,[78] calculated from equation 23, include an allowance for a small amount of covalent character in these molecules

Molecule	q_v	
	Experiment	Theory
NaF	0·014	0·011
KF	0·015	0·017
KCl	0·007	0·007
KBr	0·007	0·005
RbCl	0·020	0·018

contact preventing an expansion of the halogen radial function along the direction of the metal halogen bond. Das and Karplus[72] have investigated the effect of overlap on the field-gradient but neither of these approaches can be said to have yielded much more than to show that the problem is an extremely complex one. The actual magnitudes of the halogen coupling constants together with the success of the ionic model for the alkali metal nuclei leaves no doubt that these molecules are essentially ion pairs. The halogen ions unlike the alkali metals are however easily distorted and it is difficult to calculate exactly the effects of this distortion.

Although silver is not a member of the typical Groups it is both convenient and interesting to discuss here the coupling constants for gaseous and diatomic AgCl and AgBr, shown in Table 41. It has of course long been known that the bonding in *crystalline* silver halides involved considerable covalent character, even though the crystal structures are such that the component atoms have high coordination numbers. This covalency is very clearly and convincingly demonstrated by the halogen coupling constants which are an order of magnitude greater than those of corresponding alkali metal halides. Assuming that the halogen lone-pair pi-electrons play no part in the bonding —this assumption is of course by no means self-evident—then these couplings imply about 35–40 per cent covalent character for the metal halogen bonds. This is rather more than for the halides of the monovalent Group III elements

TABLE 12.41. Halogen coupling constants of gaseous diatomic silver halides

Compound	Nucleus	Coupling constant (Mc/s)	Ref.
AgCl	Cl^{35}	−37·32	105
AgBr	Br^{79}	307	106

(see Section C), rather surprisingly in view of the fact that for silver the valency orbital is presumably an *s*-orbital while it is a *p*-orbital for the Group III compounds.

H. THE HYDROGEN HALIDES

We will here be concerned only with the values of the halogen coupling constants; the deuterium couplings, where available, have already been discussed in Chapter 6. Table 42 shows the values for the hydrogen, deuterium and tritium halides. As is usual for covalent compounds the change of coupling constant with isotopic substitution is quite small, even here where such effects would be expected to be at a maximum. The halogen sigma orbital populations, calculated according to equation (11), Chapter 7, assuming no *sp* hybridization of the halogen sigma orbital, are also given in Table 42. As would be expected on the grounds of the relative electronegativities of the three halogens, the covalent character of the bond increases in the sense I > Br > Cl. Their actual magnitudes are also much as would be expected from the electronegativity difference between hydrogen and the halogens (see also Fig. 7.4).

TABLE 12.42. Halogen coupling constants in hydrogen, deuterium and tritium halides

Compound	Nucleus	Coupling constant (Mc/s)	Halogen sigma population	Ref.
HCl	Cl^{35}	−67·51	1·39	172
DCl	Cl^{35}	−67·41	—	173, 174
TCl	Cl^{35}	−67·21	—	175
HBr	Br^{79}	535·44	1·30	172
DBr	Br^{79}	530·58	—	173, 174, 176
TBr	Br^{79}	528·75	—	175, 177
HI	I^{127}	−1831·07	1·20	178
DI	I^{127}	−1823·38	—	173, 176 179
TI	I^{127}	−1822·61	—	177

REFERENCES

1a. R. W. Dixon and N. Bloembergen, *J. Chem. Phys.* **41**, 1720, 1739 (1964).
1b. H. C. Clark, R. J. O'Brien and J. Trotter, *Proc. Chem. Soc.* **85** (1963).
2. N. Kasai, K. Yasuda and R. Okawara, *J. Organometal. Chem.* **3**, 172 (1965).
3. M. J. Janssen, J. G. A. Luijten and G. J. M. van der Kerk, *Rec. Trav. chim. Pays-Bas* **82**, 90 (1963).
4. A. H. Muir, K. J. Ando and H. M. Cadogan, "Mössbauer Effect Data Index 1958–1965", Interscience Publishers, New York 1966.
5. C. Froese, *J. Chem. Phys.* **45**, 1417 (1966).

6. M. A. Rollier and A. Riva, *Gazz. Chim. Ital.* **77**, 361 (1947).
7. T. Chiba, *J. Phys. Soc. Japan* **13**, 860 (1958).
8. P. Perkins, *J. Chem. Soc.* A, 1218 (1967).
9. P. Love, *J. Chem. Phys.* **39**, 3044 (1963).
10. G. Wessel, *Phys. Rev.* **92**, 1581 (1953).
11. A. Bassompierre, *C.r. Lebd. Acad. Sci. Paris* **237**, 39 (1953).
12. H. K. Clark and J. L. Hoard, *J. Am. Chem. Soc.* **65**, 2115 (1943).
13. A. H. Silver and P. J. Bray, *J. Chem. Phys.* **31**, 247 (1959).
14. P. J. Bray, J. O. Edwards, J. G. O'Keefe, V. F. Ross and I. Iatsuzaki, *J. Chem. Phys.* **35**, 435 (1961).
15. B. E. Warren, H. Krutter and O. Morningstar, *J. Am. Ceram. Soc.* **19**, 202 (1936).
16. A. F. Wells, "Structural Inorganic Chemistry", 3rd Edition Oxford 1962.
17. P. A. Renes and C. H. McGillavry, *Rec. Trav. chim.* **64**, 275 (1945).
18. P. A. Casabella, P. J. Bray and R. G. Barnes, *J. Chem. Phys.* **30**, 1393 (1959).
19. H. Lew, *Phys. Rev.* **76**, 1086 (1949).
20. P. Kusch and T. G. Eck, *Phys. Rev.* **94**, 1799 (1954).
21a. A. H. Barrett and M. Mandel, *Phys. Rev.* **109**, 1572 (1958).
21b. D. R. Lide, *J. Chem. Phys.* **42**, 1013 (1965).
22. T. Rowland, *Acta Met.* **3**, 74 (1955).
23. R. Hammond and W. D. Knight, *Phys. Rev.* **120**, 762 (1960).
24. R. R. Hewitt and T. T. Taylor, *Phys. Rev.* **125**, 524 (1962).
25. S. L. Carr, B. B. Garrett and W. G. Moulton, *J. Chem. Phys.* **47**, 1170 (1967).
26. D. J. Millen and J. Pannell, *J. Chem. Soc.* 1322 (1961).
27. D. J. Millen and K. M. Sinnott, *J. Chem. Soc.* 350 (1958); Y. Morino and T. Tanaka, *J. Mol. Spec.* **16**, 179 (1965).
28. T. L. Weatherly and Q. Williams, *J. Chem. Phys.* **25**, 717 (1956).
29. K. Shimomura, *J. Phys. Soc. Japan* **12**, 1386 (1957).
30. H. Braekken, *Kong. Norsk. Vidensk. Selsk, Forhand.* **8**, 1 (1953).
31. I. Lindqvist and A. Niggli, *J. Inorg. Nucl. Chem.* **2**, 345 (1956).
32. R. G. Wyckoff, "Crystal Structures", Interscience, New York 1948.
33. Y. Abe, *J. Phys. Soc. Japan* **13**, 918 (1958).
34. I. Lindgren in "Perturbed Angular Correlations", E. Karlsson, E. Matthias and K. Siegbahn (Eds.), North Holland Publ. Co. Amsterdam 1964.
35. H. Robinson, H. G. Dehmelt and W. Gordy, *Phys. Rev.* **89**, 1305 (1953).
36. D. McCall and H. Gutowsky, *J. Chem. Phys.* **21**, 1300 (1953).
37. D. Clark, H. M. Powell and A. F. Wells, *J. Chem. Soc.* **642** (1942).
38. J. V. DiLorenzo and R. F. Schneider, *Inorg. Chem.* **6**, 766 (1967).
39. S. M. Ohlberg, *J. Am. Chem. Soc.* **81**, 811 (1959).
40. J. E. Griffiths, Reported in ref. 19 of R. R. Holmes, R. P. Carter and G. E. Peterson, *Inorg. Chem.* **3**, 1748 (1964).
41. K. Tsukada, *J. Phys. Soc. Japan* **10**, 60 (1955).
42. E. A. C. Lucken, "Proc. XIth Colloque Ampère", Eindhoven 1962.
43. R. H. Jackson and D. J. Millen, *Proc. Chem. Soc.* 10 (1959).
44. J. S. Rigden and S. S. Butcher, *J. Chem. Phys.* **40**, 2109 (1964).
45. R. F. Curl, *J. Chem. Phys.* **37**, 779 (1962).
46. J. L. Ragle, *J. Chem. Phys.* **32**, 403 (1960).
47. H. Zeldes and R. Livingston, *J. Chem. Phys.* **26**, 1102 (1957), and references therein.
48. "Interatomic Distances and Configurations of Molecules and Ions", Chemical Society Special Publications 11 and 18, London 1958 and 1965.

49. R. Livingston and H. Zeldes, *J. Chem. Phys.* **26**, 351 (1957).
50. K. Shimomura and N. Inoue, *J. Phys. Soc. Japan* **14**, 86 (1959).
51. S. Kojima, K. Tsukada, S. Ogawa and A. Shimauchi, *J. Chem. Phys.* **23**, 1963 (1955).
52. R. Kewley, K. S. R. Murty and T. M. Sugden, *Trans. Farad. Soc.* **56**, 1732 (1960).
53. E. W. Neuvar and A. W. Jache, *J. Chem. Phys.* **39**, 596 (1963).
54. A. Julg, *Tetrahedron* **19**, Suppl. 2, p. 139 (1963).
55. M. A. Whitehead and H. H. Jaffé, *J. Chem. Phys.* **34**, 2204 (1961); *J. Chem. Phys.* **36**, 3006 (1962).
56. M. A. Whitehead and H. H. Jaffé, *Theoretica Chim. Acta* **1**, 209 (1963).
57. J. Hinze and H. H. Jaffé, *J. Am. Chem. Soc.* **84**, 540 (1962).
58. R. Bersohn, *J. Chem. Phys.* **36**, 3445 (1962).
59. C. H. Townes and B. P. Dailey, *J. Chem. Phys.* **20**, 35 (1952).
60. H. Robinson, H. G. Dehmelt and W. Gordy, *J. Chem. Phys.* **22**, 511 (1954).
61. D. F. Smith, *J. Chem. Phys.* **21**, 609 (1953).
62. R. D. Burbank and F. N. Bensey, *J. Chem. Phys.* **21**, 602 (1953).
63. R. Livingston, *J. Phys. Chem.* **57**, 496 (1953).
64. D. P. Santry and G. A. Segal, *J. Chem. Phys.* **47**, 158 (1967).
65. J. S. Evans and G. Y-S. Lo, *J. phys. Chem.* **71**, 2730 (1967).
66. K. H. Boswijk and E. H. Wiebenga, *Acta Cryst.* **7**, 417 (1954).
67. P. J. Bray, *J. Chem. Phys.* **23**, 703 (1955).
68. C. D. Cornwell and R. S. Yamasaki, *J. Chem. Phys.* **27**, 1060 (1957).
69. J. C. Evans and G. Y-S. Lo, *Inorg. Chem.* **6**, 836 (1967).
70. C. G. Vonk and E. H. Wiebenga, *Acta Cryst.* **12**, 859 (1959).
71. G. Burns, *Bull. Am. Phys. Soc.* II, **3**, 147 (1958); *Phys. Rev.* **115**, 357 (1959).
72. T. P. Das and M. Karplus, *J. Chem. Phys.* **30**, 849 (1959); *J. Chem. Phys.* **42**, 2885 (1965).
73. D. R. Herschbach and V. W. Laurie, *J. Chem. Phys.* **37**, 1668 (1962).
74. H. W. de Wijn, *J. Chem. Phys.* **44**, 810 (1966).
75. A. Schlawlow, *J. Chem. Phys.* **22**, 1211 (1954).
76. J. E. Griffiths and K. B. McAfee, *Proc. Chem. Soc.* **456** (1961).
77. C. H. Townes, J. M. Mays and B. P. Dailey, *Phys. Rev.* **76**, 700 (1949).
78. V. W. Laurie, *J. Chem. Phys.* **30**, 1210 (1959).
79. J. Hoeft, *Z. Naturf.* **20a**, 1327 (1965).
80. R. H. Herber and H. A. Stöckler, unpublished data.
81. R. H. Herber, H. A. Stöckler and W. T. Reichle, *J. Chem. Phys.* **42**, 2447 (1965).
82. V. A. Bryukhanov *et al.*, *Soviet Phys. JETP*, (Engl. Trans.) **16**, 321 (1963).
83. R. H. Herber, unpublished data.
84. V. A. Bryukhanov, V. I. Goldanskii, N. N. Delyagin, E. F. Makarov and V. S. Shpinel, *Soviet Phys. JETP* **42**, 637 (1962).
85. A. Yu Aleksandrov *et al. Soviet Phys. JETP* (Eng. Trans.) **16**, 879 (1963).
86. V. I. Goldanskii, E. F. Makarov, R. A. Stukan, V. A. Trukhtanov, V. V. Khrapov, *Dokl. Akad. Nauk SSSR* **151**, 357 (1961).
87. A. Yu. Aleksandrov *et al.*, *Soviet Phys. JETP* (Eng. Trans.) **16**, 1467 (1963).
88. M. Cordey-Hayes, *J. Inorg. Nucl. Chem.* **26**, 2306 (1964).
89. R. H. Herber and G. I. Parisi (to be published) cf. "Recent Advances in Mössbauer Technology", New England Nuclear T.M.C. Symposium, Jan. 1965 (in press).
90. T. C. Gibb and N. N. Greenwood, *J. Chem. Soc.* A, **43** (1966).
91. V. I. Goldanskii, V. Ya. Rogaev and V. V. Khrapov, *Dokl. Akad. Nauk SSSR* **156**, 909 (1964).

92. M. Cordey-Hayes, *J. Inorg. Nucl. Chem.* **26**, 915 (1964).
93. M. Cordey-Hayes, "Applications of the Mössbauer Effect in Chemistry and Solid-State Physics", International Atomic Energy Agency, Technical Reports Series No. 50, Vienna 1966.
94. W. G. Laurita and W. S. Koski, *J. Am. Chem. Soc.* **81**, 3179 (1959).
95. A. H. Silver and P. J. Bray, *J. Chem. Phys.* **32**, 288 (1960).
96. M. A. Ring and W. S. Koski, *J. Chem. Phys.* **35**, 381 (1961).
97. R. G. Barnes and S. L. Segel, *J. Chem. Phys.* **25**, 180 (1956).
98. P. A. Casabella and N. C. Miller, *J. Chem. Phys.* **40**, 1363 (1964).
99. R. G. Barnes, S. L. Segel, P. J. Bray and P. A. Casabella, *J. Chem. Phys.* **26**, 1345 (1957).
100. D. Nakamura, H. Watanabe and M. Kubo, *Bull. Chem. Soc. Japan* **34**, 142 (1961).
101. H. G. Dehmelt, *J. Chem. Phys.* **21**, 380 (1953).
102. S. Segel, Previously unpublished data listed in "Catalog of Nuclear Quadrupole Interactions and Resonance Frequencies in Solids", Pt. I.
103. H. G. Robinson, *Phys. Rev.* **100**, 1731 (1955).
104. S. Kojima, K. Tsukada, S. Ogawa, A. Shimauchi and Y. Abe, *J. Phys. Soc. Japan* **9**, 805 (1954).
105. R. G. Barnes and P. J. Bray, *J. Chem. Phys.* **23**, 1177 (1955).
106. R. R. Holmes, R. P. Carter and G. E. Peterson, *Inorg. Chem.* **3**, 1748 (1964).
107. G. S. Blevins, A. W. Jache and W. Gordy, *Phys. Rev.* **97**, 684 (1955).
108. D. R. Lide, *Spectrochim. Acta* **15**, 473 (1959).
109. P. Kisliuk, *J. Chem. Phys.* **22**, 86 (1954).
110. P. Kisliuk and S. Geschwind, *J. Chem. Phys.* **21**, 828 (1953).
111. A. W. Jache, G. S. Blevins and W. Gordy, *Phys. Rev.* **97**, 680 (1955).
112. R. G. Barnes and P. J. Bray, *J. Chem. Phys.* **23**, 407 (1955).
113. S. Ogawa, *J. Phys. Soc. Japan* **13**, 618 (1958).
114. E. D. Swiger, P. J. Green, G. L. McKown and J. D. Graybeal, *J. Phys. Chem.* **69**, 949 (1965).
115. J. V. DiLorenzo and R. F. Schneider, *J. Chem. Phys.* **46**, 2434 (1967).
116. R. R. Holmes, *J. Chem. Phys.* **46**, 3718 (1967).
117. M. A. Whitehead, *Canad. J. Chem.* **42**, 1212 (1964).
118. T. Wang, *Phys. Rev.* **99**, 566 (1955).
119. T. Kushida, G. B. Benedek and N. Bloembergen, *Phys. Rev.* **104**, 1364 (1956).
120. L. Anderson, *Bull. Am. Phys. Soc.* **5**, 412 (1960).
121. K. Shimomura, T. Kushida and N. Inoue, *J. Chem. Phys.* **22**, 350 (1954).
122. G. W. Ludwig, *J. Chem. Phys.* **25**, 159 (1956).
123. H. O. Hooper and R. G. Barnes, *J. Chem. Phys.* **33**, 334 (1960).
124. A. Schmitt and W. Zeil, *Z. Naturf.* **18a**, 428 (1963).
125. J. R. Eshbach, R. E. Hillger and M. W. P. Strandberg, *Phys. Rev.* **85**, 532 (1952).
126. C. H. Townes and S. Geschwind, *Phys. Rev.* **44**, 1113 (1948).
127. R. C. Mockler and G. R. Bird, *Phys. Rev.* **98**, 1837 (1955).
128. G. C. Dousmanis, T. M. Sanders, C. H. Townes and H. J. Zeiger, *J. Chem. Phys.* **21**, 1416 (1953).
129. K. Ohno, Y. Mizuno and M. Mizushima, *J. Chem. Phys.* **28**, 691 (1958).
130. G. R. Bird and C. H. Townes, *Phys. Rev.* **94**, 1203 (1954).
131. R. Van Riet, *Ann. Soc. Sci. Bruxelles*, **76**, 56 (1962).
132. H. G. Dehmelt, *Phys. Rev.* **91**, 313 (1953).
133. T. Amano, E. Hirota and Y. Morino, *J. Phys. Soc. Japan* **22**, 399 (1967).
134. S. Geschwind, R. Gunther-Mohr and J. Silvey, *Phys. Rev.* **85**, 474 (1952).

135. B. Rosenblum and A. H. Nethercot, *J. Chem. Phys.* **27**, 828 (1957).
136. M. J. Stevenson and C. H. Townes, *Phys. Rev.* **107**, 635 (1957).
137. W. H. Flygare and J. T. Lowe, *J. Chem. Phys.* **43**, 3654 (1965).
138. W. A. Hardy, G. Silvey, C. H. Townes, B. F. Burke, M. W. F. Strandberg, G. W. Parker and V. W. Cohen, *Phys. Rev.* **92**, 1532 (1953).
139. R. Livingston, *J. Chem. Phys.* **19**, 1434 (1951).
140. S. Kojima, K. Tsukada, A. Shimauchi and Y. Hinaga, *J. Phys. Soc. Japan* **9**, 795 (1954).
141. R. V. Pound, *Phys. Rev.* **82**, 343 (1951).
142. D. A. Gilbert, A. Roberts and P. A. Griswold, *Phys. Rev.* **76**, 1723 (1949).
143. D. F. Smith, M. Tidwell and D. V. P. Williams, *Phys. Rev.* **79**, 1007 (1950).
144. C. H. Townes, F. R. Merritt and B. D. Wright, *Phys. Rev.* **73**, 1334 (1948).
145. D. F. Smith, M. Tidwell and D. V. P. Williams, *Phys. Rev.* **77**, 420 (1950).
146. T. S. Jaseja, *J. Mol. Spectroscopy* **5**, 445 (1960).
147. F. J. Adrian, *J. Chem. Phys.* **38**, 1258 (1963).
148. A. Sasane, D. Nakamura and M. Kubo, *J. Phys. Chem.* **71**, 3249 (1967).
149. R. S. Yamasaki and C. D. Cornwell, *J. Chem. Phys.* **30**, 1265 (1959).
150. G. L. Breneman and R. D. Willet, *J. Phys. Chem.* **71**, 3684 (1967).
151. R. W. G. Wyckoff, *J. Am. Chem. Soc.* **42**, 1100 (1920).
152. H. A. Tasman and K. H. Boswijk, *Acta Cryst.* **8**, 59 (1955).
153. R. C. L. Mooney, *Z. Krist.* **90**, 143 (1935).
154. J. C. Evans and G. Y-S. Lo, *J. Phys. Chem.* **71**, 3697 (1967).
155. J. C. Evans and G. Y-S. Lo, *J. Phys. Chem.* **70**, 2702 (1966).
156. G. J. Perlow and M. R. Perlow, *J. Chem. Phys.* **41**, 1157 (1964).
157. L. Wharton, L. P. Gold and W. Klemperer, *Phys. Rev.* **133**, B270 (1964).
158. R. A. Logan, R. E. Coté and P. Kusch, *Phys. Rev.* **86**, 280 (1952).
159. G. Gräff and G. Werth, *Z. Physik*, **183**, 223 (1965).
160. C. D. Hollowell, A. J. Herbert and K. Street, *J. Chem. Phys.* **41**, 3540 (1964).
161. R. Van Wachem and A. Dymanus, *J. Chem. Phys.* **46**, 3749 (1967).
162. C. A. Lee, R. O. Carlson, B. P. Fabricand and I. I. Rabi, *Phys. Rev.* **86**, 607 (1952); *Phys. Rev.* **91**, 1395 (1953); *Phys. Rev.* 1403 (1953).
163. G. Gräff, R. Schönwasser and M. Tonutti, *Z. Physik*, **199**, 157 (1967).
164. J. C. Zorn, T. C. English, J. T. Dickinson and D. A. Stephenson, *J. Chem. Phys.* **45**, 3731 (1966).
165. P. A. Bonczyk and V. W. Hughes, *Phys. Rev.* **161**, 15 (1967).
166. G. Gräff and O. Runolfsson, *Z. Physik*, **187**, 140 (1965).
167. A. Honig, M. Mandel, M. L. Stitch and C. H. Townes, *Phys. Rev.* **96**, 629 (1954).
168. D. T. F. Marple and J. W. Trischka, *Phys. Rev.* **103**, 597 (1956).
169. A. J. Herbert, F. W. Breivogel and K. Street, *J. Chem. Phys.* **41**, 2368 (1964).
170. F. W. Breivogel, A. J. Herbert and K. Street, *J. Chem. Phys.* **42**, 1555 (1965).
171. D. R. Lide, P. Cahill and L. P. Gold, *J. Chem. Phys.* **40**, 156 (1964).
172. G. Jones and W. Gordy, *Phys. Rev.* **136**, A1229 (1964).
173. M. Cowan and W. Gordy, *Phys. Rev.* **111**, 209 (1958).
174. C. A. Burrus, *J. Chem. Phys.* **31**, 1270 (1959).
175. C. A. Burrus, W. Gordy, B. Benjamin and R. Livingston, *Phys. Rev.* **97**, 1661 (1955).
176. C. A. Burrus, *J. Chem. Phys.* **30**, 976 (1959).
177. B. Rosenblum and A. H. Nethercot, *Phys. Rev.* **97**, 84 (1955).
178. M. Cowan and W. Gordy, *Phys. Rev.* **104**, 551 (1956).
179. C. A. Burrus, *J. Chem. Phys.* **28**, 427 (1958).

CHAPTER 13

COMPOUNDS OF THE TRANSITION AND RARE-EARTH ELEMENTS

A. HALIDES

1. Halogen Coupling Constants in Molecular Crystals

For the transition elements it is not convenient to discuss the results group by group since on the one hand relatively few are available and on the other hand the variable valencies exhibited by the transition elements make the group concept less significant.

In Table 1 are shown the halogen quadrupole resonances for halides forming molecular crystals. A rather wide range of coupling constants is exhibited and closer inspection of the results reveals two classes of compounds; one being typified by mercuric chloride, $HgCl_2$, where the coupling constant is much as one might expect from the Pauling electronegativity and where the metal atom has a completely filled *d*-shell. For the other class of compound the resonance frequency appears to be anomalously low, e.g. $TiCl_4$, and the metal atom has an incomplete valency *d*-shell. The results for the first group of compounds require little comment; thus $HgCl_2$ and $HgBr_2$ are linear molecules[1] and the two resonance frequencies are due to the presence of two crystallographically inequivalent halogen atoms. Two separate measurements of the ^{35}Cl asymmetry parameter in $HgCl_2$ have led to conflicting results; one, a single crystal measurement gave the value $\eta = 0{\cdot}087$[2] while the other,[3] a measurement by Morino and Toyama's method on a polycrystalline powder (Section C, Chapter 3), led to the much higher value $\eta = 0{\cdot}28 \pm 0{\cdot}1$. Of course for the isolated molecule the asymmetry parameter would be identically zero. The conflict between these two measurements has yet to be resolved. If no experimental error has been made it is possible that a large asymmetry of the magnetic resonance chemical shift tensor could give rise to anomalous results in the single crystal experiment.

TABLE 13.1. Halogen (^{35}Cl, ^{79}Br, ^{127}I) pure quadrupole resonance frequencies (Mc/s at 77°K) of the halides of miscellaneous B-subgroup elements forming molecular crystals

Compound	Nucleus	Resonance frequency (Mc/s)	Ref.
$HgCl_2$	^{35}Cl	22·050, 22·030*	27
$HgBr_2$	^{79}Br	155·47, 156·65	28
$AuCl_3$	^{35}Cl	36·116, 33·340, 23·285	29
$AuCl_3.2H_2O$	^{35}Cl	26·722, 26·809, 27·324, 27·605	29
$TiCl_4$	^{35}Cl	6·022, 6·082, 6·150	30, 31
$TiBr_4$	^{79}Br	46·309, 47·127	32
$ZrBr_4$	^{79}Br	60·09	29
$ThCl_4$	^{35}Cl	5·9181	33
$NbCl_5$	^{35}Cl	13·280	33
$TaCl_5$	^{35}Cl	13·391*	33
WCl_6	^{35}Cl	10·172	31
$VOCl_3$	^{35}Cl	11.54	30
CrO_2Cl_2	^{35}Cl	15·68	30
MoO_2Cl_2	^{35}Cl	15·40*	30
ReO_3Cl†	^{35}Cl	17·0	34

* Measured at room temperature. † Microwave measurement, the coupling constant is −34 Mc/s.

Related to the linear dihalides are the linear molecules CH_3HgCl and CH_3HgBr whose coupling constants have been determined by microwave spectroscopy to be −42 Mc/s and 350 Mc/s for the ^{35}Cl and ^{79}Br nuclei respectively.[4]

In $AuCl_3$ a crystal structure determination shows the presence of planar Au_2Cl_6 molecules where the $AuCl_3$ fragments are joined by a halogen bridge.[1] As we saw for the dimeric halides of Group IIIA the resonance frequencies of the bridging halogen atoms will normally be much lower than those of the non-bridging ones and in $AuCl_3$ this expectation is amply fulfilled. The structure of $AuCl_3 . 2H_2O$ does not appear to have been determined.

In view of the presence of low energy vacant *d*-orbitals in the second group of compounds there exists the possibility of dative π-bonding by the lone-pair electrons of the halogen atoms. Thus titanium has the atomic electron configuration Ar $3d^2 3s^2$ and while three of the *d*-orbitals are presumably used for sigma-bond formation the remaining two are available for π-bond formation. At first sight at least it seems that if the high energy 3*d*-orbitals are used for d_π-p_π bonding in the elements such as silicon then certainly they will be used here where the *d*-orbitals constitute the valence shell. For a tetrahedral molecule such as $TiCl_4$ both of the halogen atom p_π lone-pair electrons can interact with the vacant *d*-orbitals leading therefore to a partial triple bond character for the titanium-halogen bond, and the coupling constants are given by equation (1), Chapter 12.

From the electronegativity of titanium we may estimate a chlorine sigma-bond population of about 1·60 corresponding to a resonance frequency of 22 Mc/s and hence from the observed frequency of 6·0 Mc/s estimate that the population of the chlorine π-orbitals is only 1·7 (each orbital) corresponding to 30 percent triple bond character for the Ti—Cl bond. The same argument leads to a similar value of the pi-character of the Ti—Br bond in $TiBr_4$ and indeed all the compounds for which the results are given above may be treated in the same way.

A further piece of evidence supporting the pi-bond hypothesis lies in the temperature coefficient of the resonance frequency. As we have seen almost all compounds studied have a negative temperature coefficient and this phenomenon is well explained by the Bayer theory. The compounds WCl_6, $TiBr_4$ and $ThCl_4$ all exhibit a positive temperature coefficient and a reasonably satisfactory explanation of this phenomenon has been given in terms of the effect of molecular vibrations on pi-bonding. The quantitative aspects of this theory have been developed only for the hexahalides K_2ReCl_6, K_2ReBr_6 and K_2WCl_6 which also exhibit this phenomenon and for this reason we delay its presentation until the next section.

In some cases, e.g. $VOCl_3$, CrO_2Cl_2, etc. d_π-p_π bonding could give rise to appreciable asymmetry of the quadrupole coupling tensor but no measurements of this kind appear to have been made.

2. Halogen Coupling Constants in Polymeric Halogen-bridged Crystals

Many of the halides of B subgroup elements crystallize not as molecular crystals but as systems exhibiting infinite one-, two- or three-dimensional polymeric structures where the halogen atoms are shared between two or more metal centres. As we have seen for bridged dimeric molecules halogen atoms in such a situation will almost always have a low coupling constant and reference to the results collected in Table 2 shows that this is also certainly the case here. Barnes *et al.*[5] have calculated theoretical coupling constants for many of the systems shown in the table using both ionic point charge and partial covalent models and they concluded that only the latter gives a reasonable description of the bonding. Since the coordination number of the halogen atoms in most of these compounds is three or more they did not consider the possible effects of multiple-bonding. The structure of mercuric iodide consists of infinite chains of HgI_4 tetrahedra linked by sharing of edges.[6] It will in general be expected therefore that the halogen field-gradient tensor will be asymmetric as is observed here ($e^2Qq = 799{\cdot}0$ Mc/s, $\eta = 0{\cdot}527$) and we may calculate the $Hg\widehat{I}Hg$ interorbital angle by

TABLE 13.2. Halogen (^{35}Cl, ^{81}Br or ^{127}I) pure quadrupole resonance frequencies or quadrupole coupling constants (Mc/s at room temperature) of halides of subgroup B metals forming infinite layer or chain type lattices. The crystal structures of all these compounds are described in Ref. 1

Compound	Crystal structure type	Halogen pure quadrupole resonance frequency (Mc/s)	Halogen quadrupole coupling constant (Mc/s)	Ref.
$CdBr_2$	$CdCl_2$	14·341	—	5
$FeCl_2$	$CdCl_2$	—	4·74	35
CdI_2	CdI_2	(1) 14·488, 14·793 (2) 29·018, 29·472	—	5
ZnI_2	CdI_2	(1) 182·18, 183·75, 188·88	—	36
$TiCl_2$	CdI_2	4·17	—	5
$CrCl_2$	?	8·52	—	5
HgI_2	HgI_2	(1) 151·12 (2) 229·47	—	36
$TiCl_3$	BiI_3	7·39	—	5
$CrBr_3$	BiI_3	85·73	—	5
VCl_3	BiI_3	9·40	—	5
$CrCl_3$	$CrCl_3$	12·812, 12·848	—	5
$FeCl_3$	$CrCl_3$	10·026	—	37
$PrCl_3$	UCl_3	—	4·5667, $\eta = 0·58$*	38
$CeCl_3$	UCl_3	—	4·387*	38
$NdCl_3$	UCl_3	—	4·729*	38
$SmCl_3$	UCl_3	—	5·033*	38
$CuCl_2.2H_2O$	$CuCl_2.2H_2O$	—	17·94*	39
$CoCl_2.2H_2O$	$CoCl_2.2H_2O$	—	5·227†	40

* at 4°K ; † at 77°K.

equation (5), Chapter 12, to obtain 100° which is in reasonable agreement with the interbond angle (103°) determined by X-ray crystallography.

3. Halogen Resonances in Complex Ionic Halides

Almost entirely through the efforts of Kubo and Nakamura[7] there is available an extensive series of halogen quadrupole coupling constants of complex ionic halides of the general formula R_2MX_6. In the cases where the salts of different monovalent cations were studied the frequencies were not markedly affected by the nature of the cation. In Table 3 are shown the resonance frequencies for the potassium salts, except for the $[TeCl_6]^{2-}$ and $[PbCl_6]^{2-}$ anions where the results for the ammonium and salts are given, and for the $[SnI_6]^{2-}$ anion where only the result for the rubidium salt is available.

TABLE 13.3. Halogen (^{35}Cl, ^{79}Br and ^{127}I) pure quadrupole resonances of the potassium salts of the complex halide dianions K_2MX_6. For $[SnI_6]$ the frequency is for the rubidium salt and for $[TeCl_6]^{2-}$ and $[PbCl_6]^{2-}$ for the ammonium salts (Ref. 41 and references therein)

X in $(MX_6)^{2-}$	Pure quadrupole resonance frequencies (Mc/s at room temperature)			
	^{35}Cl	^{79}Br	^{127}I	
M in $[MX_6]^{2-}$			$\pm\frac{1}{2} \leftrightarrow \pm\frac{3}{2}$	$\pm\frac{3}{2} \leftrightarrow \pm\frac{5}{2}$
Sn	15·065	128·71	130·48	—
		128·02		
		127·12		
Pb	15·60	—	—	—
Se	20·431	169·036	—	—
Te	14·993	133·477	151·991	302·915
		133·001	151·579	302·502
		132·520	150·571	300·753
Pd	26·55	201·81	—	—
W	10·22	—	—	—
Re	13·887	112·70	121·91	243·68
			122·99	245·16
			123·13	245·34
Os	16·839	—	—	—
Ir	20·732	—	—	—
Pt	25·813	200·18	202·20	403·80
			201·99	403·56
			198·90	397·68

Nakamura and Kubo[7] have shown that the coupling constants fall into three classes illustrated in Fig. 1. Ionic characters were calculated from the coupling constants in the usual way with the exception that the halogen sigma bond was assumed to have 15 per cent *s*-character, and as we shall see below, certain assumptions were made as to the extent of pi-bonding in the transition metal derivatives. It is clear that in cases where only outer *d*-orbitals are present, i.e. for Se, Te, Sn and Pb compounds, the ionic character is very high, consistent with a formulation in terms of bonds involving only valence *s*- and *p*-orbitals, i.e. electron-rich compounds similar to the chloro-iodides (Section F, Chapter 12). The higher frequencies of the Se/Te group of compounds compared with the Sn/Pb series is a natural consequence of the two extra electrons in the Group IV derivatives.

The transition metal complexes, where the *d*-orbitals are partially occupied and form part of the valency shell, constitute yet a third class having much higher resonance frequencies and thus exhibiting much more covalent metal-halogen bonds. In this series moreover the presence of the low-energy and partially filled *d*-orbitals can lead to considerable dative bonding by the lone-pair electrons on the halogen atoms with a consequent reduction in coupling constant and an increase in the apparent ionic character of the metal–halogen bond.

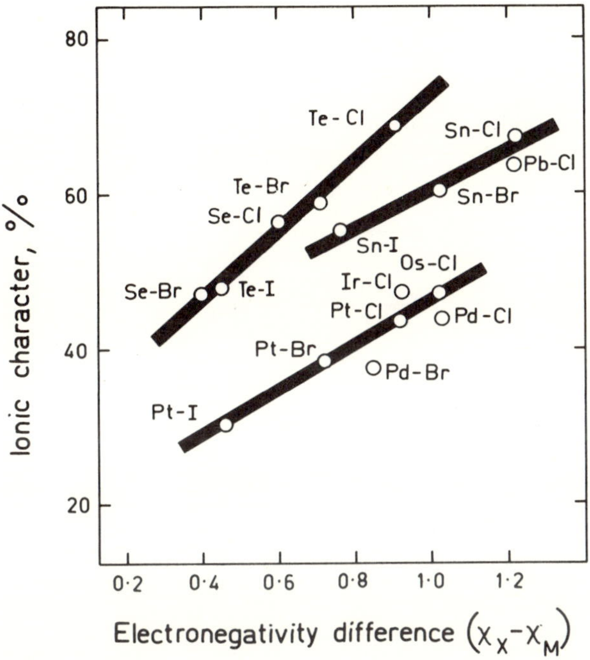

FIG. 13.1. Dependence of the ionic character of metal-ligand bonds in hexahalo complexes on the difference between the electronegativities of ligands and the central atom.

A scheme of one-electron molecular orbitals for octahedral transition metal halide complexes is shown in Fig. 2. Of particular interest in the present context is the question of pi-bonding between the ligand and the central metal atom, and as shown in the figure the twelve p_π orbitals on the ligands span the irreducible representations of the octahedral group (O_h),

$$T_{1u} + T_{1g} + T_{2u} + T_{2g}.$$

The t_{1g} and t_{2u} orbitals are essentially non-bonding, the t_{1u} orbitals can interact weakly with the vacant *p*-orbitals on the central atom. The t_{2g} orbitals, however, interact rather more strongly with three of the central atom d-orbitals to give rise to molecular orbitals of the form:

$$\pi_{xy} = (1 - \beta^2)^{\frac{1}{2}}\, d_{xy} \pm \frac{\beta}{2}[p_y^1 - p_y^2 + p_x^3 - p_x^4]$$

$$\pi_{yz} = (1 - \beta^2)^{\frac{1}{2}}\, d_{yz} \pm \frac{\beta}{2}[p_z^3 - p_z^4 + p_y^5 - p_y^6]$$

$$\pi_{xz} = (1 - \beta^2)^{\frac{1}{2}}\, d_{xz} \pm \frac{\beta}{2}[p_z^1 - p_z^2 + p_x^5 - p_x^6]. \tag{1}$$

The bonding orbitals, π, correspond to the positive sign while the corresponding antibonding orbitals have the negative sign. The symbols p_x^i, p_y^i refer of course to the p_x or p_y orbitals of ligand atom i, numbered so that atoms 1 and 4 lie on the x axis, atoms 2 and 5 on the y axis and 3 and 6 on the z axis.

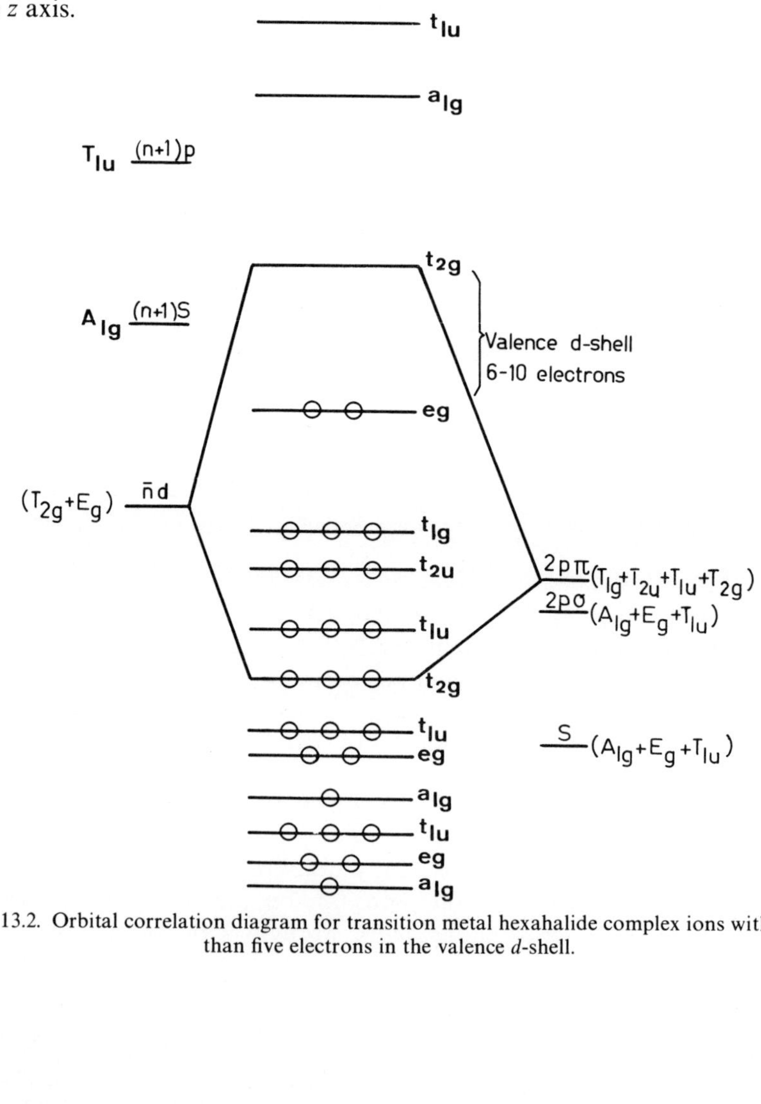

FIG. 13.2. Orbital correlation diagram for transition metal hexahalide complex ions with more than five electrons in the valence d-shell.

The considerable interest of the hexachloride complexes of the metals Pt, Ir, Os, Re and W lies in the fact that their electronic structure is very similar with the exception that the number of electrons in the antibonding π^* orbital decreases by one as the series is descended. In K_2PtCl_6 the π^* orbital is just completely filled and it is clear from the form of the orbitals that no net π-delocalization of the halogen lone-pair electrons can occur here. For K_2IrCl_6 however this effect should become apparent and increase in a regular manner from osmium to tungsten. Any π-delocalization due to the t_{1u} orbitals will probably be approximately constant and indeed most discussions of quadrupole couplings in these compounds have, to date, ignored them.

In octahedral complexes π-bonding does not result in the development of a halogen asymmetry parameter. The hexachloroiridate ion is, however, paramagnetic and Griffiths and Owen [8,9] have estimated from the hyperfine splitting data in its electron spin resonance spectrum that the chlorine-halogen bond has 5·4 per cent π-character. Nakamura and Kubo have supposed that the partial π-character of the metal halogen bond increases linearly with the valence shell d-electron deficiency thus leading to the results shown in Table 4. It is interesting to note in this series how the charge on the central atom remains approximately constant. We remind the reader, however, that the numerical analysis neglects the t_{1u} π-bonding.

TABLE 13.4. Analysis of ^{35}Cl pure quadrupole resonance frequencies for the transition metal hexachloride anions $[MCl_6]^-$

Compound	Ionic character	σ-bond character	π-bond character	Net charge on metal ion
K_2PtCl_6	0·44	0·56	0·0	0·64
K_2IrCl_6	0·47	0·48	0·054	0·80
K_2OsCl_6	0·47	0·43	0·108	0·80
K_2ReCl_6	0·45	0·39	0·162	0·70
K_2WCl_6	0·43	0·35	0·216	0·58

Support for the π-bonding hypothesis in general and for some of the details of the above analysis in particular comes from a consideration of the temperature coefficient of the quadrupole coupling constant in these hexa-halides. In the discussion of the quadrupole resonance frequencies of the B-sub group halides it was mentioned that a few of these compounds exhibited an unusual positive coefficient of resonance frequency with temperature and that it was thought that this phenomenon was connected with π-bonding. It is certainly true that only compounds in which such pi-bonding is likely is a positive temperature coefficient exhibited and in the present series the only likely are K_2ReCl_6, K_2ReBr_6 and K_2WCl_6 where, as Table 2 shows, the pi-character is probably fairly considerable. Perhaps even more significant is the extremely good linear correlation shown in Fig. 3 between the temperature coefficient of the quadrupole resonance frequency and the electron-deficit in the valency $5d$-shell for the series K_2WCl_6, K_2ReCl_6, K_2OsCl_6, K_2IrCl_6 and K_2PtCl_6 which indicates that the temperature coefficient may be represented as the sum of a constant lattice term (negative) and a term proportional to the extent of pi-bonding (positive). Both Haas and Marram[10] and the Japanese group[11] have shown that in a perfectly general manner, metal halogen bonding vibrations would reduce the average pi-bond character of the metal halogen bond and thus increase the coupling constant. This effect therefore would give rise to the positive term required by Fig. 3 and it was shown that indeed the magnitude of the term should be approximately proportional to the extent of pi-bonding and thus to the $5d$ shell deficiency. In order to put the theory on a more quantitative basis the pi-bonding is expressed in the usual molecular orbital terms

$$\Psi_0 = C_1\phi_1 + C_2\phi_2$$

$$\Psi_0^* = C_2\phi_1 - C_1\phi_2 \tag{2}$$

where Ψ_0 and Ψ_0^* are the bonding and antibonding π-orbitals respectively while ϕ_1 and ϕ_2 are one of the chlorine p_π-orbitals and ϕ_2 is one of the metal d_π-orbitals. This expression is of course, for simplicity, only for a single metal chlorine bond, but the orbitals of equation (1) may be reduced to linear combinations of such two-centre bonds and no loss of generality results. These two wave-functions are of course solutions of the usual secular equation

$$\begin{vmatrix} H_{11} - E & H_{12} \\ H_{12} & H_{22} - E \end{vmatrix} = 0 \tag{3}$$

where $H_{ij} = \int \phi_i H \phi_j \, d\tau$. If the chlorine atom is displaced sideways from its equilibrium position the off-diagonal element is reduced by a factor which we may approximate by $\cos\theta$ where θ is the angular displacement (see Fig. 4). Noting that $\cos\theta = \sqrt{1 - \sin^2\theta} \simeq 1 - \frac{1}{2}\sin^2\theta$ this displacement may be conveniently treated as a perturbation to yield the perturbed wave functions:

$$\Psi = \left[C_1 + \left(\frac{C_2 H_{12}\sin^2\theta}{2(E_1 - E_2)}\right)C_2\right]\phi_1 + \left[C_2 - \frac{C_1 H_{12}\sin^2\theta}{2(E_1 - E_2)}\right]\phi_2 \tag{4}$$

$$\Psi^* = \left[C_2 - \frac{C_1 H_{12}\sin^2\theta}{2(E_1 - E_2)}\right]\phi_1 - \left[C_1 + \frac{C_2 H_{12}\sin^2\theta}{2(E_1 - E_2)}\right]\phi_2 \tag{5}$$

where E_1 and E_2 are the energies of the bonding and antibonding orbitals.

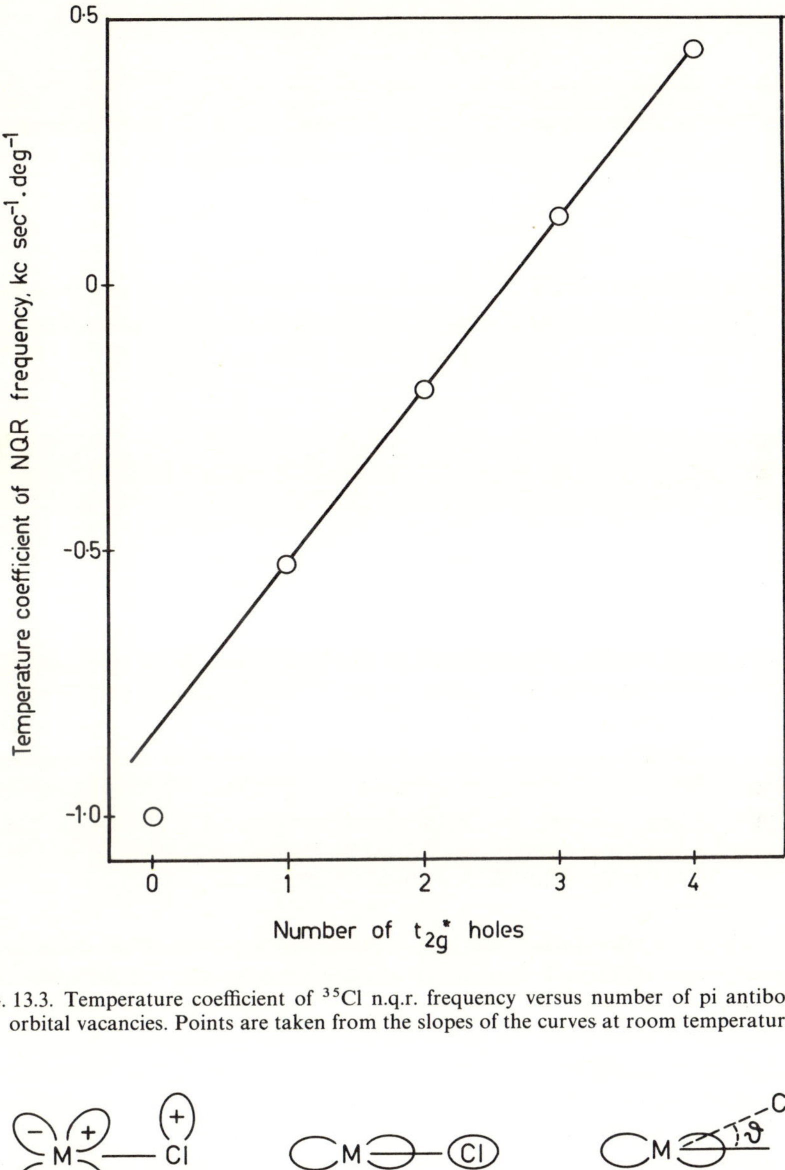

FIG. 13.3. Temperature coefficient of ^{35}Cl n.q.r. frequency versus number of pi antibonding orbital vacancies. Points are taken from the slopes of the curves at room temperature.

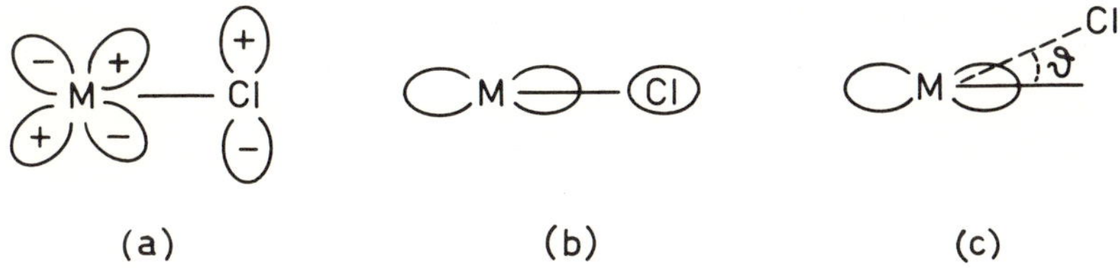

FIG. 13.4. d_π-p_π bonding in MCl_6 : (a) equilibrium position, nodal plane perpendicular to paper, (b) equilibrium position, nodal plane in paper, (c) displaced position, nodal plane in paper.

The orbital populations of interest are of course those averaged over the vibrational motion and $\sin^2\theta$ is replaced by $\langle\sin^2\theta\rangle \simeq \langle\theta^2\rangle$. Carrying out a similar analysis for all the different bonding vibrations, they obtain for the average number of electrons, N_π, in one of the chlorine atom p_π orbitals in K_2IrCl_6 where there is one electron short in the antibonding orbital:

$$N_\pi = 2C_1^2 + C_2^2 + \frac{5 \cdot C_1C_2 \cdot H_{12}\langle\theta^2\rangle}{E_2 - E_1}. \tag{6}$$

If the bending vibrations are assumed to be due to a harmonic oscillator potential $\langle\theta^2\rangle$ may be readily expressed as a function of temperature as

$$\langle\theta^2\rangle = \frac{\hbar}{I\omega}\left[\frac{1}{2} + \frac{1}{\exp(\hbar\omega/kT) - 1}\right] \tag{7}$$

where ω is the vibrational frequency, I the moment of inertion and T the absolute temperature. Substitution in (1) yields:

$$N_\pi = 2C_1^2 + C_2^2 + \frac{5C_1C_2H_{12}}{E_2 - E_1}\cdot\frac{\hbar}{I\omega}\left[\frac{1}{2} + \frac{1}{\exp(\hbar\omega/kT) - 1}\right]. \tag{8}$$

Hass and Marram[10] showed that the stretching vibrations did not result in a reduction in coupling constant; it is also clear from Fig. 4 that owing to the cylindrical symmetry of p orbitals the t_{1u} pi-bonding which was not taken into account is not much affected by bonding vibrations. The last term in the above expression therefore shows the temperature dependence of N_π in K_2IrCl_6. According to equation (1), Chapter 12, this is just the temperature dependence of the coupling constant.

Since both the bonding and antibonding orbitals are filled in K_2PtCl_6 there is no term of this kind in this compound and the observed (negative) temperature coefficient is just that due to the lattice. Moreover these compounds are all isomorphous having, within 1·5 per cent, identical lattice parameters and metal-halogen bond lengths so that the temperature coefficient due to the lattice vibrations may be assumed constant. The difference in temperature coefficient between K_2PtCl_6 and K_2IrCl_6 is then just due to the temperature dependent term of equation (8). Furthermore if the reasonable assumption is made that the term $C_1C_2H_{12}/(E_2 - E_1)$ is constant throughout the series then equation (8) may be extended to yield the experimentally observed result that the temperature coefficient is be proportional to the deficiency of $5d$ electrons. The magnitude of the term

$$\frac{C_1C_2H_{12}}{(E_2 - E_1)}$$

is of course unknown and this difficulty was overcome by using one of the experimental vibration frequencies of the $[PtCl_6]^=$ ion, ($\nu_5 = 162\ cm^{-1}$) and calculating a value for it to produce agreement with experiment for the frequency change due to π-bonding at room temperature. The π-bonding term may then be calculated as a function of temperature for K_2IrCl_6 and K_2OsCl_6 with the very satisfactory result shown in Fig. 5.

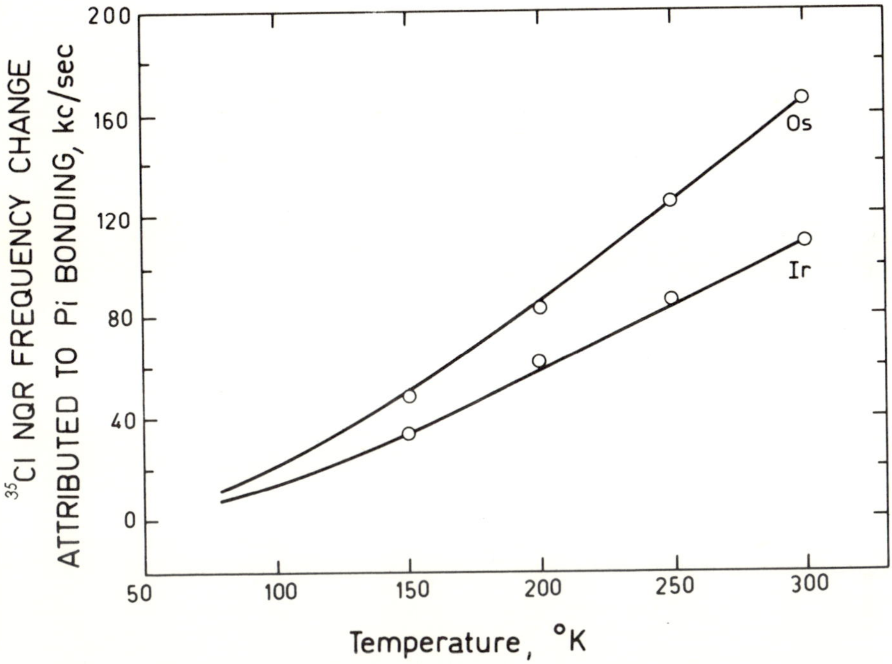

FIG. 13.5. Frequency change attributed to pi-bonding effects as a function of temperature. Solid lines are theoretical curves. Experimental points are from smoothed experimental frequency-temperature plots.

This theory has been given in some detail since it is an extremely nice example of the way in which the maximum information may be obtained from quadrupole resonance frequencies. It is unfortunate that it is only rarely that such a similar series of compounds exists.

There remains the quadrupole coupling constants of a few ionic complexes which do not fall into the previous category. They are listed in Table 5. For the tetrabromopalladate(II) and tetrabromo- and tetrachloroplatinate(II) anions only one resonance was observed in agreement with the known crystal structures. The frequency is very markedly less than for the corresponding tetravalent hexabromides and this could well be due in part to a large amount of dative pi-bonding by the lone-pair electrons of the halogen.

TABLE 13.5. Halogen (^{35}Cl, ^{79}Br, ^{127}I) pure quadrupole resonance frequencies (Mc/s) at room temperature in complex anionic halides

Compound	Resonance frequency (Mc/s)	Ref.
K_2PtCl_4	17·925	42
K_2PdBr_4	129·34	43
K_2PtBr_4	138·70	43
$KAuCl_4$*	27·30, 27·57, 27·89, 28·20	44
$NaAuCl_4.2H_2O$	27·37, 28·34	44
$NH_4AuCl_4.\frac{1}{3}H_2O$	26·59, 26·94, 27·39, 27·58	44
K_2HgI_4	123·46, 129·30, 148·49	28
trans-[$CO(H_2N.CH_2.CH_2NH_2)_2Cl_2$]$Cl.HCl.XH_2O$	16·058	45

* At 80°K.

This could of course now be revealed by measurements of the asymmetry parameter (the complexes have the square planar configuration) but this measurement has not so far been reported, no doubt due to difficulties in obtaining single crystals of suitable dimensions.

Although this section is nominally devoted to halogen coupling constants it is convenient to include herein the results of measurements of N^{14} couplings in complex cyanides (Table 6). As we saw in Chapter 11, it is rather difficult to correlate the N^{14} coupling constant in the nitrile group with the nature

TABLE 13.6. N^{14} pure quadrupole coupling constants of complex cyanides (Ref. 46)

Compound	e^2Qq (Mc/s)	η
$K_2Zn(CN)_4$	4·1390	0·0000
$K_2Cd(CN)_4$	4·1988	0·0000
$K_2Hg(CN)_4$	4·0475	0·0000
	4·0739	0·0137
$K_3Cu(CN)_4$	3·9639	0·0294
$K_2Pt(CN)_4.3H_2O$	3·467	0·032
$K_3Co(CN)_6$	3·684	0·030

of the group to which it is bonded but even so the values given in the Table fall clearly into two groups. The tetracyano complexes of Zn^{II}, Cd^{II}, Hg^{II} and Cu^{I} all have tetrahedral configurations and have the central atom in formally a d^{10} configuration. The N^{14} coupling constants are all about 4·0 Mc/s. For the isolated ions the asymmetry parameters are identically zero and the small values observed in the mercury and copper complexes are presumably due to crystalline electrostatic fields. The tetracyano platinum complex on the other hand has a d^8 configuration and has a square-planar molecular structure while the octahedral $K_3Co(CN)_6$ has a central atom d^6 configuration. The coupling constants of these two compounds with incomplete *d*-shells are notably lower than those with a complete *d*-shell.

Equation (22), Chapter 11, states that the coupling constants for linear cyanides are given by the relationship

$$\frac{e^2Qq}{e^2Qq_0} = 1 + \frac{b}{2} - a \qquad (9)$$

where *b* and *a* are the effective sigma and pi orbital populations of the nitrogen atom. Provided that the coupling constants of the cyano complexes are negative as in covalent cyanides the lower coupling constants of the complexes with the incomplete *d*-shell could be due either to *a* being greater or *b* being smaller than in the complexes with the complete *d*-shell. In other words, this may mean that the multiple bond order of the metal–ligand bond is greater in the d^{10} complexes than in the d^8 and d^6 complexes. This, of course, is the opposite trend to that observed in the octahedral chloro complexes above but is in fact what would be expected since whereas the halide ions are pi-electron donors the cyano group is a pi-electron acceptor.

If extensive metal–ligand pi-bonding is occurring then it is surprising that in the square planar $Pt(CN)_4$ ion the N^{14} asymmetry parameter is so small. This could conceivably be accidental and a more extensive study of such square planar systems would be welcome.

B. Fe^{57m} COUPLING CONSTANTS

More than four hundred papers concerning Fe^{57m} Mössbauer spectroscopy have appeared and in the majority of these quadrupole splitting effects have been observed. It is convenient therefore to discuss the data for Fe^{57} separately from those for other transition metal nuclei.

The electron-configuration of atomic iron is $Ar3d^64s^2$ and its usual oxidation states are 2 or 3 with configurations $Ar3d^6$ and $Ar3d^5$ respectively. The intraatomic field-gradient thus arises essentially from the 3*d*-electron shell. In many of the compounds of iron the central atom is surrounded by six electron-donor ligands in an octahedral configuration. In these circumstances it is well-known that the five-levels of the *d*-shell split up into two groups: a triply degenerate t_{2g} set and a doubly degenerate e_g set. The symbols t_{2g} and e_g are derived from the corresponding irreducible representations of the octahedral group. Depending upon the separation between the t_{2g} and the e_g levels, i.e. the strength of the interaction between the central atom and the ligands, the spins of the 3*d* electrons in ferrous and ferric compounds may be either as paired as possible (low total spin) or as unpaired as possible (high total spin).

It is a straightforward matter in principle to calculate the field-gradients for the four configurations shown in Fig. 6 provided that care is taken to label our axes correctly. Just as in polycoordinate nitrogen compounds we must be sure that the field-gradient axes and the orbital axes coincide or that if not we perform the necessary rotations and calculate the appropriate

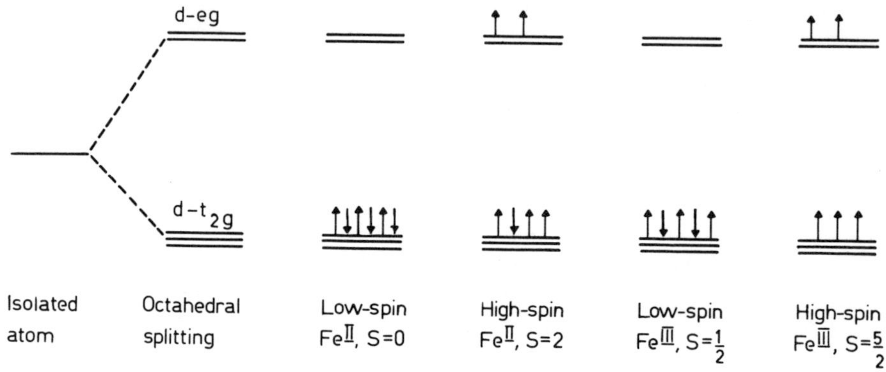

FIG. 13.6. Splitting of *d*-electrons in an octahedral environment and electron configurations of octahedral high- and low-spin ferrous and ferric complexes.

off-diagonal elements. In Table 7 are shown the contributions of each of the *d*-orbitals, labelled in an axis system going through the six ligands, to the field-gradient along the *z*-axis (equation 6, Chapter 4). Provided that the complex has the field-gradient axes oriented in this way then the total

TABLE 13.7. *d*-orbital contributions to field-gradients in octahedral molecules

Orbitals, real form	Magnetic quantum number, $\|m\|$	Irreducible representation in octahedral symmetry	eq_{zz}
d_{xy}	2	t_{2g}	$\frac{4}{7}r^{-3}$
d_{yz}	1	t_{2g}	$-\frac{2}{7}r^{-3}$
d_{xz}	1	t_{2g}	$-\frac{2}{7}r^{-3}$
d_{z^2}	0	e_g	$-\frac{4}{7}r^{-3}$
$d_{x^2-y^2}$	2	e_g	$\frac{4}{7}r^{-3}$

field-gradient may be obtained by adding the individual contributions. It is thus easy to see that high-spin Fe^{III} and low-spin Fe^{II} have a zero intra-atomic field-gradient. This could have been seen directly from the fact that these are both *S*-state ions. With the other two systems, however, there is a problem. These are both orbitally degenerate and according to the Jahn–Teller Theorem cannot be octahedrally symmetric but will distort in such a way as to remove the orbital degeneracy. In both cases this orbital degeneracy essentially concerns the t_{2g} level and we may concentrate our attention on this for the moment. One way in which this degeneracy can be lifted is for the complex to change its dimensions along one of the axes, *z* let us say. The field-gradient axes still coincide with the orbital axes but in low-spin Fe^{III} for example the unpaired *d*-electron is moving in an orbital which is a linear combination of d_{xy}, d_{xz} and d_{yz}. We can treat this more simply if we use the hole formalism and think of a positive hole in the complete t_{2g} shell. The wave function of this hole is thus also a linear combination of d_{xy}, d_{xz} and d_{yz}.

$$\Psi = \frac{a}{\sqrt{2}}\{d_{xz} + d_{yz}\} + (1 - a^2)^{\frac{1}{2}} d_{xy}. \tag{10}$$

This gives rise to a field-gradient

$$(q_0)_z = \frac{2e}{2}(3a^2 - 2)\left\langle \frac{1}{r^3} \right\rangle \tag{11}$$

(Remember that the field-gradient of a hole has the opposite sign to that of an electron.)

The mixing coefficient *a* depends on the details of the distortion and the nature of the metal-ligand interaction. The positive hole can exist in two other states of higher energy with wave functions given by

$$\Psi_1 = \frac{1}{\sqrt{2}}(d_{xz} - d_{yz}) \tag{12}$$

$$\Psi_2 = \frac{(1 - a^2)^{\frac{1}{2}}}{\sqrt{2}}(d_{xz} + d_{yz}) - a d_{xy}. \tag{13}$$

These two states have field-gradients of, respectively

$$(q_1)_z = \frac{2e}{7}\left\langle \frac{1}{r^3} \right\rangle \tag{14}$$

$$(q_2)_z = \frac{2e}{7}(1 - 3a^2)\left\langle \frac{1}{r^3} \right\rangle. \tag{15}$$

If the energy separations between these three states are comparable with kT then all will be thermally populated and the quadrupole coupling constant will be the weighted average of the couplings given by equations (11), (14) and (15). This will of course give rise to a strong temperature dependence of the coupling constant. When the separations are all much smaller than kT the average coupling constant is zero. Note that according to the value of a^2 the field-gradient and hence the coupling constant can be either of positive or negative sign.

The Jahn–Teller splitting can occur in other ways besides a tetragonal distortion and in this case the calculation of the coupling constant becomes more complex. In particular the field-gradient axes may no longer coincide with the orbital axes given in Table 7. The general problem of *d*-orbital field-gradients in such complexes, with particular reference to high-spin ferrous complexes, has been discussed by Ingalls.[12]

A similar treatment may be applied to the iron nucleus in other than a six-coordinate environment. A case of particular interest is the $[FeCl_4]^=$ ion which is tetrahedral. The *d*-orbital splitting in this case is the same as in the octahedral environment but now the e_g orbitals have the lower energy. The electron configuration of these high-spin species is thus $e_g^3 t_g^3$ and the Jahn–Teller theorem predicts that the system will be distorted. In this case an extremely marked temperature dependence was observed,[13] typically going from less than 1·0 mm/sec at room temperature to more than 3·0 mm/sec at 1·5°K. It is interesting to note that in this case X-ray crystallography was unable to detect the distortion and it had been concluded that the $[FeCl_4]^=$ ion was a regular tetrahedron.

It will be appreciated that a quantitative understanding of the actual experimental values of the coupling constants in iron complexes is not easily arrived at. The situation is further complicated by the relatively low field-gradient produced by *d*-electrons and the consequent importance of interionic terms. Much effort has indeed been devoted to this problem and to the consequent evaluation of the Sternheimer shielding parameter (see Chapter 5). For this reason no attempt is given here to discuss any of the results in detail or to attempt a comprehensive survey of the data. Instead Table 8 shows a few representative data for ferrous and ferric complexes, illustrating the points discussed above. For three of the ferrous compounds the sign of the coupling constant has been determined by means of the method described in Chapter 3, and they illustrate the point that the sign is by no means self-evident.

The coupling constant of the low-spin Fe^{III} complexes is in general much lower than those of high-spin Fe^{II}. This is probably because in the low-spin complexes the metal ligand bonds are highly covalent and the effective *d*-orbital population drops below the implicitly-assumed value of unity.

Up to the present we have been implicitly discussing the spectra of systems in which the six ligands are identical or nearly so. In mixed complexes the situation increases once more in complexity, since the populations of all the different *d*-orbitals may no longer be anything like equivalent, particularly

TABLE 13.8. Quadrupole splittings of Fe^{57m} in high and low spin ferrous and ferric complexes

Compound	Temp. (°K)	Oxidation state	Spin state	Quadrupole splitting (mm/sec)	Ref.
$FeSO_4.7H_2O$	166	II	high	3·47 ± 0·01	47, 49
	333			+3·10 ± 0·001	
$FeC_2O_4.2H_2O$	166	II	high	−1·942 ± 0·01	47, 49
	331			−1·668 ± 0·01	
$FeCl_2.2H_2O$	158	II	high	+3·090	47
	333			+2·927	
$NH_4Fe(SO_4)_2.12H_2O$	165	III	high	0·6	47
$Fe_2(SO_4).x.H_2O$	298	III	high	0·282 ± 0·010	47
$FePO_4.4H_2O$	163	III	high	0·657 ± 0·007	47
	333	III	high	0·622 ± 0·007	
$Ag_4Fe(CN)_6$	298	II	low		48
$Ag_3Fe(CN)_6$	147	III	low	0·858 ± 0·006	48
	298	III	low	0·767 ± 0·006	
$(NMe_4)_2FeCl_4$	1·5	II	high	3·30	13
	363			0·58	

in the strongly-bonded low-spin complexes. Without a knowledge of the sign of the coupling constant, its temperature dependence and the direction of the field-gradient axes, an analysis of the data in terms of molecular electronic structure becomes almost impossible. If these data are available however and if in addition magnetic hyperfine splitting is present as well as the quadrupole effects, an analysis of all the Mössbauer data together can throw considerable light on the compounds electronic structure. However such a discussion is really outside the scope of this book and it will be clear to the reader that the correlation of Fe^{57m} quadrupole coupling constants in different compounds is nothing like as straightforward as for a nucleus such as chlorine.

In addition to the compounds discussed above, iron forms an extensive series of compounds with carbon monoxide and unsaturated organic compounds in which the iron atom is more often than not in a low oxidation state (the assignment of oxidation state to these compounds is often a matter of taste!). The results for these compounds have been surveyed by several authors[14,15,16]: it is probably fair to say that they provide disappointingly, but as we have seen, understandably, little information. An exception to this remark is the well-studied compound, ferrocene $[(C_5H_5)_2Fe]$ and the corresponding ferricinium ion. Related to ferrocene is the isoelectronic cobaltocene for which the Co^{59} coupling constants have been determined and which will be discussed with them. Ferrocene is of course the compound whose discovery sparked off the present intense study of organometallic compounds and has the structure illustrated in Fig. 7. The other compounds are isostructural with this.

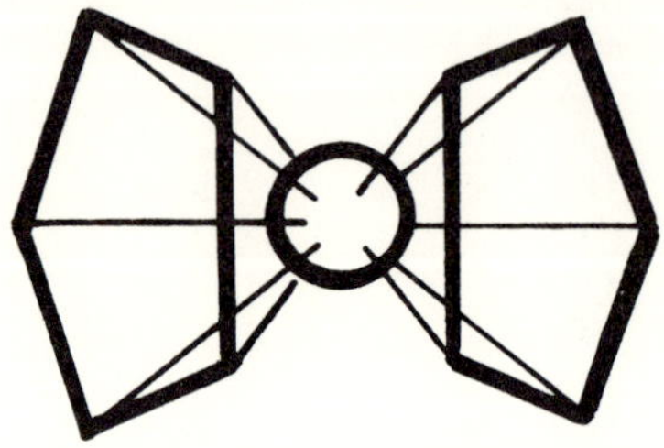

FIG. 13.7 Schematic drawing of the molecular structure of ferrocene.

From the present point of view these molecules have the advantage that the field-gradient principal axis is symmetry-determined and the field-gradient tensor has cylindrical symmetry. Furthermore the sign of the coupling constant in this ferrocene has been determined by the magnetic perturbation method (see Chapter 3). The results are shown in Table 9.

TABLE 13.9. Quadrupole coupling constants (Fe^{57}, mm/sec; Co^{59} Mc/sec) for transition-metal cyclopentadienyl complexes

Compound	Nucleus	Quadrupole coupling constant	Ref.
$(C_5H_5)_2Fe$, ferrocene	Fe^{57m}	+2·4 mm/sec	50, 21
$[(C_5H_5)_2Fe]^+$, ferricinium	Fe^{57m}	1·08 mm/sec	51
$[(C_5H_5)_2Co]^+$, cobaltocenium	Co^{59}	1·70 Mc/s	52

Ferrocene has been the object of considerable theoretical study of its electronic structure. The problem has been approached both from the extreme view point of considering ferrocene to be an ionic Fe^{++} ion with two $C_5H_5^-$ ligands and from the more realistic molecular orbital point of view. In the former case the calculation of the intraatomic field-gradient due to the *d*-electrons is straightforward provided that one decides in which orbitals to place the electrons. Under the D_{5h} symmetry of the molecule the

d-levels split into the subsets $A'_1(d_z2)$, $E'_2(d_{x^2-y^2}, d_{xy})$ and $E''_1(d_{xz}, d_{yz})$. Matsen[17] proposed that the levels $A'_1 + E''_1$ are occupied. From Table 7 this yields a field-gradient $-\frac{16}{7}\langle r^{-3}\rangle$ and implies contrary to observation a negative coupling constant. On the other hand Voitländer *et al.*[18,19,20] proposed that the $A'_1 + E'_2$ levels were occupied, leading to a predicted positive field-gradient of $\frac{8}{7}\langle r^{-3}\rangle$. Both these assignments were made before the experimental determination of the sign of the field-gradient. Thus the Mössbauer spectrum directs the choice of the correct ionic model. Covalent bonding of course enters considerably into the structure of ferrocene and in fact the value $\frac{8}{7}\langle r^{-3}\rangle$ is about twice as great as the observed field-gradient. Collins[21] has used the molecular orbitals of Shustorovich and Dyatkina[22] and of Dahl and Ballhausen[23] to calculate a fairly satisfactory value of the field-gradient in ferrocene while Voitländer *et al.*[18,19,20] have made independent calculations of this quantity, both for ferrocene and for the cobaltocenium ion.

The qualitatively much lower coupling constant of the ferricinium ion can be easily understood on the basis of Voitländer's ionic model for in this case the electron is removed from the E'_2 level so that the field-gradient is now $\frac{4}{7}\langle r^{-3}\rangle$. The actual magnitude of the quadrupole splitting in the ferricinium ion depends somewhat on the counter ion but the qualitative predictions of this simple model are excellent.

C. COUPLING CONSTANTS OF TRANSITION METAL AND RARE-EARTH NUCLEI OTHER THAN Fe^{57m}

Although many of the transition metal and rare-earth elements have naturally occurring quadrupolar isotopes present in reasonable concentration, rather few values of their nuclear quadrupole coupling constants are available. Although a large number of the paramagnetic compounds of this series have been studied by means of their electron spin resonance spectra, in many cases the nucleus is in such a highly symmetric environment that the quadrupole coupling constant is small or identically zero. Even when appreciable quadrupole coupling is present investigators often do not do anything other than note the fact, since of course these effects only influence the "forbidden" transitions $\Delta m = \pm 1, \pm 2$ (see Chapter 3). For diamagnetic compounds a few values are available from pure quadrupole resonance or from quadrupole satellites in nuclear magnetic resonance. Table 10 shows a selection of coupling constants for diamagnetic transition metal complexes. The only nucleus which is at all well-studied is Co^{59} and in particular in unsymmetrical octahedral Co^{III} complexes. In addition to those values given in Table 10, Hartmann and Sillescu[24] have estimated the Co^{59} coupling constants for a number of such complexes from their Co^{59}-n.m.r. line widths

TABLE 13.10. Quadrupole coupling constants of transition metal nuclei in diamagnetic compounds

Compound	Temp. (°K)	Nucleus	Resonance frequencies (Mc/s)	e^2Qq (Mc/s)	η	Ref.
V_2O_5	Room	V^{51}	—	$0{\cdot}735 \pm 0{\cdot}1$	—	53
NH_4VO_3	Room	V^{51}	—	$3{\cdot}04 \pm 0{\cdot}1$	—	53
$Na_3VO_4.14H_2O$	Room	V^{51}	—	$3{\cdot}4 \pm 0{\cdot}05$	—	53
$NaVO_3$	Room	V^{51}	—	$3{\cdot}1 \pm 0{\cdot}02$	—	53
$Na_4V_2O_7.16H_2O$	Room	V^{51}	—	$3{\cdot}1 \pm 0{\cdot}02$	—	53
$C_5H_5Mn(CO)_3$	293	Mn^{55}	$9{\cdot}6(\nu_1)$, $19{\cdot}28(\nu_2)$	64·3	0	52, 25
$Mn_2(CO)_{10}$	298	Mn^{55}	—	3·05	0·35	54
$[(C_5H_5)_2Co]ClO_4$	78	Co^{59}	$12{\cdot}25(\nu_1)$, 24·08, $24{\cdot}50(\nu_2)$, 36·12, $36{\cdot}74(\nu_3)$	171·6	0	52, 55
$Co_2(CO)_8$	78	Co^{59}	18·625, 37·270	—	—	52
trans$[Co\{C_2H_4(NH_2)_2\}_2Cl_2]Cl.HCl.xH_2O$	Room	Co^{59}	$5{\cdot}979(\nu_1)$ $10{\cdot}001(\nu_2)$ $15{\cdot}296(\nu_3)$	71·73	$0{\cdot}222 \pm 0{\cdot}002$	24, 45
$[Co(NH_3)_5H_2O](ClO_4)_3$	298	Co^{59}	—	54·8	0	24
$NH_4[Co(NO_2)_4(NH_3)_2]$	298	Co^{59}	—	32·35, 33·05	0	24
trans$[CoCl_2(NH_3)_4]Cl$	Room	Co^{59}	$8{\cdot}370(\nu_2)$ $12{\cdot}665(\nu_3)$	59·23	0·136	56

trans[Co{$C_2H_4(NH_2)_2$}$_2Cl_2$]Cl	Room	Co^{59}	5·367, 8·368, 12·893	60·63	0·272	56
trans[Co{$C_2H_4(NH_2)_2$}$_2Cl_2$]NO_3		Co^{59}	4·777, 8·877, 13·426	62·78	0·132	56
trans[Co{$C_2H_4(NH_2)_2$}$_2Br_2$]Br		Co^{59}	8·392(ν_2), 12·860(ν_3)	60·36	0·235	56
$KCu(CN)_2$	Room	Cu^{63}	32·647	60·9	0·78 ± 0·07	57, 58
$K_3Cu(CN)_4$	87	Cu^{63}	—	1·113	—	59
Cu_2O	20	Cu^{63}	26·697	—	—	57
$Re_2(CO)_{10}$	Room	Re^{185}	28·836(ν_1) 39.650(ν_2)	111·5	0·63	60
		Re^{187}	27·191(ν_1) 37·730(ν_2)	106·1	0·63	
AuCl	300	Au^{197}	257·075, 257·173, 257·377	—	—	61
$NbCl_5$	298	Nb^{93}	3·030(ν_1), 2·085(ν_2), 2·527(ν_3), 3·648(ν_4)	78·26	0·34974 ± 0·00002	33
$KNbO_3$ (rhombohedral)	77		1·335(ν_2), 2·004(ν_3), 2·674(ν_4)	16·0	0·00	62
$KNbO_3$ (orthorhombic)	298		3·030(ν_1), 2·085(ν_2) 2·527(ν_3), 3·648(ν_4)	23·12	0·80	62
$LiNbO_3$	Room	Nb^{93}	—	22·02	—	63

in solution. All these complexes are in the low-spin state, thus being iso-electronic with Fe^{II}, and as we have seen would be expected to have, in the electrostatic approximation, a vanishingly small coupling constant. In order to explain the observed values it is necessary to use the more realistic covalent model for the bonding in these complexes. In this case the coupling constants are a measure of the population differences between the various metal ligand bonds. The results for the cobaltocenium ion have already been discussed in connection with the Fe^{57m} coupling in ferrocene. Semin[25] has studied the coupling constants of substituted cyclopentadienyl manganese tricarbonyls and shown that the substituent mainly acts to alter the asymmetry parameter from the value zero present in the unsubstituted molecule.

In addition to the values shown in Table 10 the volatile manganese and rhenium oxyhalides have been studied by microwave spectroscopy (Table 11). In one case (ReO_3F) a remarkable dependence of the coupling constant on the vibrational state was noticed. The *s-p* orbitals used for the essentially tetrahedral framework of this compound are given by equation (16). If the population of the ReO bonds (ψ_2, ψ_3, ψ_4) is given the symbol b while that of the Re-F bond is a then the coupling constant is given by

$$\left|\frac{e^2Qq}{e^2Qq_0}\right| = \left|\frac{3(b - a)\cos\gamma}{1 - \cos\gamma}\right| \tag{16}$$

where γ is the $\widehat{\mathrm{OReO}}$ angle. It will be remembered that the value of the angular term is quite sensitive to the value of γ. This alone would not be enough to account for the large variation in coupling constant with vibrational state but if $|b - a|$ is small compared to the individual values of b and a then small changes in average electronic structure will produce large changes in the coupling constant. That $(b - a)$ is small is consistent with the relative electronegativities of oxygen and fluorine and is substantiated by the much higher coupling constant of ReO_3Cl.

The only transition metal nucleus for which a large number of quadrupole coupling constants is available is Cu^{63} in distorted octahedral divalent copper complexes. Here the coupling constant has a fairly uniform value of around 130 Mc/s.[26]* The configuration of Cu^{++} is Ar $3d^9$ and the field gradient would then be due to a single hole in the $3d$-shell. In this case the coupling constant would be expected to be largely independent of the nature of the ligands surrounding the Cu^{63} nucleus.

Coupling constants for a few rare-earth nuclei are available from electron spin resonance and Mössbauer spectra. The intraatomic contribution to

* The constant P, given in wave-numbers, shown therein is equal to e^2Qq (4I(2I − 1)/3).

TABLE 13.11. Mn^{55} and $Re^{185,187}$ coupling constants from microwave spectroscopy

Molecule	Nucleus	Coupling constant (Mc/s)	Ref.
MnO_3F	Mn^{55}	$+16{\cdot}8$	34
ReO_3F	Re^{182}	$-48{\cdot}4 \pm 1{\cdot}3\ (v = 0)$	64
		$-27{\cdot}0 \pm 1{\cdot}5\ (v_3 = 1)$	
		$-34{\cdot}9 \pm 2{\cdot}9\ (v_5 = 1)$	
		$-58{\cdot}2 \pm 1{\cdot}3\ (v_6 = 1)$	
		$-37{\cdot}8 \pm 3{\cdot}6\ (v_3 = 1, v_6 = 1)$	
		$-17{\cdot}0 \pm 2{\cdot}2\ (v_3 = 2)$	
		$-25{\cdot}7 \pm 4{\cdot}3\ (v_6 = 1)$	
ReO_3Cl	Re^{185}	270 ± 6	34
	Re^{187}	253 ± 6	

the field-gradient arise from *f* electrons and the small value of $\langle 1/r^3 \rangle$ of these means that the interionic contribution is considerable. For this reason considerable efforts have been made to calculate the Sternheimer shielding factor of these species (see Chapter 5). It is probably true to say that the coupling constants of these nuclei present the theoretician with problems rather than information.

REFERENCES

1. A. F. Wells, "Structural Inorganic Chemistry," 3rd Edition, Oxford 1962.
2. H. Negita, T. Tanaka, T. Okuda and H. Shimada, *Inorg. Chem.* **5**, 2126 (1966).
3. Dinesh and P. T. Narasimhan, *J. Chem. Phys.* **45**, 2170 (1966).
4. W. Gordy and J. Sheridan, *J. Chem. Phys.* **22**, 92 (1954).
5. R. G. Barnes, S. L. Segel and W. H. Jones, *J. Appl. Phys.* **33**, 296 (1962).
6. R. G. Wyckoff, "Crystal Structures," Interscience, New York, 1948.
7. M. Kubo and D. Nakamura, "Advances in Inorganic and Radiochemistry", **8**, 257 (1966), and references contained therein.
8. J. H. E. Griffiths and J. Owen, *Proc. R. Soc.* A**226**, 96 (1954).
9. J. H. E. Griffiths, J. Owen and I. M. Ward, *Proc. R. Soc.* A**219**, 526 (1953).
10. T. E. Haas and E. P. Marram, *J. Chem. Phys.* **43**, 3985 (1965).
11. R. Ikeda, D. Nakamura and M. Kubo, *J. Phys. Chem.* **69**, 2101 (1965).
12. R. Ingalls, *Phys. Rev.* **133**, A787 (1964).
13. P. R. Edwards, C. E. Johnson and R. J. P. Williams, *J. Chem. Phys.* **47**, 2074 (1967).
14. G. K. Wertheim, R. B. King and R. H. Herber, *Inorg. Chem.* **3**, 101 (1964).
15. R. L. Collins and R. Pettit, *J. Chem. Phys.* **39**, 3433 (1963).
16. R. L. Collins and R. Pettit, *J. Am. Chem. Soc.* **85**, 2332 (1963).
17. F. A. Matsen, *J. Am. Chem. Soc.* **81**, 2023 (1959).

18. J. Schraml and J. Voitländer, *Z. Naturf.* **20**a, 1424 (1965).
19. B. Höfflinger and J. Voitländer, *Z. Naturf.* **18**a, 1185 (1963).
20. B. Höfflinger and J. Voitländer, *Z. Naturf.* **18**a, 1074 (1963).
21. R. L. Collins, *J. Chem. Phys.* **42**, 1072 (1965).
22. E. M. Shustorovich and M. E. Dyatkina, *Dokl. Akad. Nauk. SSSR* **128**, 1234 (1959).
23. J. P. Dahl and C. F. Ballhausen, *Mat. Fys. Medd. Dan Vid. Selsk.* **33**, No. 5 (1961).
24. H. Hartmann and H. Sillescu, *Theor. Chim. Acta* **2**, 371 (1964).
25. G. K. Semin, *Acta Crystallog.* **21**, A151 (1967).
26. E. König, *Magnetische Eigenschaften der Koordinations—und metallorganischen Verbindungen der Uebergangselemente*, Springer Verlag, Berlin 1966.
27. A. Bassompierre, *C.r. hebd. Acad. Sci. Paris*, **236**, 596 (1953).
28. D. Nakamura, Y. Uehara, Y. Kurita and M. Kubo, *J. Chem. Phys.* **31**, 1433 (1959).
29. S. Segel, Unreported data quoted in *Catalogue of Nuclear Quadrupole Interactions and Resonance Frequencies in Solids*, Pt. I.
30. H. G. Dehmelt, *J. Chem. Phys.* **21**, 380 (1953).
31. R. Hamlen and W. S. Koski, *J. Chem. Phys.* **25**, 360 (1956).
32. R. G. Barnes and R. D. Engardt, *J. Chem. Phys.* **29**, 248 (1958).
33. A. H. Reddoch, *J. Chem. Phys.* **35**, 1085 (1961).
34. A. Javan and A. Engelbrecht, *Phys. Rev.* **96**, 649 (1954).
35. R. G. Barnes and S. L. Segel, *J. Chem. Phys.* **37**, 1895 (1962).
36. S. Kojima, K. Tsukada, S. Ogawa and A. Shimauchi, *J. Chem. Phys.* **23**, 1963 (1955).
37. A. Narath, *J. Chem. Phys.* **40**, 1169 (1964).
38. W. E. Hughes, C. G. Montgomery, W. G. Moulton and E. H. Carlson, *J. Chem. Phys.* **41**, 3470 (1964).
39. W. J. O'Sullivan, W. W. Simmons and W. A. Robinson, *Phys. Rev.* **140**A, 1759 (1965).
40. A. Narath, B. Morosin and A. T. Fromhold, *Bull. Am. Phys. Soc.* **8**, 359 (1963).
41. D. Nakamura, *Bull. Chem. Soc. Japan* **36**, 1662 (1963), and references therein.
42. E. P. Marram, E. J. McNiff and J. L. Ragle, *J. Phys. Chem.* **67**, 1719 (1963).
43. K. Ito, D. Nakamura, Y. Kurita, K. Ito and M. Kubo, *J. Am. Chem. Soc.* **83**, 4526 (1961).
44. V. Caglioti, C. Furlani, F. Orestano and F. M. Capece, *Atti Accad. Nazl. Lincei, Rend., Classe Sci. Fis., Mat. Nat.* **35**, 10 (1963).
45. H. Hartmann, M. Fleissner and H. Sillescu, *Naturwiss.* **50**, 591 (1963).
46. R. Ikeda, D. Nakamura and M. Kubo, Proc. Xth International Conference on Coordination Chemistry, Japan, p. 110 (1967).
47. W. Kerler and W. Neuwirth, *Z. Phys.* **167**, 176 (1962).
48. W. Kerler, W. Neuwirth, E. Fluck, P. Kuhn and B. Zimmermann, *Z. Phys.* **173**, 321 (1963).
49. R. W. Grant, H. Wiedersich, A. H. Muir, U. Gonser and W. N. Delgass, *J. Chem. Phys.* **45**, 1015 (1966).
50. L. M. Epstein, *J. Chem. Phys.* **36**, 2731 (1962).
51. U. Zahn, P. Kienle and H. Eicher, *Z. Phys.* **166**, 220 (1962).
52. J. Voitländer, H. Klocke, R. Longino and H. Thieme, *Naturwiss.* **49**, 491 (1962).
53. V. Saraswati, *J. Phys. Soc. Japan*, **23**, 761 (1967).
54. E. A. C. Lucken and D. F. Williams, Proc. XV. Colloque Ampère, Ljubljana (1966).
55. J. Voitländer and R. Longino, *Naturwiss.* **46**, 664 (1959).

56. I. Watanabe and Y. Yamagata, *J. Chem. Phys.* **46**, 407 (1967).
57. H. Kruger and U. Meyer-Berkhout, *Z. Phys.* **132**, 171 (1952).
58. G. L. McKown and J. D. Graybeal, *J. Chem. Phys.* **44**, 610 (1966).
59. G. Becker, *Z. Phys.* **130**, 415 (1951).
60. S. L. Segel and R. G. Barnes, *Phys. Rev.* **107**, 638 (1957).
61. P. Machmer, M. Read and P. Cornil, *C.r. hebd. Acad. Sci., Paris* **262**, 650 (1966).
62. R. M. Cotts and W. D. Knight, *Phys. Rev.* **96**, 1285 (1954).
63. G. Petersen, P. M. Bridenbaugh and P. Green, *J. Chem. Phys.* **46**, 4009 (1967), and personal communication.
64. J. F. Lotspeich, A. Javan and A. Engelbrecht, *J. Chem. Phys.* **31**, 633 (1959).

CHAPTER 14

CHARGE TRANSFER COMPLEXES

The phenomenon of the formation of well-defined complexes between molecules is well-known but, at least in the details of the bonding, cannot be said to be well-understood. These compounds range from straightforward donor-acceptor complexes such as that between diethylether and boron trifluoride whose structure is well-represented by the dative bond formulation

$$(C_2H_5)_2O \rightarrow BF_3$$

to the complexes between aromatic molecules, such as the hydrocarbon picrates so familiar to the organic chemist. In a series of papers mostly concerned with the electronic absorption spectra of these complexes, Mulliken has proposed that in their ground state these complexes may in general be formulated as donor-acceptor compounds in which, at least for the complexes such as the picrates, only a small amount of actual charge-transfer takes place from the π-electrons of one molecule to vacant π-orbitals of the other. In some cases there is indeed clear evidence from electron spin resonance spectroscopy that complete transfer of one electron, resulting in two paramagnetic ions, does indeed take place, in many others the complexes remain diamagnetic, and it is with these that we will be concerned here. The crystal structure of many such compounds has been determined, particularly by Hassel and his co-workers.[1] The advantage of nuclear quadrupole resonance spectroscopy in this field is that, particularly when the crystal structure is known, the changes in n.q.r. frequency produced by complex formation can in principle provide very good evidence of the direction and extent of electron transfer.

The results of pure quadrupole resonance investigations of molecular complexes are summarized in Tables 1 to 4. Some of the earliest attempts at observing pure quadrupole resonance in such complexes were unsuccessful and it transpires that, particularly for the weak π-π complexes, considerable care must be taken to ensure the presence of a well-ordered crystal structure. Thus the freshly prepared complex between naphthalene and picryl chloride yielded no ^{35}Cl resonance signal, and this appeared only after the compound had been annealed for 48 hours at a temperature just below its melting point.

The crystal structure[2] of the $SnCl_4 . 2POCl_3$ complex, measured at room temperature, shows that, like the analogous $SnCl_4 . 2SeOCl_2$, the tin atom is in the octahedral configuration with the $POCl_3$ groups occupying *cis*

TABLE 14.1. ^{35}Cl, $^{79,81}Br$ pure quadrupole resonance frequencies (Mc/s at 77°K) of miscellaneous complexes

Component A	Component B	Nucleus	Pure quadrupole resonance frequency (Mc/s at 77°K)	Ref.
CBr_4 Carbon tetrabromide	—	^{81}Br	265·45 266·06 266·4–267·06 267·97 268·30 269·25	11
CBr_4	p-$C_6H_4(CH_3)_2$ p-xylene	^{79}Br	317·00 322·09	11
		^{81}Br	264·95 269·18	
Br_2	—	^{81}Br	319·46	11
Br_2	p-$C_6H_4(CH_3)_2$ p-xylene	^{81}Br	321·83	11
CCl_4 Carbon tetrachloride	—	^{35}Cl	40·465 to 40·817 (15 lines)	11
CCl_4	p-$C_6H_4(CH_3)_2$ p-xylene	^{35}Cl	40·472 40·689	11
$SnCl_4$	—	^{35}Cl	24·294 24·226 24·140 23·719	12
$SnCl_4$	2.$(C_2H_5)_2O$ Diethylether	^{35}Cl	19·473 19·435	12
$SnCl_4$	$C_6H_5NO_2$ Nitrobenzene	^{35}Cl	23·08 (average of six lines)	13
$POCl_3$ Phosphorus oxychloride	—	^{35}Cl	28·9835 28·9378	12
$SnCl_4$	2.$POCl_3$	^{35}Cl	30·213 30·117 21·146 19·807 19·035	12
$AsCl_3$ Arsenic trichloride	—	^{35}Cl	25·406 25·058 24·960	12
$AsCl_3$	$POCl_3$	35Cl	29·208 29·168 29·154 25·481 25·125 24·799	12

TABLE 14.1—*Continued*

Component A	Component B	Nucleus	Pure quadrupole resonance frequency (Mc/s at 77°K)	Ref.
$HgCl_2$ Mercuric chloride	—	^{35}Cl	22·873	12
			22·521	
$HgCl_2$	$C_4H_8O_2$ Dioxan		21·196	
$SbCl_5$	—	^{35}Cl	18·76 ± 0·03	14
			27·42 ± 0·03	
			27·85 ± 0·02	
			28·01 ± 0·02	
			29·80 ± 0·03	
			29·98 ± 0·03	
			30·42 ± 0·03	
			30·50 ± 0·03	
			27·93 ± 0·02	
		^{121}Sb or ^{123}Sb	28·69 ± 0·03	
			31·09 ± 0·04	
			38·48 ± 0·03	
			39·63 ± 0·04	
			42·89 ± 0·03	
			54·22 ± 0·06	
$SbCl_5$	$POCl_3$	^{35}Cl	24·44 ± 0·02	14
			25·85 ± 0·02	
			26·22 ± 0·03	
			27·33 ± 0·03	
			30·58 ± 0·02	
			30·64 ± 0·02	
		^{121}Sb	34·05 ± 0·03	
			58·37 ± 0·03	
		^{123}Sb	25·30 ± 0·06	
			34·48 ± 0·03	
			53·76 ± 0·05	
$SbCl_5$	CH_3CN	^{35}Cl	24·99 ± 0·02	14
			26·37 ± 0·02	
		^{121}Sb	32·47 ± 0·02	
			64·87 ± 0·04	
		^{123}Sb	19·70 ± 0·02	
			39·43 ± 0·03	
			59·06 ± 0·04	

positions and bonded to the tin by oxygen. The ^{35}Cl n.q.r. resonance, however, while being perfectly visible at 77°K, vanishes suddenly at 183°K and it was concluded that a phase-change had occurred. The details of the crystal structure may not therefore be relevant to the low-temperature n.q.r. results. It is unlikely, however, that any very fundamental change in structure occurs.

TABLE 14.2. ^{35}Cl pure quadrupole resonance frequencies (Mc/s) of complexes between aromatic molecules

Component A	Component B	^{35}Cl n.q.r. frequency	Temperature	Ref.
$C_6Cl_4O_2$	—	37·5851		
Chloranil		37·5148	77°K	15
		37·4698		
		37·4417		
Chloranil	Hexamethyl-benzene	37·5042	77°K	
		37·7161		
$C_6H_2(NO_2)_3Cl$ Picrylchloride	—	38·795	291°K	16
Picrylchloride	Naphthalene	39·130		
C_6F_5Cl Pentafluorochlorobenzene	—	39·410	77°K	17
Pentafluorochlorobenzene	Benzene	38·717	77°K	
Pentafluorochlorobenzene	Mesitylene	38·684	77°K	
Pentafluorochlorobenzene	Aniline	38·640	77°K	
Pentafluorochlorobenzene	Triethylamine	39·284	77°K	
Pentafluorochlorobenzene	Toluene	39·158	77°K	
1,3,5-$C_6F_3Cl_3$ 1,3,5 trichloro 2,4,6 trifluorobenzene	—	39·312	77°K	
1,3,5 trichloro 2,4,6 trifluorobenzene	Mesitylene	38·850		
1,3,5 trichloro 2,4,6 trifluorobenzene	Hexamethyl-benzene	38·830		

If possible π-bonding effects in the P—Cl and Sn—Cl bonds are neglected for the moment, both the magnitude and the sign in the shifts of the resonance of the chlorine atom attached to tin (-3 to -4 Mc/s) are in accordance with the formulation of the bonding as a coordinate link, $(Cl_3P{=}O)_2 \rightarrow SnCl_4$, which leads to a greater ionicity of the Sn—Cl bond. The positive shift ($+1$ Mc/s) in the phosphorus oxychloride ^{35}Cl resonance frequency is likewise in accordance with this hypothesis, the smaller magnitude in the shift presumably being due to the fact that the chlorine atom is one bond further removed from the site of complex formation.

The crystal structure of $SbCl_5.POCl_3$ has likewise been determined[3] and is similar to that of the $SnCl_4.POCl_3$ complex in that the oxygen atom of the $POCl_3$ group forms a sixth ligand to the antimony atom. In this case the number of different resonances agrees exactly with the room temperature crystal structure. An interpretation of the results in this case and an assignment of the ^{35}Cl coupling constants to the different chlorine atoms in the complex is rendered difficult, without a single crystal study, by the complexity

TABLE 14.3. ^{35}Cl, $^{121,123}Sb$ pure quadrupole resonance frequencies (Mc/s at 77°K) of complexes of $SbCl_3$ with organic compounds. For the antimony isotopes the frequency of the $(\pm\frac{1}{2} \leftrightarrow \pm\frac{3}{2})$ transition is given.

Component A	Component B	Pure quadrupole resonance frequency (Mc/s) ^{35}Cl	^{123}Sb	^{121}Sb	$\left(\frac{e^2Qq}{\eta}\right)Sb^{121}$	(η)Sb, per cent	Ref.
—	$SbCl_3$	20·9077	59·728	39·108	383·108	18·8	18, 12
		19·3047					
C_6H_6*	$2.SbCl_3$	20·509	59·774	38·566	392·678	10·3	12, 7
		20·459	59·602	37·117			
		20·349					
		20·068					
		18·749					
		18·655					
$C_6H_5.CH_3$ Toluene	$2.SbCl_3$	—	59·998	—	—	—	7
			59·306				
			58·639				
$C_6H_5.C_2H_5$ Ethylbenzene	$2.SbCl_3$	—	58·761	—	—	—	7
o-$C_6H_4(CH_3)_2$ *o*-xylene	$2.SbCl_3$	—	58·839	—	—	—	
m-$C_6H_4(CH_3)_2$ *m*-xylene	$2.SbCl_3$	—	58·503	—	—	—	
p-$C_6H_4(CH_3)_2$ *p*-xylene	$2.SbCl_3$	20·631	58·268	37·990	—	—	12
		19·609					
		18·361					

$C_6H_5.C_6H_5$ Biphenyl	$2.SbCl_3$	20·671	58·384	36·358	—	—	12, 7
		20·378	55·598	35·260	385·815	9·9	
		19·667					
		19·229					
		19·160					
		18·873					
$C_{10}H_8$ Naphthalene	$2.SbCl_3$	20·810	59·427	36·058	396·180	0·0	12, 7
		19·893					
		19·187					
$C_6H_5OCH_3$ Anisole	$2.SbCl_3$		61·200	38·521	385·137	10·4	7
C_6H_5OH Phenol	$2.SbCl_3$	—	58·960	36·074	391·715	5·6	7
$C_6H_5CH_3$ Toluene	$SbCl_3$	—	57·312	36·860	371·597	16·1	7
$C_6H_5C_2H_5$ Ethylbenzene	$SbCl_3$	20·076	58·689	37·988	379·191	17·0	7
o-$C_6H_5CH_3$ *o*-xylene	$SbCl_3$	—	57·074	—	379·191	17·07	7
m-$C_6H_5CH_3$ *m*-xylene	$SbCl_3$	—	58·495	—	—	—	7
p-$C_6H_5CH_3$ *p*-xylene	$SbCl_3$	—	58·257	—	—	—	7
1,3,5-$C_6H_3(CH_3)_3$ Mesitylene	$SbCl_3$	—	56·374	—	—	—	
$C_6H_5OCH_3$ Anisole	$SbCl_3$	—	58·457	36·426	385·137	10·4	7
$C_6H_5OC_2H_5$ Phenetole	$SbCl_3$	—	56·063	35·167	368·190	11·8	7
$C_6H_5CO.CH_3$ Acetophenone	$SbCl_3$	—	62·042	41·801	391·012	23·0	7
C_6H_5Cl Chlorobenzene	$SbCl_3$	34·57†	59·662	—	—	—	7

* Measured at 133°K; † Chlorine atom in chlorobenzene.

TABLE 14.4. $^{121,123}Sb$ and ^{81}Br pure quadrupole resonance frequencies (Mc/s) at 77°K of complexes of $SbBr_3$ and organic compounds

Component A	Component B	Pure quadrupole resonance frequency (Mc/s)			$(e^2Qq)Sb^{121}$	η	Ref.
		Sb^{121}	Sb^{123}	Br^{81}			
—	$SbBr_3$	49·291	—	137·429 (α phase) 144·488 136·885 (β phase) 137·621 143·346	—	—	7, 19
C_6H_6 Benzene	$2SbBr_3$	50·717	33·568	141·899 142·614	323·395	20·41	7, 19
1,3,5-$C_6H_3(CH_3)_3$ Mesitylene	$2SbBr_3$	48·459 51·279	—	—	—	—	7, 19
$C_6H_5.C_6H_5$ Biphenyl	$2SbBr_3$	47·208	28·717	—	314·412	3·0	7
$C_{10}H_8$ Naphthalene	$2SbBr_3$	50·301	30·654	141·885 143·257 144·340	344·724	4·1	7, 19
$C_6H_5.CH_3$ Toluene	$SbBr_3$	49·500		—	—	—	7
$C_6H_5.C_2H_5$ Ethylbenzene	$SbBr_3$	48·992	31·817	135·880 138·842 140·330	316·024	17·5	7, 19
o-$C_6H_4(CH_3)_2$ o-xylene	$SbBr_3$	46·323	28·627	—	306·457	8·4	7
m-$C_6H_4(CH_3)_2$ m-xylene	$SbBr_3$	46·574	—	—	—	—	7
$C_6H_5.OCH_3$ Anisole	$SbBr_3$	50·194	31·342		330·338	10·9	7
$C_6H_5OC_2H_5$ Phenetole	$SbBr_3$	47·610	30·189		310·746	13·8	7

of the spectrum of $SbCl_5$ itself. By analogy with the stannic chloride complex, however, and in agreement with the crystal structure it is likely that the two highest frequencies arise from the chlorine atoms of the $POCl_3$ group. In this case the ^{35}Cl shift would be around $+1{\cdot}5$ Mc/s.

The structure of the ether-tin tetrachloride complex is unknown, again the tin atom is almost certainly in an octahedral environment with the ether molecules, possibly in the *cis* configuration, linked to the tin by a coordinate bond involving the unshared electrons on the oxygen atom. The large negative shift (-5 Mc/s) in the ^{35}Cl resonance is again in accordance with this formulation.

The $HgCl_2$.Dioxan complex has been shown to contain square-planar coordinated mercury atoms (I)

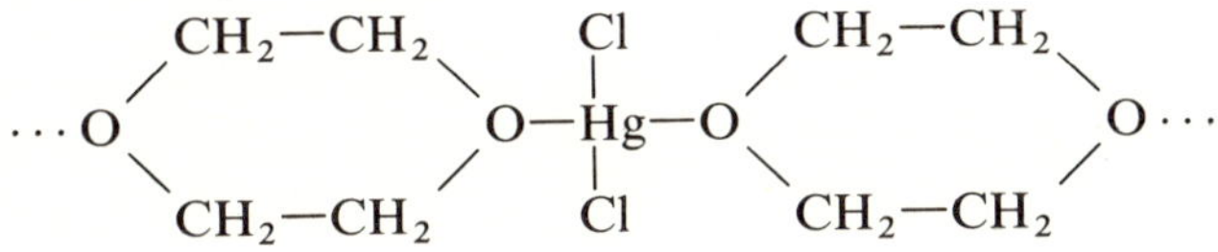

(I)

and in agreement with the crystal structure,[4] only one ^{35}Cl resonance frequently was observed. The shift in ^{35}Cl resonance frequency on complex formation is however much less in this case.

The structure of the $POCl_3.AsCl_3$ complex is unknown. In this case the shifts in ^{35}Cl resonance frequency are small for both molecules although again positive for the $POCl_3$ fragment. It is regrettable that so far no studies of the Zeeman effect have been made on complexes of the above type. The directions of the field-gradient tensors and the value of the asymmetry parameter would be most interesting. The shift in pure quadrupole resonance frequency can, by itself, be misleading, since if appreciable π-bonding is present an increase in electron density in the sigma bond will probably be accompanied by a similar increase in the pi-bond resulting in only a small resultant shift in resonant frequency.

In contrast to the results for the above compounds, where complex formation is probably the result of the formation of a coordinate dative bond from the unshared electrons on one atom to the vacant atomic orbitals on another, the shifts in resonant frequency when one or more of the partners in the complex uses delocalized *molecular* orbitals are in general rather small. Thus the crystal structure[5] of the bromine-benzene complex consists of infinite chains of alternate bromine and benzene molecules with the Br—Br axis perpendicular to the aromatic plane (II).

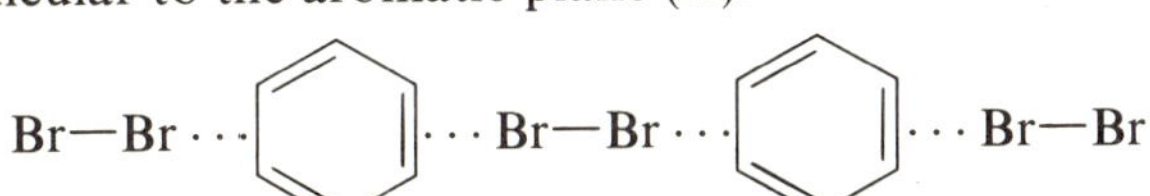

(II)

Similarly the crystal structure[6] of the *p*-xyiene-carbon tetrabromide complex consists of infinite chains of alternate xylene and carbon tetrabromide molecules (III)

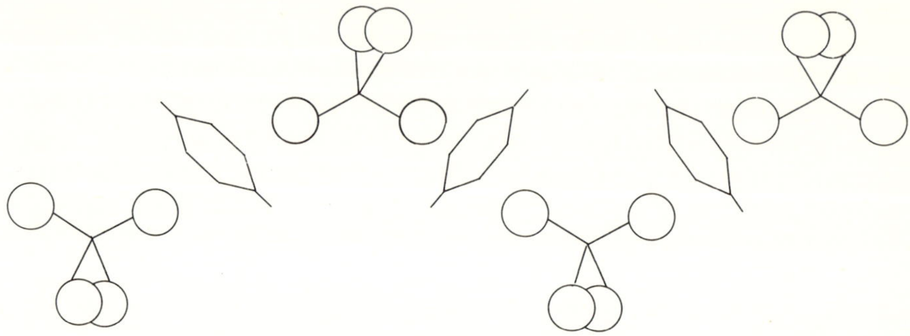

(III)

This last structure is interesting in that only two of the bromine atoms on CBr_4 participate in bond-formation. The ^{81}Br resonance frequencies of both of these complexes are almost insignificantly different from those of the free parent molecule and in particular for the CBr_4/*p*-xylene complex very little difference exists between the complexed and the uncomplexed bromine atoms. The shifts are if anything smaller when both partners are π-acceptors or donors (Table 2) and these results have led to the statement that there is no charge transfer in charge-transfer complexes.[11] It certainly seems clear that, whatever the nature of the bonding, it is very little different from that which binds the molecules of the pure individual components together in their respective crystalline states. Once again, however, it is unfortunate that so far only quadrupole resonance frequencies have been determined and not all the components of the field-gradient tensor and their orientation in the molecular coordinate system.

Grechishkin and Kyuntsel[7] have made an extensive investigation of the antimony trichloride and tribromide complexes of various organic compounds (Tables 3 and 4) where the resonance frequencies of both the halogen and antimony nuclei could be determined. In addition, either by making use of the relationship

$$\frac{(\nu)\mathrm{Sb}^{123}}{(\nu)\mathrm{Sb}^{121}} = \frac{12{\cdot}747}{22} \cdot \frac{1 + 3{\cdot}6333\eta^2 - 7{\cdot}2607\eta^4}{1 + 1{\cdot}0925\eta^2 - 0{\cdot}6340\eta^4}$$

or by direct observation of the $(\pm\frac{1}{2} \leftrightarrow \pm\frac{3}{2})$ and $(\pm\frac{3}{2} \leftrightarrow \pm\frac{5}{2})$ transitions for ^{123}Sb, the asymmetry parameters and coupling constants of the antimony nuclei were determined. Here it is clear that the effect on η for Sb is much more pronounced than on the coupling constants themselves. If these results were combined with crystal structure determinations they would certainly add much to our knowledge of the bonding in molecular complexes. As it is unfortunately none of these compounds appears to be of known crystal structure.

In addition to the results shown in Tables 1–4 Biedenkapp and Weiss[8] have carried out an interesting study of over thirty molecular complexes formed between trichloroacetic acid and a variety of oxygen containing organic compounds (aldehydes, ketones, acids, esters, phenols, alcohols and ethers). They have also studied the ^{35}Cl frequencies of a number of trichloroacetates. Their results are summarized in Fig. 1 and illustrate again the apparently small changes in electron distribution attendant on complex formation.

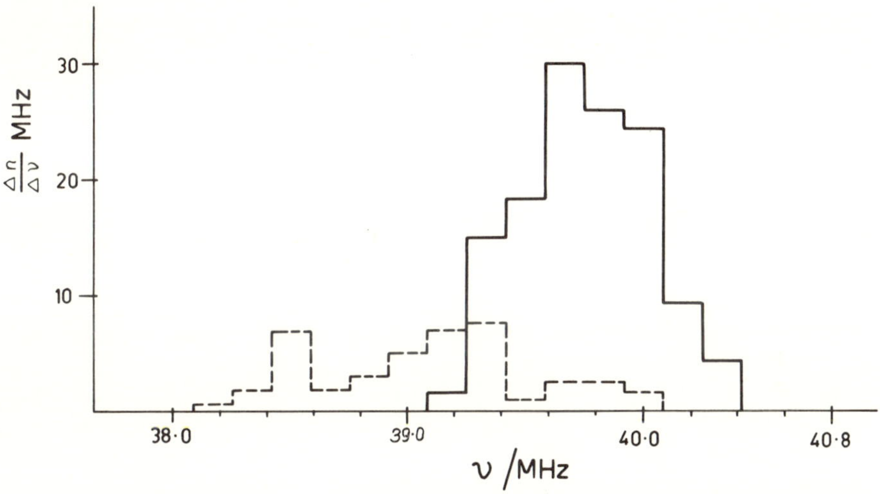

FIG. 14.1. Number of ^{35}Cl resonance frequencies lying in a 200 kc/s interval as a function of frequency. —— Trichloroacetic acid addition compounds; – – – – Trichloroacetates.

Finally sulphur, in the form of cyclic S_8 molecules, forms complexes with certain covalent iodides and the pure quadrupole resonance data for these complexes are shown in Table 5. For the complexes of arsenic and antimony triiodides the changes in coupling constants on complex-formation are large. However, as discussed in Section D, Chapter 12, these triiodides crystallize as halogen bridged macromolecules while the crystal structure[9] of $AsI_3.3S_8$ reveals the presence of isolated AsI_3 molecules. The frequencies of the parent molecule and the complex are not therefore comparable and little information as to the nature of the bonding can be gleaned from the quadrupole resonance results. A Zeeman single crystal study of the complex[10] shows that the field-gradients at both the ^{75}As and ^{127}I nuclei have essentially axial symmetry. The angle between the ^{127}I field-gradient tensors, and hence presumably between the AsI bonds, is determined to be 101·7° ± 0·5°.

TABLE 14.5. Pure quadrupole resonance frequencies (Mc/s at 77°K) of complexes between covalent iodides and sulphur (S_8) (Ref. 20)

Compound	Nucleus	ν_1	ν_2	e^2Qq	η
AsI_3	As^{75}	29·338	—	—	—
	I^{127}	207·023	395·979	1330·23	0·1891
$AsI_3.3S_8$	As^{75}	49·501	—	—	—
	I^{127}	227·543	455·059	1516·86	0·007
SbI_3	Sb^{121}	25·406	—	—	—
	I^{127}	174·356	254·637	895·83	0·565
$SbI_3.3S_8$*	Sb^{121}	37.461	—	—	—
	I^{127}	184·151	367·023	1226·25	0·03
SnI_4*	I^{127}	203·470	406·902	1356·36	0·008
		204·490	408·988	1363·28	0·000
$SnI_4.2S_8$*	I^{127}	204·019	408·008	1360·04	0·002
		206·784	413·153	1377·40	0·028
$SnI_4.4S_8$*	I^{127}	201·961	403·897	1346·3	0·00
		202·036	404·081	1346·9	0·00
		206·853	413·238	1377·7	0·03
		213·148	—	—	—

* Measured at room temperature.

REFERENCES

1. O. Hassel and C. Rømming, *Q. Rev.* **16**, 1 (1962).
2. C. I. Brändén, *Acta Chem. Scand.* **17**, 759 (1963).
3. I. Lindqvist and C. I. Brändén, *Acta Chem. Scand.* **12**, 134 (1958); *Acta Chem. Scand.* **17**, 353 (1963); *Acta Cryst.* **12**, 642 (1959).
4. O. Hassel and J. Hvoslef, *Acta Chem. Scand.* **8**, 1953 (1954).
5. O. Hassel and K. O. Stromme, *Acta Chem. Scand.* **12**, 1146 (1958).
6. F. J. Strieter and D. H. Templeton, *J. Chem. Phys.* **37**, 161 (1962).
7. V. Grechishkin and I. A. Kyuntsel, *Zh. Strukt. Khim.* **5**, 53 (1964); *Optika Spek.* **16**, 161 (1964).
8. D. Biedenkapp and A. Weiss, *Ber. Bunsengesell. Phys. Chemie* **70**, 788 (1966).
9. H. Hertel, *Z. Phys. Chem.* **15**, 51 (1932).
10. Y. Abe, *J. Phys. Soc. Japan* **13**, 918 (1958).
11. H. O. Hooper, *J. Chem. Phys.* **41**, 599 (1964).
12. D. Biedenkapp and A. Weiss, *Z. Naturf.* **19**a, 1518 (1964).
13. H. G. Dehmelt, *J. Chem. Phys.* **21**, 380 (1953).
14. R. F. Schneider and J. V. DiLorenzo, *J. Chem. Phys.* **47**, 2343 (1967).

15. D. Douglass, *J. Chem. Phys.* **32**, 1882 (1960).
16. A. I. Kitaigorodskii and E. I. Fedin, *Zh. Strukt. Khim.* **2**, 216 (1961).
17. G. K. Semin, V. I. Robas, V. D. Shteingartz and G. G. Yakobson, *Zh. Strukt. Khim.* **6**, 143 (1965).
18. T. C. Wang, *Phys. Rev.* **99**, 566 (1955).
19. H. Negita, T. Okuda and M. Kashima, *J. Chem. Phys.* **45**, 1076 (1966); *J. Chem. Phys.* **46**, 2450 (1967).
20. S. Ogawa, *J. Phys. Soc. Japan* **13**, 618 (1958).

AUTHOR INDEX

Numbers in parentheses are reference numbers and are included to assist in locating references when the authors' names are not mentioned in the text. Numbers in italics refer to the page on which the reference is listed.

H

I

J

K

L

T

X-Y-Z

SUBJECT INDEX

O

P

R

S

T

V

W